Adverse Meteorological Phenomena in Northeast Brazil

Adverse Meteorological Phenomena in Northeast Brazil

Edited by

Natalia Fedorova and Vladimir Levit

Cambridge Scholars Publishing

Adverse Meteorological Phenomena in Northeast Brazil

Edited by Natalia Fedorova and Vladimir Levit

This book first published 2023

Cambridge Scholars Publishing

Lady Stephenson Library, Newcastle upon Tyne, NE6 2PA, UK

British Library Cataloguing in Publication Data
A catalogue record for this book is available from the British Library

ISBN (10): 1-5275-3115-5
ISBN (13): 978-1-5275-3115-4

CONTENTS

Part II
Short Term Forecasting in Brazilian Northeast

List of Abbreviations and Acronyms

BNE – Brazilian Northeast
EBNE – Eastern BNE
NH – Northern Hemisphere
SH – Southern Hemisphere

Names of the BNE states:

RN – Rio Grande do Norte,
PB – Paraiba,
PE – Pernambuco,
AL – Alagoas,
SE – Sergipe,
BA – Bahia.

Cities in the BNE.

south of the BNE:
Salvador city (13.01°S, 38.51°W) (Bahia State).
middle of the BNE:
Maceio city (9.55°S, 5.77°W) (Alagoas State).

Meteorological Centers

INPE (Instituto Nacional de Pesquisas Espaciais) – Nacional Institute of Space Research
CPTEC (Centro de Previsão de Tempo e Estudos Climáticos) – Center for Weather Forecasting and Climate Studies
INMET (Instituto Nacional de Meteorologia) – National Institute of Meteorology
ECMWF – European Centre for Medium-range Weather Forecast
NCEP – National Centers for Environmental Prediction

Synoptic systems

L – Low/cyclone
H – High/Anticyclone
ITCZ – InterTropical Convergence Zone
UTCV – Upper Tropospheric Cyclonic Vortex
UTT – Upper Tropospheric Trough,
MTCV – Middle Tropospheric Cyclonic Vortexes
SACZ- South Atlantic Convergence Zone

JS – Jet Stream
SJS – Subtropical Jet Stream
PJS – Polar Jet Stream
PJSSH – polar jet stream of Southern Hemisphere
PJSNH – polar jet stream of Northern Hemisphere
STJSSH – subtropical jet streams of Southern Hemispheres
STJSNH – subtropical jet streams of Northern Hemispheres
BNEJS – Brazilian Northeast Jet Stream

FZ – Frontal Zone
TW – Trade Winds
WDTW – Wave Disturbance in Trade Winds
MCC – Mesoscale Convective Complexes
EW – Easterly Waves
Wave – cyclonic curvature at streamlines
WDTW – Wave Disturbances on the Trade Winds
SASH – South Atlantic Subtropical High
SASR – South Atlantic Subtropical Ridge
SAD – South Atlantic Dipole

Variables

u – zonal wind component ($m.s^{-1}$)
v – meridional wind component ($m.s^{-1}$)
u_g – **x** component of geostrophic wind
v_g – **y** component of geostrophic wind
ω – **p** component of wind (vertical motion in p- coordinates; ($hPa\ s^{-1}$)
w – vertical component of wind ($m\ s^{-1}$)
g – gravitational acceleration
ς – vorticity
RV – relative vorticity

CRV – cyclonic relative vorticity
ρ – air density
t – time
Z – geopotential (height in geopotential meters, gpm)
Thickness 500-1000 hPa (dam)
f – Coriolis parameter
SLP – surface level pressure (hPa)
MSLP – Mean Sea Level Pressure (hPa)
SST – Sea Surface Temperature (°C)
Stl – streamlines
Q – omega (Pas^{-1})
Conv – convergence of air current
Div – divergence of air current
T – temperature (°C)
AdvT – temperature advection
Ttr – temperature transformation
OLR – Outgoing Long Wave Radiation (Wm^2)

γ – vertical temperature gradient, lapse rate of temperature
γ_d – vertical temperature gradient for dry adiabatic processes, adiabatic lapse rate of temperature
Tmax – maximum temperature
Hmax – layer heights with Tmax (hPa)
H – diabatic term
R – universal gas constant
Cp – heat capacity of air at constant pressure
RH – relative humidity
P – precipitation
Pw – precipitable water (mm)
SH – Specific humidity (gkg^{-1})
Mdiv – moisture flux divergence ($10^{-5}gkg^{-1}s^{-1}$)

Sp – simulated profile
Fp – forecasted or predicted profile.

Satellite

IR – infrared satellite imagery
WV – water vapor satellite imagery
VIS – visible imagery

INTRODUCTION

Geographical conditions of the Brazilian Northeast

The continent of South America consists of four regions (Figure 1) with different weather conditions and processes of formation (Cavalcanti, 2012). These regions are: 1) Amazonia, 2) Northeast Brazil or Brazilian Northeast (BNE), 3) Southeast Brazil and 4) La Plata Basin. This book is dedicated to region 2, the Brazilian Northeast (BNE).

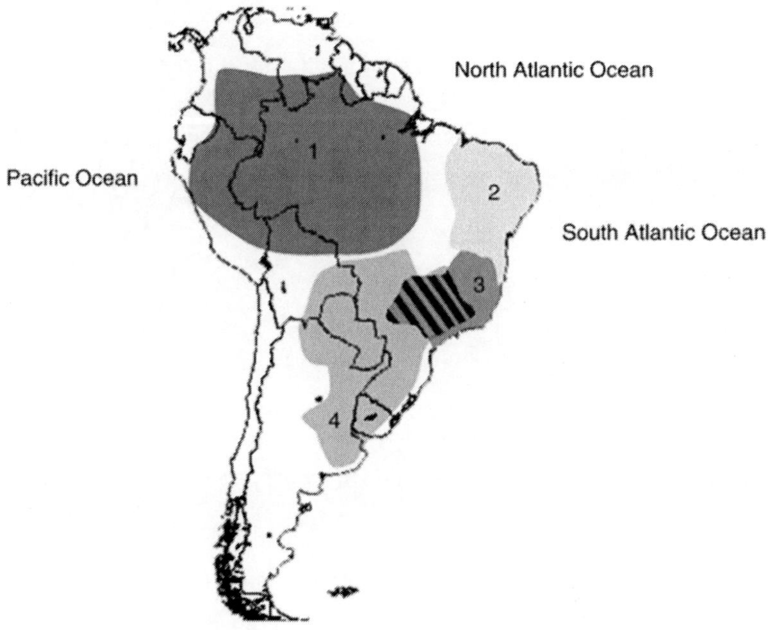

Figure 1- Regions of South America:
1 — Amazonia, 2 — Brazilian Northeast, 3 — Southeast Brazil, 4 — La Plata Basin (stripes indicate common areas of La Plata Basin and Southeast Brazil).
Source: Cavalcanti (2012)

Northeast South America is a very important agricultural region and one of the most important tourist regions in Brazil. Correct weather forecasting plays a key role in the economic success of this region. The location of this region can be seen on the map of Brazil (Figure 2). The BNE includes the following states (from north to south along the coast): Maranhao, Piaui, Ceara, Rio Grande do Norte, Paraiba, Pernambuco, Alagoas, Sergipe and Bahia.

a)

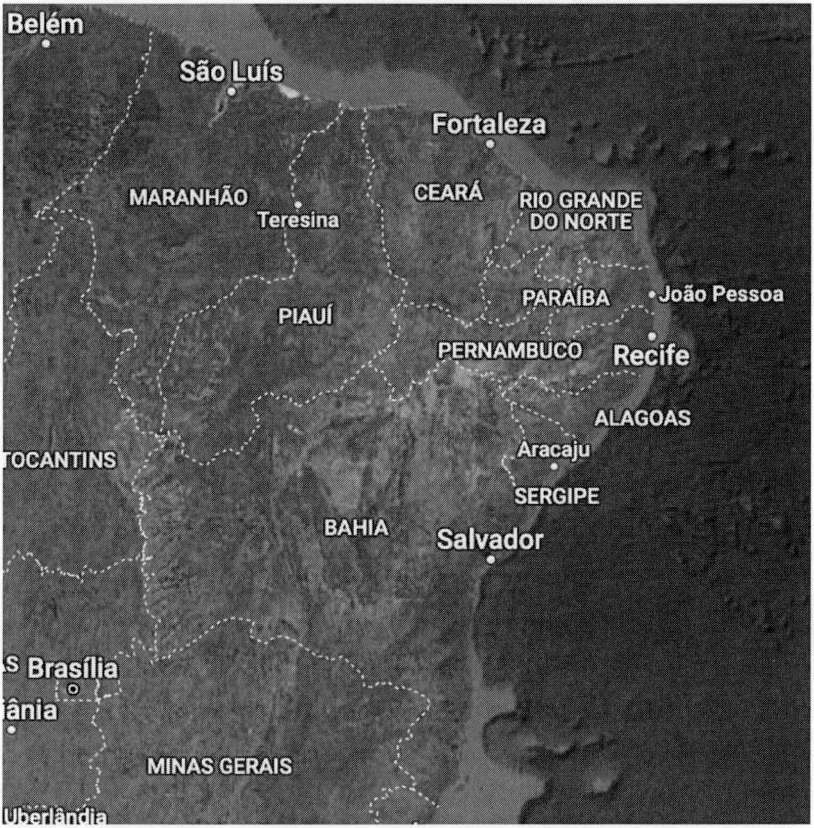

b)

Figure 2- Location of the BNE region (dark grey) on the map of Brazil (a) and the BNE States with capitals (b).
Source: wikipedia.org; Google, maps

Relief of the BNE

The BNE relief (Figure 3) is characterized by the presence of 1) coastal plains close to the northeastern beaches, 2) the Borborema Plateau, located in the interior of the region in the east, 3) the Parnaiba River Basin in the west, and 4) some uplands such as the plateau, in the south.

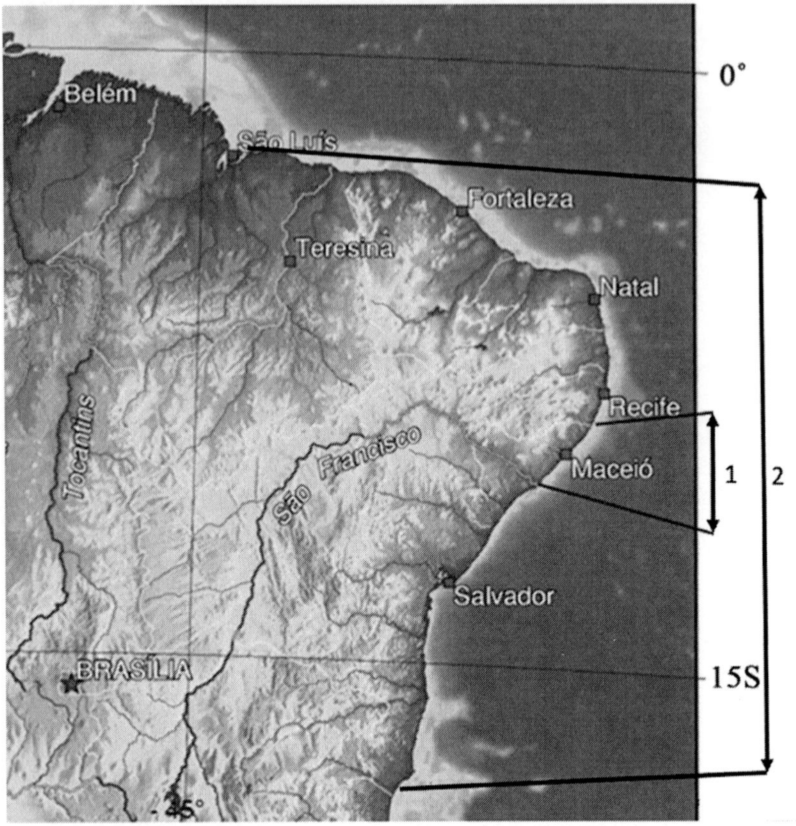

Figure 3- Relief in the BNE: 1- central region and 2 - all BNE
Source: IBGE (2006); Kulikova et al. (2014)

Geoclimatic regions and precipitation in the BNE

Four geoclimatic regions with meridional distribution are distinguished in the BNE: 1– *Meio-Norte*; 2 – *Sertao*, 3 –*Agreste,* 4 – *Zona da Mata* (Figure 4). Region 1, Meio-Norte, furthest from the Atlantic Ocean. Region 2, *Sertao*, is located in the highest regions. Two regions stretch along the coast of the Atlantic Ocean, one close to the coast, *4- Zona da Mata*, and the other *3 –Agreste* on the slope.

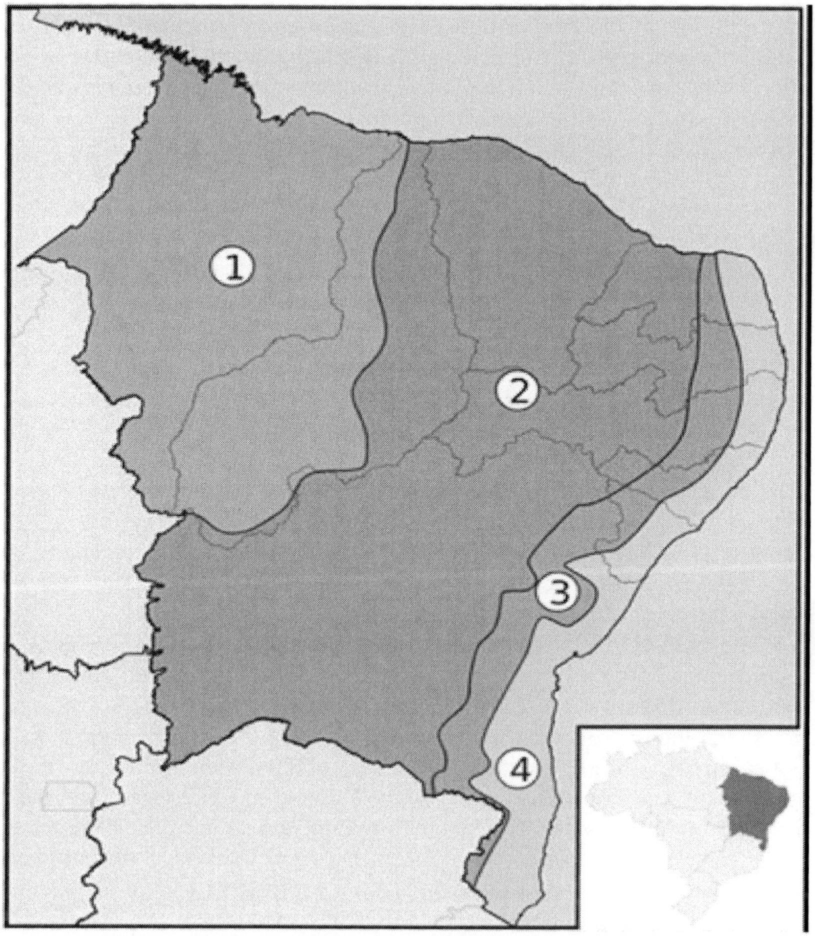

Figure 4 – Map of the geoclimatic subregions of the BNE:
1– Meio-Norte; 2 – Sertao, 3 –Agreste, 4 – Zona da Mata
Source: wikipedia.org; Matheus (2018); Fedorova et al. (2019)

Three precipitation patterns are described in the BNE (Molion, Bernardo, 2002): 1) The northern region has a humid season from February to May 2) The southern region has a humid season from November to February and 3) the coastal region has a rainy season from April to July. The rainy season in the central part of the coastal region usually occurs in May-July (precipitation $\geq$ 150 mm per month) or April-August (precipitation $\geq$ 100 mm per month) (Pontes Da Silva, 2008, 2011). Heavy precipitation in the central part of the coastal region usually fell in the autumn/winter period, and convective rains in January (Pontes Da Silva, 2011). The spatial distribution of precipitation in the central part showed the presence of two regions: semi-arid in the west (500 - 700 mm per year) and humid along the coast (2000 mm per year) (Molion & Bernardo, 2002, Pontes Da Silva, 2011).

The spatial and temporal characteristics of precipitation make their diagnostics and forecasting extremely difficult. The precipitation formation mechanisms and their evolution are associated with dynamic and thermodynamic processes interacting from large scales to cloud microphysics scale (Schwerdtfeger, 1976). This book will describe the dynamic and thermodynamic process in the BNE.

Objectives

The book analyzes the formation and forecast of adverse meteorological phenomena in the tropical region of the *Brazilian Northeast* (*BNE*). This book is a *Short-Range Weather Forecasting Guide* for the Tropical Region of Northeast Brazil. In addition, the main methods of forecasting can be used, after special additional verification in other tropical regions.

The meteorological systems of the tropical and extra-tropical regions of both hemispheres influence the phenomena formation in the BNE. The physical processes of these phenomena formations differ from the processes in other regions. Adverse phenomena (heavy and weak rain, thunderstorms, low visibility, fog, and mist) create many problems for the safety of the population and for transport (road and aviation). Therefore, the main purpose of this book is to combine information about the physical processes of adverse phenomena formation with practical methods for short-term forecasting. Synoptic and thermodynamic methods will be used together with the results of numerical models. This information is essential to improve weather forecasting and thus save lives.

The book contains information on both well-known and recently described synoptic-scale systems such as the Middle Tropospheric Cyclonic Vortexes, Brazilian Northeast Jet Stream, and Frontal Zones in the tropical region. The interactions between the hemispheres are also described, as well as the influence of tropical cyclones of the Northern Hemisphere on the BNE. In addition, the book presents practical methods for short-term weather forecasting in Northeast Brazil using thermodynamic analysis, satellite data and numerical models.

This book is intended primarily for undergraduate, graduate, and postgraduate students of meteorology. Also, this book will be useful to professionals in meteorology, such as professors, researchers, and operational meteorologists. In addition, even experts from another field can use it if they want to better understand the topic. This book will form part of the main text of courses in synoptic meteorology and operational meteorology at Brazilian Federal Universities.

Bibliography

Cavalcanti I. F.A. *"Large scale and synoptic features associated with extreme precipitation over South America: A review and case studies for the first decade of the 21st century."* Atmospheric Research 118, 27–40, (2012).

Fedorova Natalia, Pontes Da Silva Bruce F. P. Levit Vladimir *"Complexos Convectivos de Mesoescala sobre o Nordeste do Brasil.*" Maceio: Ef. Clube dos Autores, 248 p., 2019. ISBN: 978-65-900634-0-3.

Kulikova, N. Fedorova, V. Levit, E. S. Cordeiro *"Sea surface temperature anomaly and precipitation distribution in the Alagoas state of the Brazilian northeast"*, Natural Science 6, 1159-1178, (2014), DOI 10.4236/ns.2014.614104.

Molion, L.C.B. and Bernardo, S.O. *"Uma revisão da dinâmica das chuvas no Nordeste Brasileiro."* Revista Brasileira de Meteorologia 17, 1-10, (2002).

Matheus Nelson P. A.*"Desenvolvimento e deslocamento dos complexos convectivos de mesoescala sobre o oceano e continente no Nordeste do Brasil.* Tese. UFAL-ICAT, (2018).

Pontes Da Silva, B.F., Fedorova, N., Levit, V., Cavalcanti, J.V. and Rocha, A.C. *"El Niño-Oscilação Sul e sua relação com as precipitações observadas no período chuvoso de Alagoas entre 2003 e 2006".* In: XV Congresso Brasileiro de Meteorologia, San Paulo, (2008).

Pontes Da Silva B.F., Fedorova N., Levit V, Peresetsky A. *"Sistemas sinóticos associados às precipitações intensas no Estado de Alagoas."* Revista Brasileira de Meteorologia 26, 295-310, (2011).

Schwerdtfeger, W. *"Climates of Central and South America."* Elsevier Scientific Publishing Company, New York, 1976.

PART I

SYNOPTIC AND MESOSCALE SYSTEMS AFFECTING THE WEATHER FORMATION IN BRAZILIAN NORTHEAST

CHAPTER I.1

INTERTROPICAL CONVERGENCE ZONE

NATALIA FEDOROVA,
DAVIDSON LIMA DE MELO AND
VLADIMIR LEVIT

1. Identification

The Intertropical Convergence Zone (ITCZ) is the most important weather system in the tropical region that can be seen around the globe. The ITCZ is characterized by a band of low pressure, convergence of the trade winds (TW, Chapter I-2), and a cloudiness band on satellite imagery. TW convergence defined the classical method of ITCZ identification: TW confluence at low levels and difluence at high levels. This difluence/convergence creates air lifting and therefore the ITCZ controls precipitation formation (Schneider, 2014). In some books, the ITCZ is also referred to as the *Intertropical Front* (Djuric, 1994).

Intense instability creates convective clouds along the ITCZ. Therefore, the second method of ITCZ determining is based on the identification of the band of the convective cloud on satellite imagery (Xavier et al., 2000; Uvo, 1989).

The ITCZ is located over the oceans with positive anomalies of Sea Surface Temperature (SST) and negative anomalies of the surface level pressure (SLP) (Hastenrath, 1991). ITCZ oscillated on average between ~8 °N during summer of the Northern Hemisphere (July) and ~1 °N during summer of the Southern Hemisphere (January) (Citeau et al., 1988).

2. Annual and seasonal frequencies.

ITCZ around the globe

The annual variation of the ITCZ location is different around the globe (Figure I-1) (Lockwood, 1974; Hastenrath, 1984, 1991; Weninger, 2014).

The ITCZ usually migrates towards the warming hemisphere (Schneider et al., 2014). The average ITCZ location is north of the Equator because the Northern Hemisphere is warmer than the Southern Hemisphere. Over the Atlantic, the ITCZ is located north of the equator from July to December and near the equator from January to May (Xavier et al., 2000; Hastenrath and Lamb, 1977).

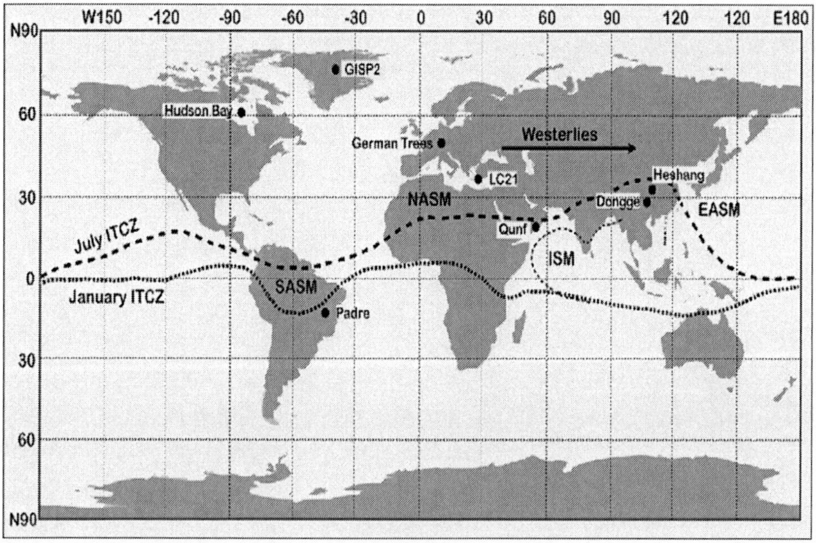

Figure I-1-1 – ITCZ location in July and January
Source: Weninger, 2014

The mechanisms of interannual climate variability, global circulation and, hence the ITCZ location are determined by annual cycles (Hastenrath, 1984). Its position coincides with areas of the ocean with positive SST anomalies, which indicates the relationship between the ocean and the atmosphere (Hastenrath, 1984; Asnasi, 1993).

ITCZ over the Atlantic

In the north of South America, the change in ITCZ location is small, from 2°N in July to 2°S in January (Weninger, 2014). More detailed information shows that ITCZ annual variation over the Atlantic is approximately 7° of latitude (Figure I-2a) (Carvalho & Oyama, 2013). Its displacement ranged from 1°N in April to 8°N in August. The width varied from 3° in March to 6° in October (Figure I-2b). The ITCZ reaches its southernmost position from February to April and tends to decrease in width and intensity during this period. The ITCZ position coincides with the region of a positive SST anomaly (Hastenrath, 1991; Asnani, 1993). On average, positive width anomalies are related with negative SST anomalies and positive convergence of surface winds.

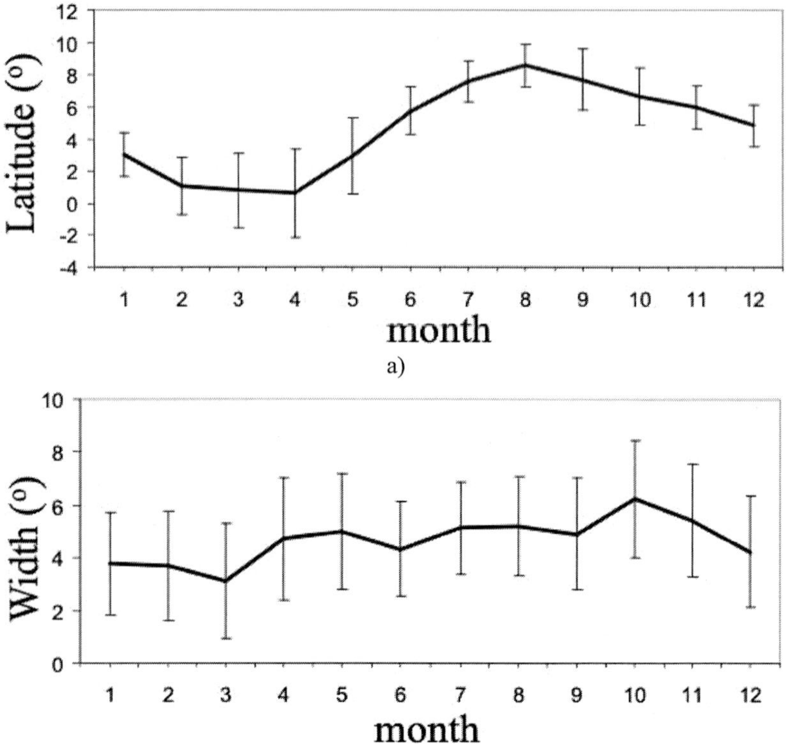

b)

Figure I-1-2 – ITCZ over the Atlantic: a) location per latitude and b) width. Source: Carvalho & Oyama, 2013

The sea surface temperature (SST) with a negative anomaly or cooling SST in North Atlantic (28-12°N) triggered strong meridional processes (Foltz et al., 2012). During one intense event, SST was 18° C colder than normal, and Equatorial Atlantic was 0.5° C warmer than normal. This very intense SST gradient created anomalous northern trade winds and an anomalous southward shift of the ITCZ. This cooling was initiated by stronger trade winds in January and February 2009. All these processes were associated with an anomalous intense Subtropical High (Chapter I-2). Unusually strong trade winds cooled the ocean due to wind-induced evaporation and deepened the mixed layer by 5-20 m. Near the equator, the wind responded to the anomalous SST gradient between the hemispheres and became northwesterly.

3. Vertical and horizontal structure

Idealized ITCZ

The idealized ITCZ model represents a warm core at the low and upper levels, while a cold core at the middle levels (Figure I-3, Estoque, 1975). Large moisture content is observed through all levels in the central core. Relatively low humidity was found near 700 hPa (Figure I-4). The wind was relatively calm at low levels and significantly disturbed at middle levels. Similar results were obtained for two sources of information: Atlantic Trade Wind Experiment (ATEX) and Line Island Experiment (LIE).

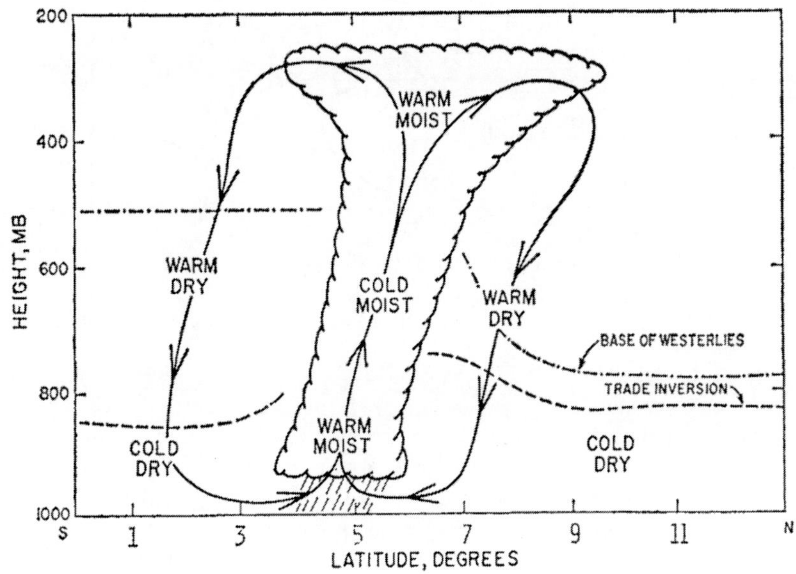

Figure I-1-3 – Schematic diagram of an idealized ITCZ model.
ITCZ: cloud-shaped outline
Source: Estoque, 1975

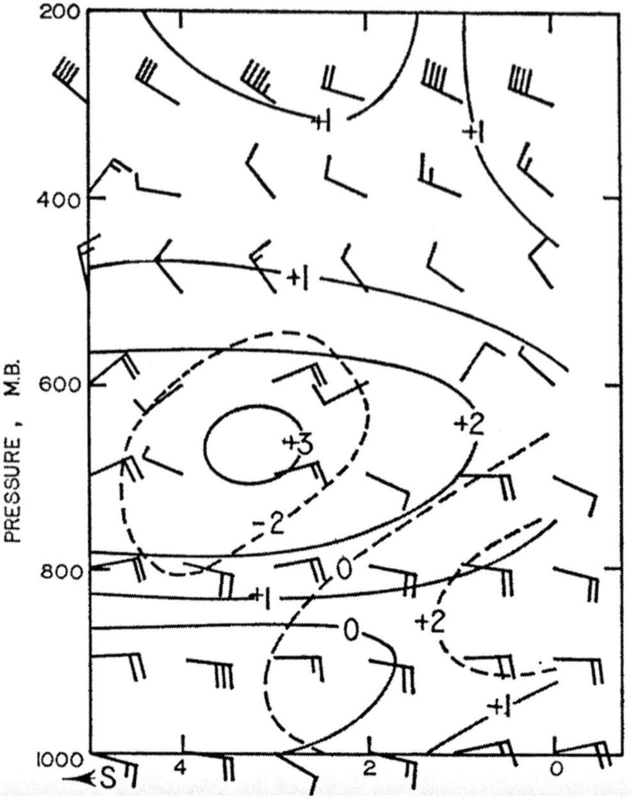

Figure I-1-4 – Vertical structure of the ITCZ along a meridional cross-section based on the Line Island Experiment (LIE) observations.
Temperature (ºC) - full lines, maxing ratio (g/kg) - dashed lines, wind - convectional form (full barb=10 knots)
Source: Estoque, 1975

The absence of any significant temperature maximum at any ITCZ levels was confirmed by various studies by Estoque (1975) and Estoque and Douglas (1978). This conclusion shows that condensational heating is a secondary mechanism for the development and maintenance of the ITCZ. Conversely, the same authors found that the ITCZ of the Central Atlantic presents a warm core at high levels (Estoque & Douglas, 1978). The main mechanisms of ITCZ formation are barotropic instability or "pressure energy flux" from extratropical regions (Estoque & Douglas, 1978).

Double Band

Double bands in the Atlantic were more frequent in March and April (Teodoro et al., 2019). The climatological position of the single ITCZ bands is along 2ºN, and the double bands are along 2ºS. The two bands are 4º latitude apart from each other. Low-level mass convergence and outgoing longwave radiation (OLR) represent the double bands better than the other variables.

When a cold-water current enters the warm ocean around the ITCZ, dividing it, two ITCZ bands or Double Band are formed (Lietzke et al., 2001). The cloudiness band bifurcates, forming two bands visible on the satellite image. The longer its duration, the more precipitation is recorded in the affected region (Uvo, 1989).

The *Eastern Pacific* ITCZ pattern shows a synoptic-scale precipitable water (PW) dual structure, in which "*the southern and northern bands straddled at the ITCZ*" (Chen et al., 2014). This structure produces meridional dipoles, which propagate zonally. These dipoles are observed in summer and autumn. The vertical structure of the meridional wind is baroclinic, and the vertical motion has two peaks at 850 and 300 hPa. Shallow convection within deep convection forms east of the dipoles. Deep convection tends to be suppressed by the intrusion to the west of the dipole of dry cold air at middle levels. The dual band can be related to wind divergence and humidity advection. Strong meridional dipoles are closely associated with ITCZ distortion and breakdown. For the Atlantic region, such information has not yet been received.

4. ITCZ formation

Thermic theory

The scheme of the global circulation and, as a result, the ITCZ formation, was originally based on climatological maps. Thermal convection within the equatorial depression was defined as the main force of the Global Circulation (many books are cited in Pedelaborde, 1958). The example shows a map of the radiation balance per year; almost identical values are visible on this map in a wide band between 20ºN and 20ºS, approximately (Figure I-5). Many other maps show similar results, such as global radiation (Lockwood, 1974), surface temperature (Fedorova, 2001, 2008). All these maps show a contradiction with thermic theory, since the highest values are observed not only along the equator, but also in a wide band around the globe. Therefore, according to this theory, thermal convection should be observed in this large band, but this is not confirmed by observational data.

On the other hand, according to Pedelaborde (1958, with reference to Rossby), there is no thermal lifting in the equatorial region. This is due to strong cloud cover in the equatorial region, which reduces surface heating. Compared to anticyclones in the subtropics, where the highest temperature and subsidence of air masses are observed in cloudless areas (Chapter I-2, SASH). As a result, the lifting in the equatorial region is the result of dynamic processes due to trade winds confluence. Additional evidence of the predominance of the dynamic process is the frequent absence of the antitrade wind at the upper levels. But in the thermal theory, the presence of the antitrade wind at the upper levels is necessary for the SASH formation (Pedelaborde, 1958).

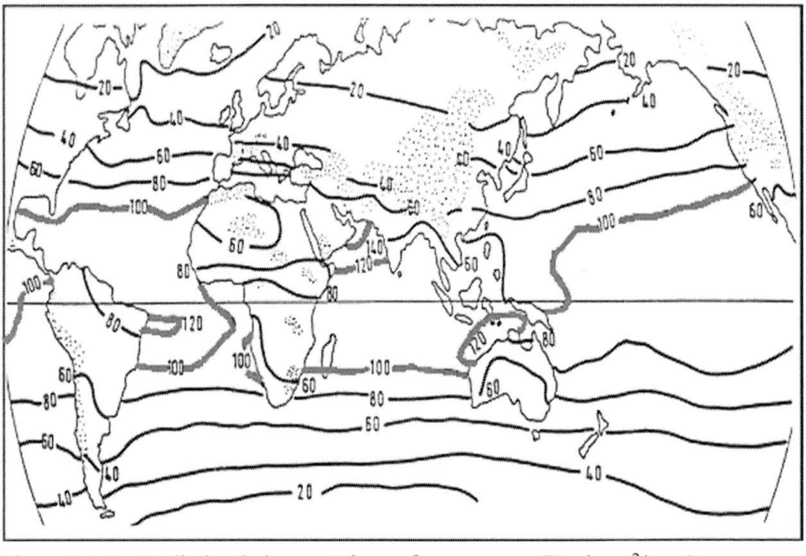

Figure I-1-5 – Radiation balance at the surface per year (Kcal cm^{-2}/year)
Source: Lockwood, 1974

Many authors (for example, Djuric, 1994) refer to the ITCZ as the *Intertropical Front* (Figure I-6). It is located between the trade winds of both hemispheres, usually within 10° of latitude, but may cross the equator. Often the ITCZ is difficult to distinguish, and it expands into a large transition zone between the northern and southern trade winds.

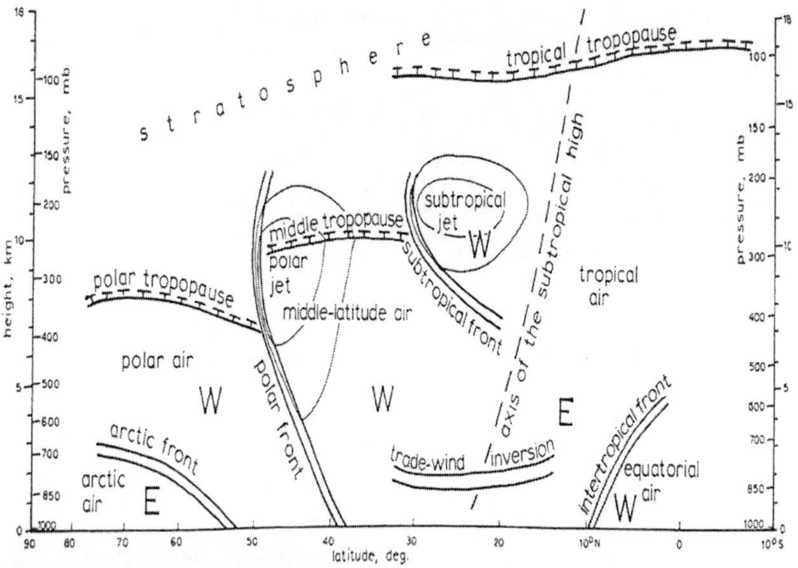

Figure I-1-6 – Main structural parts of the atmosphere in a vertical meridional section.
E and W- prevalent wind directions.
Source: Djuric, 1994

Dynamic theory

Synoptic-scale processes in the polar regions, such as the Rossby wave, cause Global Circulation on the planet and, in particular, in tropical regions. As a result of the displacement of the Rossby wave, an ITCZ is formed (Figure I-7) (Pedelaborde, 1958). That is, the primary source of energy circulation on the planet is in the polar regions (Rossby, 1949).

Rossby waves are synoptic-scale waves that are generated by the conservation of potential vorticity (Rossby, 1939). The Rossby wave displacement can be observed using satellite images (Fedorova 2001, 2008; Yamashita, 2012). Due to the intrusion of cold air masses from the south or cold advection, together with cold sea surface temperature (SST), the propagation of the Rossby wave is created (Philander & Pacanowski, 1981; Hastenrath & Lamb, 1977; Fedorova 2001, 2008).

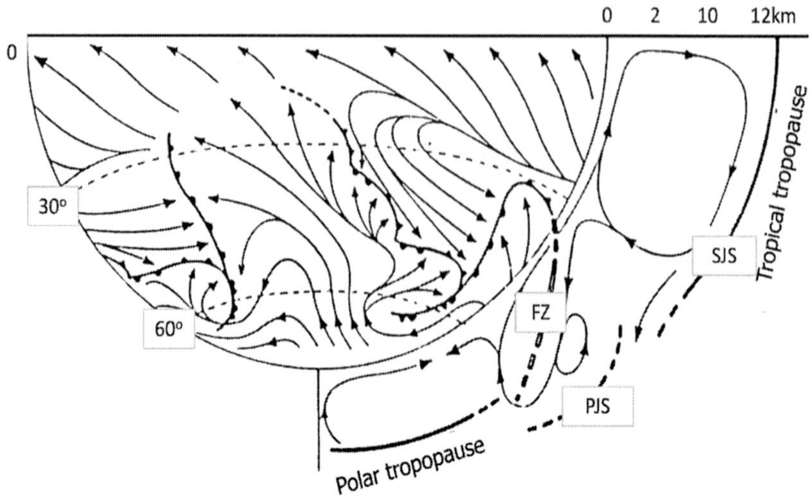

Figure I-1-7 – Main circulation systems in the Southern Hemisphere.
The Rossby wave is represented by the series of baroclinic cyclones and anticyclones.
FZ - Frontal Zone
PJS - Polar Jet Stream
SJS -Subtropical Jet Stream
Source: Palmen, Newton, 1969 (adapted for the Southern Hemisphere)

Examples of the meridional entry of the Rossby wave will be presented in Chapters I-2, I-7 and I-9. The four examples below show different types of Rossby waves. The propagation of these waves from the northwest to the southeast is seen in Figures I-8*a* and I-8*c*, and more meridional in figure I-8*b* and I-8*d*.

In all figures Rossby wave cloudiness is combined with ITCZ; but this union took place in the western region of South America (Figures I-8*a,* I-8*b*) or in the eastern region, in the BNE (Figure I-8*c* and I-8*d*). These examples also show the different types of association of the ITCZ with the Rossby wave and their influence on the BNE weather: 1) only ITCZ have a direct impact on the BNE weather (Figure I-8*c*), 2) only the Rossby wave affect the BNE (Figures I-8 *a,* I-8*b*), and 3) rarely both systems jointly form the BNE weather (ITCZ united with Rossby wave) (Figure I-8*c* and I-8*d*).

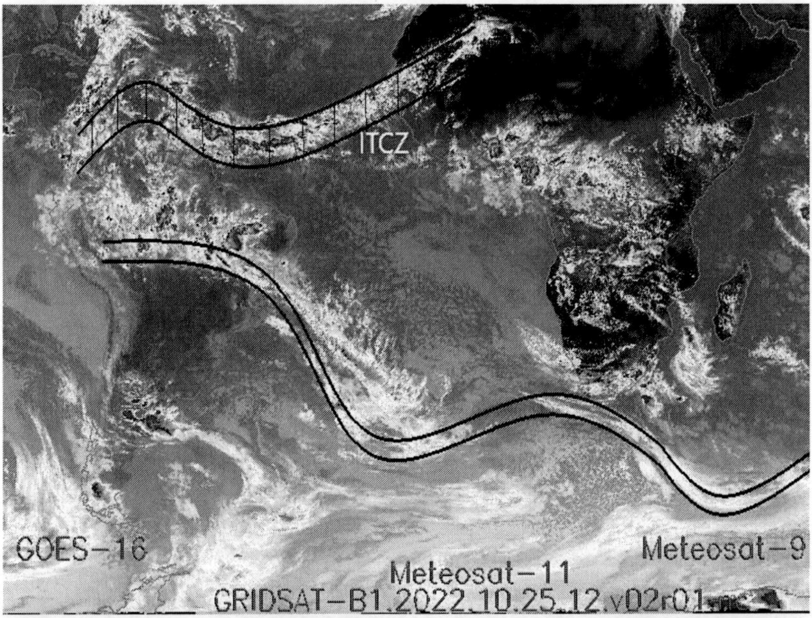
ITCZ
GOES-16
Meteosat-9
Meteosat-11
GRIDSAT-B1.2022.10.25.12.v02r01

a)

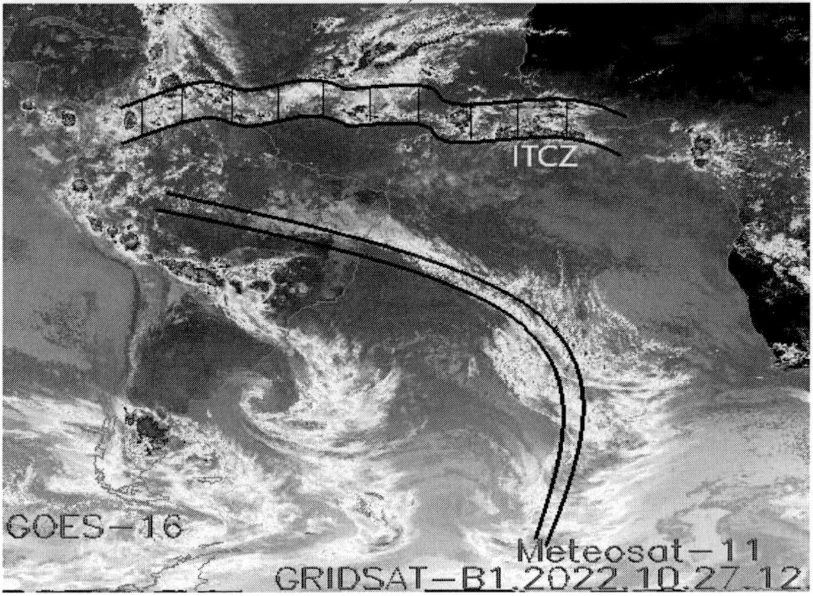
ITCZ
GOES-16
Meteosat-11
GRIDSAT-B1.2022.10.27.12

b)

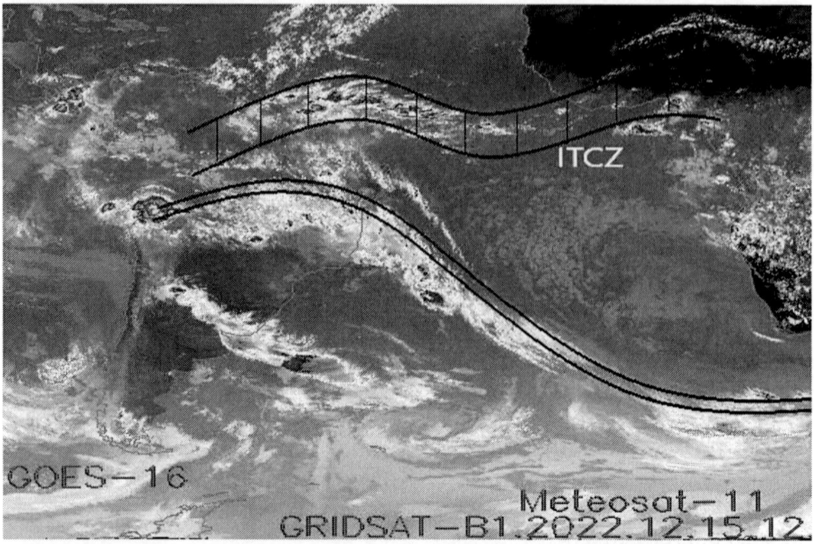

c)

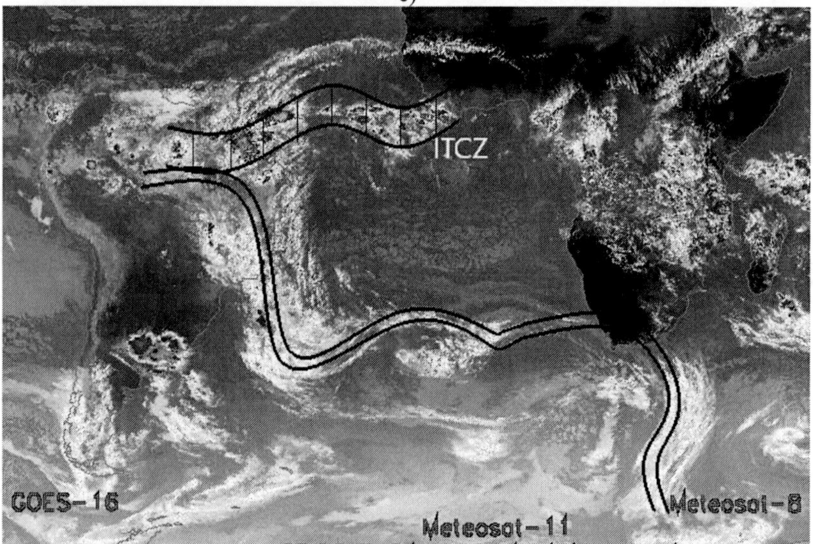

d)

Figure I-1-8 – Association of ITCZ with Rossby wave:
a) 25 October 2022, b) 27 October 2022, c) 15 December 2022, d) 01 January 2023
Double line - Rossby wave; double line with crossbars - ITCZ
Source: CRIDSAT, Meteosat

5. Precipitation and meteorological phenomena

The most intense precipitation on Earth is observed in the ITCZ (Schneider et al., 2014). Precipitation in the north of the BNE are observed in March and April (Hastenrath & Heller, 1977). Precipitation during these months creates the following general circulation factors: 1) streamline convergence or the ITCZ location reaches its southernmost position, 2) the equatorial region of the South Atlantic is warmer than the North Atlantic (Moura & Shukla, 1981); this leads to a decrease in the sinking and an increase in the lifting above the BNE, 3) relatively warm equatorial waters contribute to increased humidity and instability at low levels.

Precipitation intensity in ITCZ ranged from 10 mm/day in March to 13 mm/day in July (Figure I-9) (Carvalho & Oyama, 2013). Precipitation in ITCZ bands of small width (43% of all events) was zoned between 2-4°N and varied from 5 to 12.5 mm/day with a maximum at 1°N (Figure I-10). Precipitation in large ITCZ bands (57% of all events) shows the distribution of SW-NE from the north of the BNE to the African continent. Precipitation of 5-20 mm/day was observed between 4°S-4°N.

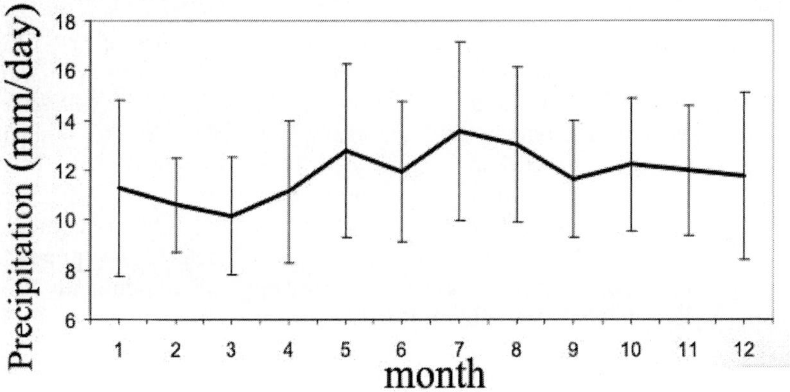

Figure I-1-9 – Precipitation (mm/day) associated with ITCZ in the Atlantic
Source: Carvalho & Oyama, 2013

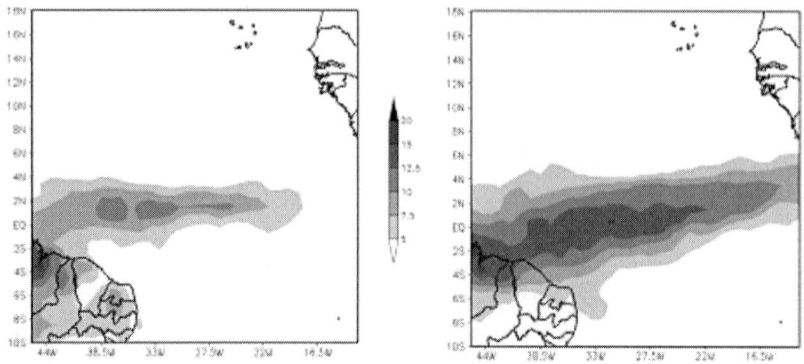

a) b)

Figure I-1-10 – Precipitation (mm/day) in ITCZ events in the Atlantic with small (a) and large (b) widths of band in March and April for 10 years (1999-2008).
Source: Carvalho & Oyama, 2013

In the equatorial Atlantic, cloudiness pattern, water vapor, and precipitation exhibit a vertical circulation where lifting occurs 100 km south of line of wind confluence at surface level and sinking 300 km north of this line (Hastenrath, 1984).

An intense negative SST anomaly in the North Atlantic (28-12°N) creates a very intense temperature gradient, anomalous northerly winds, and an anomalous ITCZ shift to the south (Foltz et al., 2012). As a result, strong flooding was recorded in the BNE. For example, in 2009 a strong event with a negative Atlantic meridional mode (AMM) was observed (Foltz et al., 2012). It was initiated by anomalous cooling in the tropical part of the North Atlantic (12-25°N). Cold SST anomalies in January and February coincided with a moderate La Niña in the equatorial Pacific Ocean. These processes create an anomalously strong North Atlantic High and consequently more intense trade winds (TW) (Figure I-11*a*).

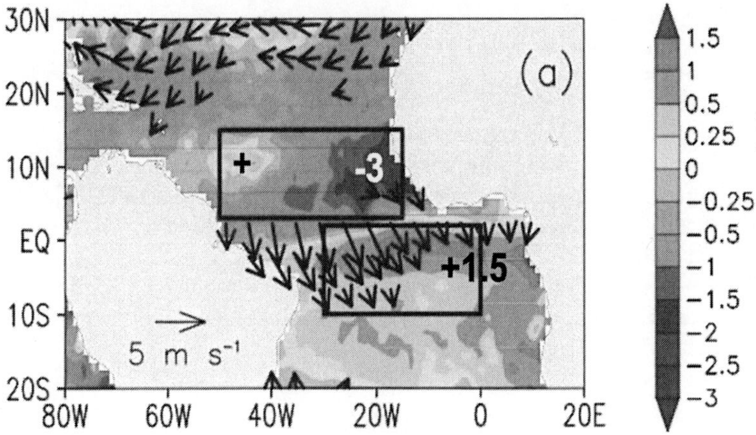

a)

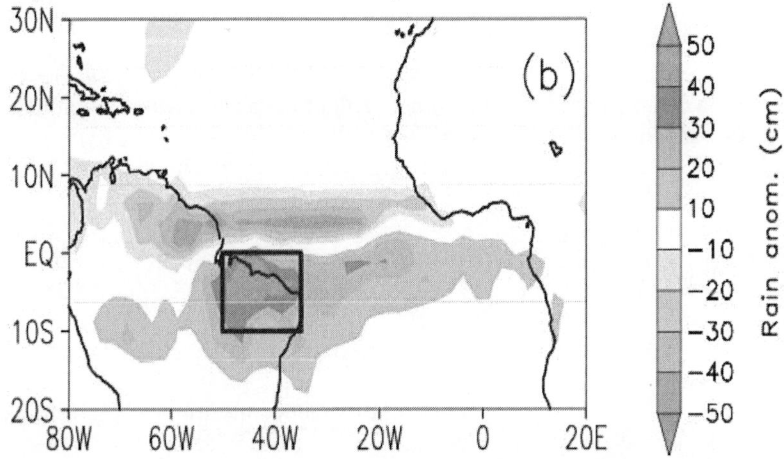

b)

Figure I-1-11 – The conditions of the ocean and atmosphere for 2009 strong flooding in the BNE.

Interannual anomalies of TMI-AMSR-E* SST (shaded) and QuikSCAT** wind speed (vectors, for wind speed anomaly >1m/s) averaged during April-May 2009. GPCP rainfall anomaly (shaded) with respect to the 2003-08 monthly mean seasonal cycle.

**TMI-AMSR-E:*

TMI: Microwave Imager; AMSR-E: Advanced Microwave Scanning Radiometer for Earth Observing System.

***Surface wind velocity from the SeaWinds instrument on the Quick Scatterometer (QuikSCAT) satellite is available from Institut Franxcais de Recherche pour l'exploitation de la Mer (IFREMER)/CentreERSd'ArchivageetdeTraitement*

(CERSAT) on a 0.58 3 0.58 daily grid from July 1999 to November 2009 (ftp://ftp.ifremer.fr/ifremer/cersat/ products/gridded/mwf-quikscat).
Source: Foltz et al., 2012

The coldest SST anomalies shifted southward in February and March 2009. The anomalous meridional SST gradient in the spring of 2009 was the strongest since the start of satellite SST measurements in 1982. The SST gradient and associated surface wind anomalies shifted the ITCZ to the south and contributed to severe flooding in the BNE (Figure I-11*b*).

Precipitation and extreme events are formed due to the connection of the ITCZ with the frontal zone (Muanza, 2018). This process will be described in Chapter I-7.

6. References

Asnani G. C. *Tropical Meteorology.* Indian Institute of Tropical Meteorology, 1202p, ISBN: 8190040006, 1993.

Carvalho M.A.V., Oyama M.D., "Variabilidade da largura e intensidade da zona de convergência intertropical Atlântica: aspectos observacionais." *Revista Brasileira de Meteorologia*, 28, n.3, 305 – 316. (2013).

Citeua, J.; Bergés, J. C.; Demarcq, H.; Mahé, G., "Position de la Zone de Convergence a 28°N et température de surface de l'océan." *Veille Climatique Satellitaire*, 21: 3-7. (1988).

Chen G., Takayabu Y., N., Yokoyama C. "Synoptic-Scale Dual Structure of Precipitable Water along the Eastern Pacific ITCZ." *Journal of Climate*, 27, 6288-6304. (2014).

Djuric Dusan, *Weather Analysis*. New Jersey: Prentice Hall, 304p, ISBN: 0-13-501149-3, (1994).

Estoque, M. A. "Structure of the mid-oceanic intertropical convergence zone." *J. Meteorol. Soc. Japan* 53, 317-322. (1975).

Estoque, M. A; Douglas. M. "Structure of the Intertropical Convergence Zone Over the GATE area." *Tellus* 30 (1): 55-61. (1978).

Fedorova Natalia. *Meteorologia Sinótica II.* Pelotas: Ed. Universitária/UFPel, 242p, ISBN: 85-7192-154-7, 2001.

Fedorova Natalia. *Sinótica IV: Sistemas e processos sinóticos atuantes na América do Sul.* Material didático: sinopses, figuras, equações Maceió: Ed. Universitária/ UFAL, 180 p, ISBN: 978-85-7177-422-3, 2008.

Foltz G.R., McPhaden M.J., Lumpkin R. "A Strong Atlantic Meridional Mode Event in 2009: The Role of Mixed Layer Dynamics." *Journal of Climate* 25, 363-380. (2012).

Hastenrath Stefan, Lamb P. "Climatic Atlas of the Tropical Atlantic and Eastern Pacific Oceans." *University of Wisconsin Press.* 97p. (1977).

Hastenrath S., Heller L. "Dynamics of climatic hazards in northeast Brazil." *Quarterly Journal of the Royal Meteorological Society* 103, n. 435, p. 77-92, (1977).

Hastenrath S. "Interannual variability and annual cycle: Mechanisms of circulation and climate in the tropical Atlantic sector." *Monthly Weather Review* 112, n. 6,1097-1107. (1984).

Hastenrath Stefan. *Climate Dynamics of Tropics*. Dordrecht, Kluwer, 488p. (1991).

Lockwood John. G. *World climatology. An environmental approach.* New York: St. Martin's Press. 330p. (1974).

Lietzke, C.E.; Deser, C., Haar, T.H.V. "Evolutionary Structure of the Eastern Pacific Double ITCZ Based on Satellite Moisture Profile Retrievals." *Journal of Climate* 14: 743–751. (2001).

Moura A. D., Shukla J. "On the dynamics of droughts in northeast Brazil: Observations, theory and numerical experiments with a general circulation model." *Journal of the atmospheric sciences*, 38, n. 12, p. 2653-2675, (1981).

Muanza, G. A. "*Padrões de circulação atmosférico associados à ligação da extremidade frontal com a ZCIT*". MSc. Thesis. Federal University of Alagoas, Maceió, Brazil. 73p. 2018.

Palmén, E., Newton C.W. *Atmospheric circulation systems: their structure and physical interpretation.* Academic press, New York. 603p. 1969.

Pedelaborde P., *Les Moussons*. Paris, Colin, "Coll. Armand Colin", n 325, 208p. 1958.

Philander S.G.H., Pacanowski R.C. "The oceanic response to cross-equatorial winds (with application to upwelling in low latitudes)". *Tellus* 33: 201-210. (1981).

Rossby, C. G. "Relation between variations in the intensity of the zonal circulation of the atmosphere and the displacements of the semi-permanent centers of action." *J. mar. Res.* 2, 38-55. (1939).

Rossby C.G. *On the Dispersion of Planetary Waves in a Barotropic Atmosphere*. 1949. https://doi.org/10.1111/j.2153-3490.1949.tb01928.x

Schneider, T.; Bischoff, T.; Haug, G. H. "Migrations and dynamics of the intertropical convergence zone." *Nature* 513, n. 7516, 45-53. (2014).

Teodoro T.A., Reboita M.S., Escobar G.C.J., "Characterization of the Double Band of the Intertropical Convergence Zone (ITCZ) over the Atlantic Ocean." *Anuário do Instituto de Geociências – UFRJ*, 42-2, 282-298. (2019).

Uvo, C. R. B., Nobre, C. A. A. "Zona de Convergência Intertropical (ZCIT) e a precipitação no norte do Nordeste do Brasil. Parte I: A posição da ZCIT no Atlântico Equatorial." *Climanálise* 4, n. 07, 34-40. (1989).

Weninger, B., Clare, L., Gerritsen, F., Horejs, B., Krauß, R., Linstädter, J., Özbal, R., & Rohling, E. J. " Neolithisation of the Aegean and Southeast Europe during the 6600–6000 calBC period of Rapid Climate Change." *Documenta Praehistorica* 41, 1–31. https://doi.org/10.4312/dp.41.1 (2014).

Xavier, T. M. B. S.; Xavier, A. F. S; Dias, P. L. S.; Dias, M. A. S. "A Zona de Convergência Intertropical e Suas Relações Com a Chuva do Ceará (1964-98)." *Revista Brasileira de Meteorologia* 15, 27-43. (2000).

Yamashita, M. K. "Análise da influência vertical de ondas de Rossby longas no Atlântico Sul." Dissertação de Mestrado, Instituto Oceanográfico, Universidade de São Paulo, São Paulo. doi:10.11606/D.21.2012.tde-11122012-160207. (2012).

CHAPTER I.2

SOUTH ATLANTIC SUBTROPICAL HIGH

NATALIA FEDOROVA
BRUCE FRANCISCO PONTES DA SILVA
VLADIMIR LEVIT
MARCIO HENRIQUE DOS SANTOS SILVEIRA

1. Identification

Subtropical Highs are defined as regions of anticyclonic flow in the lower troposphere (Figure I-2-1). They are associated with downward movements of the global circulation pattern, the incursion of anticyclonic vorticity from extratropical anticyclones, and the exchange of heat and humidity with the surface of the oceans (Hastenrath, 1985). These Highs are barotropic systems with a warm core (Vasquez, 1994; Fedorova 2008 a,b) and cause the northward air movement, i.e. trade winds (see Chapter I-3)

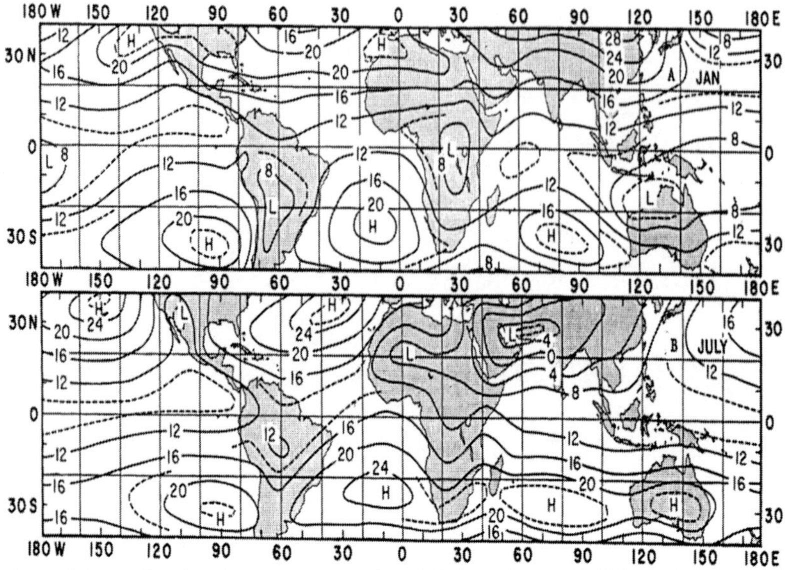

Figure I-2-1 – Sea level pressure over the global tropics at 1000 hPa. *January (top), July (bottom).*
Source: Hastenrath, 1985

Annual variations in the position of the centers of the Subtropical High are observed between 30-40°N (Figure I-2-2). In the Southern Hemisphere, this variation in the Pacific and Indian Oceans is almost the same, approximately between 25-35°S. The smallest displacement of the Subtropical High centers is found in the South Atlantic, between 27-33° S.

Subtropical High over the South Atlantic Ocean, namely the South Atlantic Subtropical High (SASH), affects the weather throughout Brazil.

2. Annual and seasonal frequencies

Subtropical Highs are closest to the equator in the austral winter (27°S in August) and move poleward in the austral summer (33°S in February) (Figures I-2-2, I-2-3). It is located further East in austral summer (0°W in February) and shifted further west in austral winter (14°W in August).

Observation of the SASH annual cycle is essential for understanding the regional climate in the Northeast of Brazil, because changes on its intensity or position will change the wind pattern in this region.

The SASH annual cycle has two distinct peaks in both intensity and size. It is strongest during the cold season, peaking in July (Hastenrath & Lamb, 1977).

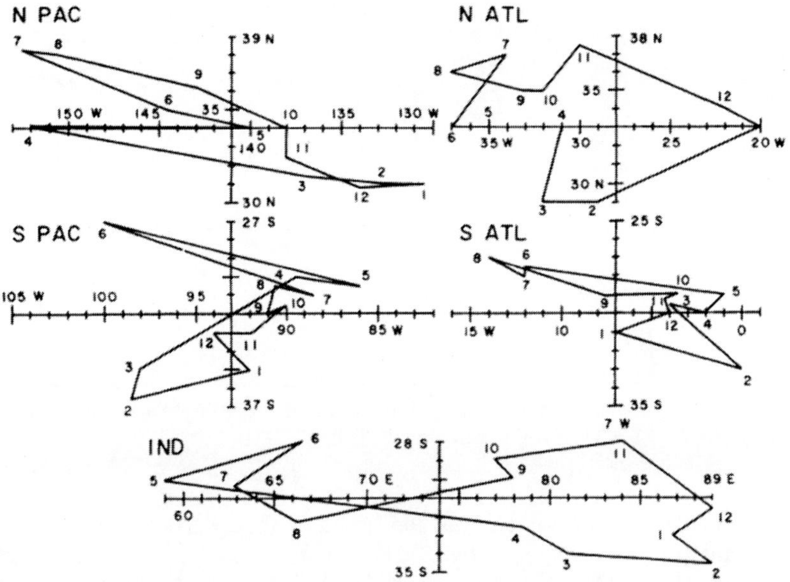

Figure I-2-2 - Annual variation of latitude and longitude of the position of Subtropical High centers.
Numbers from 1 to 12 represent calendar months.
N and S means southern and northern, respectively. PAC is Pacific Ocean, ATL means Atlantic Ocean and IND is Indian Ocean.
Source: U.S. Navy, 1974, 1976, 1977, 1978, 1979

It is very important to note that in the BNE near the coast at a latitude of 12°S according to climatological data, in winter can be seen a trough above the ocean and a ridge over the mainland (Reboita et al., 2019). Since this trough/ridge is visible from the climatological information, this means that these systems (or wave disturbances in the trade winds – WDTW) can have a high formation frequency. This will be discussed in Chapter I-3.

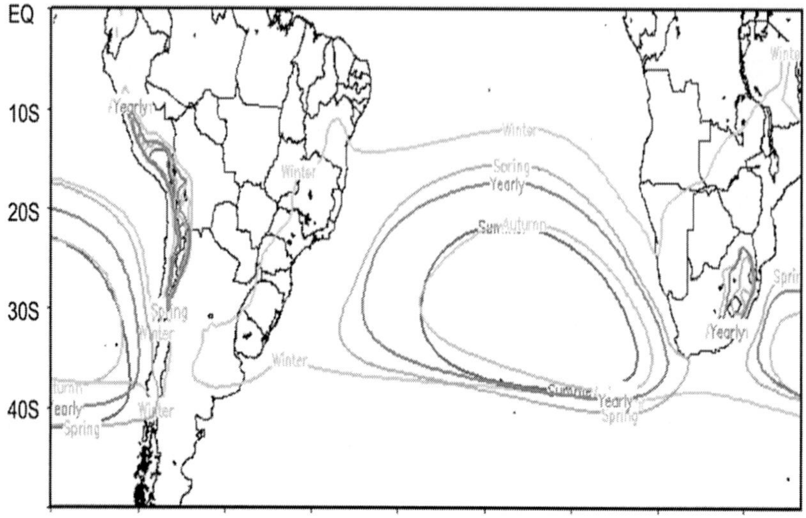

Figure I-2-3 – Seasonal and annual climatology of the 1018 hPa isobar based on climatic (1979-2005) ERA-Interim reanalysis data.
Source: Reboita et al., 2019

This SASH annual cycle was confirmed by the ERA-Interim and JRA-55 reanalysis products (Sun et al., 2017) – Figure I-2-4. Values were scaled by removing the annual mean of the area-averaged geopotential height (1517.16 gpm) over the South Atlantic region bounded by 50°S-07.5°N and 67.5°W-37.5°E.

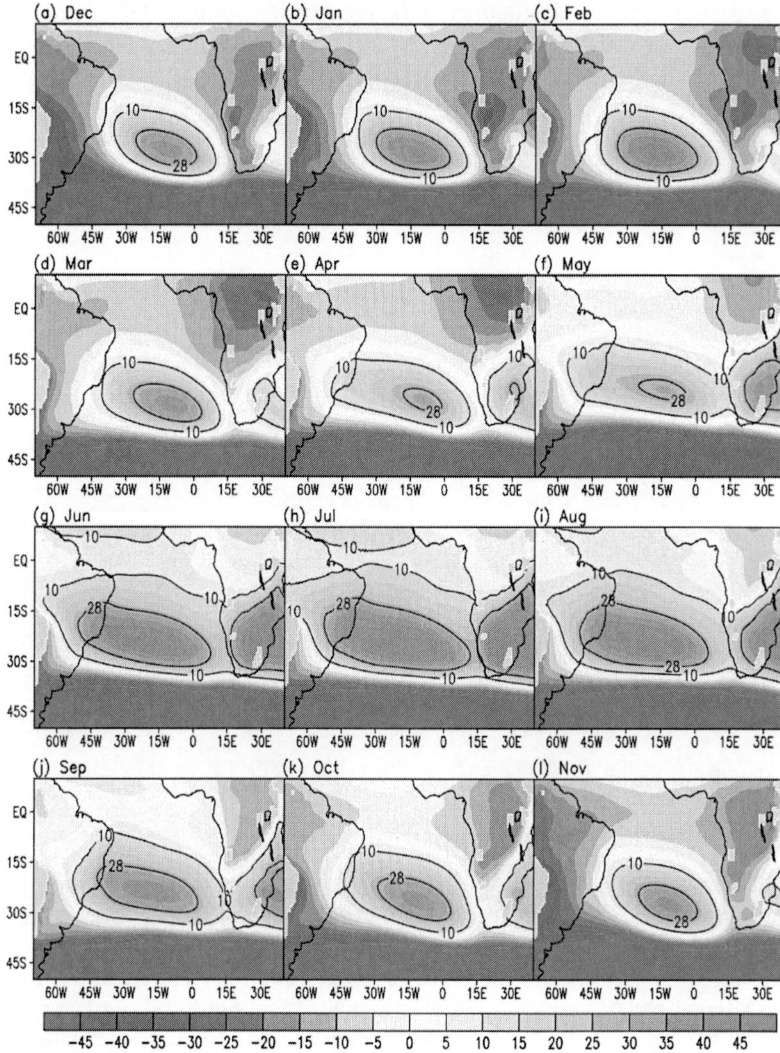

Figure I-2-4 - ERAI 1979-2015 climatological monthly 850-hPa scaled geopotential height (gpm).
The 10- and 28-gpm values are marked by black contours.
Source: Sun et al., 2017

SASHs are strongest and largest during the solstice months in austral winter, when its center is either closest to the equator (Figure I-2-5) or on the western side of the South Atlantic (Bryson & Layey, 1958; Sun et al., 2017; Reboita et al., 2019). SASH is located furthest poleward in the center of the South Atlantic in late austral summer.

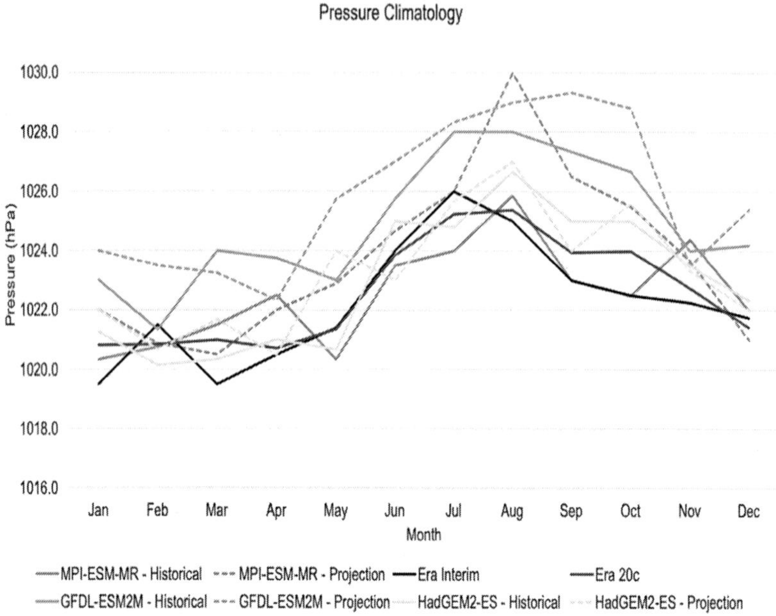

Figure I-2-5 – Monthly pressure climatology (hPa) of the SASH center according to different models.
Source: Reboita et al., 2019

Climatological analysis of SASH average pressure, location, and intensity provide more detailed information (Degola, 2013). A decrease in central pressure values is observed from November to April. In the summer months, the stability of its spatial coverage, due to continental warming, stands out. Such an annual cycle is explained by the SASH regeneration process, which often occurs due the intrusion into the SASH circulation of baroclinic anticyclones (migrating Highs) of the rear side of the Rossby wave. This process will be described in detail in section 5 of this Chapter.

As a result of the change in the position of the SASH during the year, prevailing winds are formed in the BNE: from the east/northeast in summer and from the southeast in winter (Servain & Lukas, 1990). The wind thus blows perpendicular to the coast, favoring the nighttime convergence zone associated with the land breeze.

3. Vertical structure

SASH is a warm-core barotropic High (Vasquez, 1994; Fedorova 2008 a,b). High contains warm air at the center through the entire troposphere. The anticyclonic circulation is strong near the surface and increases with height. Convergence causes an increase in mass and an air sinking.

Climatological data show the SASH location in January between 20-30°S at sea level and at 500 hPa to the north, about 20°S (Figures I-2-6 and I-2-7, Taljaard et al., 1969 quoted in Loon 1972). In July, the *steering current* (*main current*) is more intense (Fedorova, 2001), and its northern side reaches 30°S, while High cannot be identified on climatological charts (Figures I-2-8).

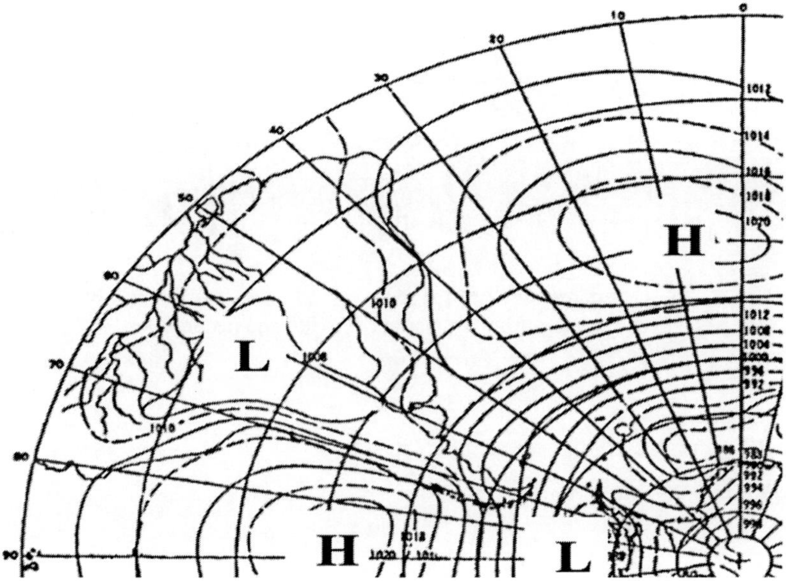

a)

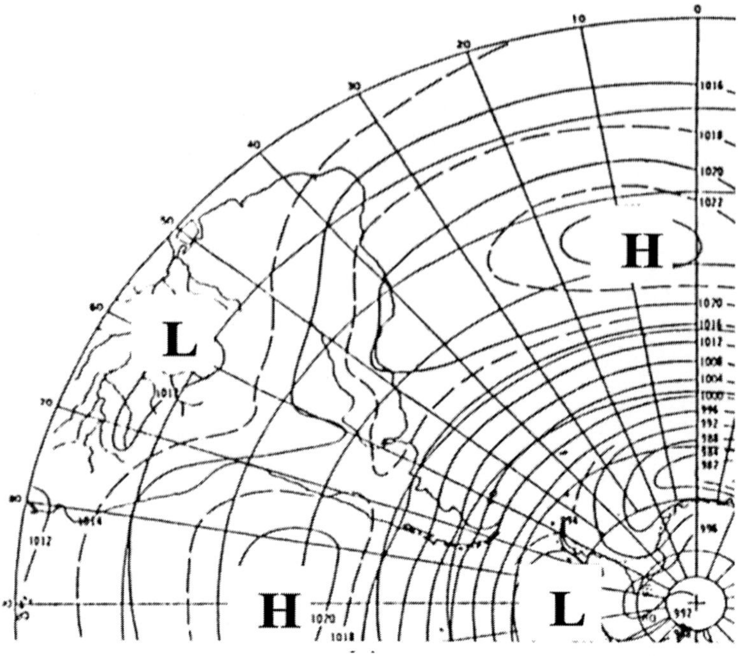

b)

Figure I-2-6 – Mean pressure (mb) at sea level in January (a) and in July (b)
Source: adapted fragment from Taljaard et al., 1969, quoted in Loon, 1972

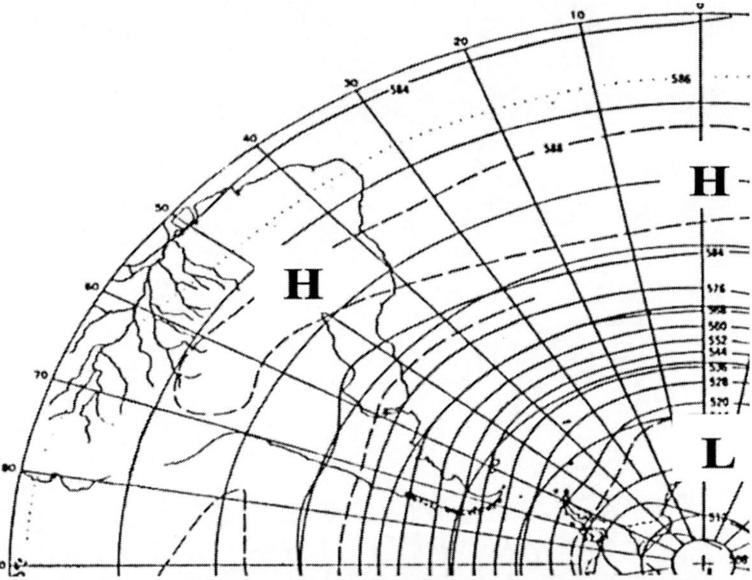

a)

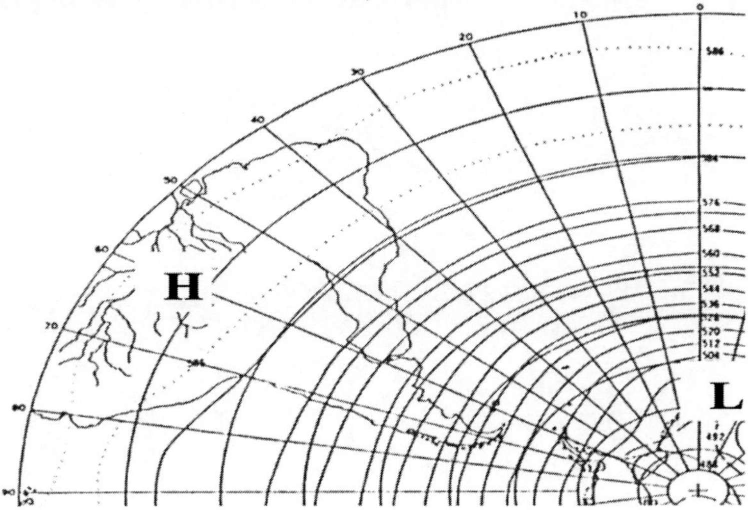

b)

Figure I-2-7 – Mean height (dkm) at 500 hPa in January (a) and in July (b)
Source: adapted fragment from Taljaard et al., 1969, quoted in Loon, 1972

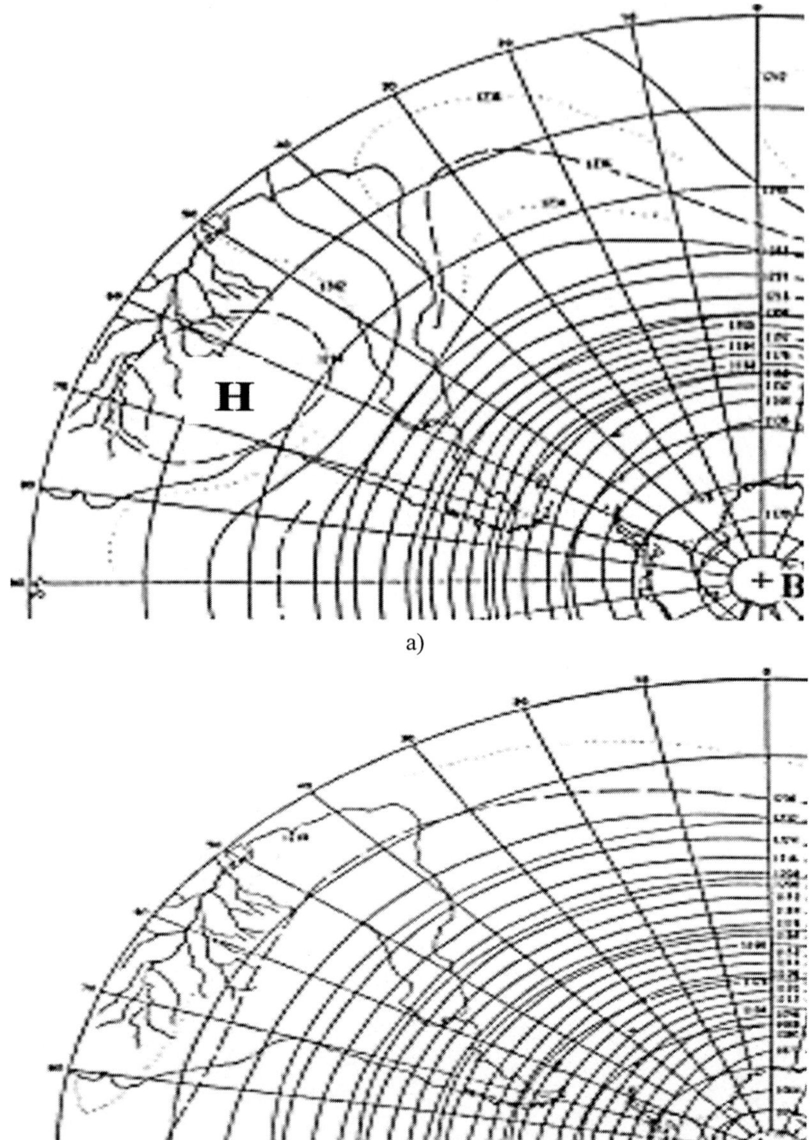

Figure I-2-8 – Mean height (dkm) at 200 hPa in January (a) and in July (b)
Source: adapted fragment from Taljaard et al., 1969, quoted in Loon, 1972

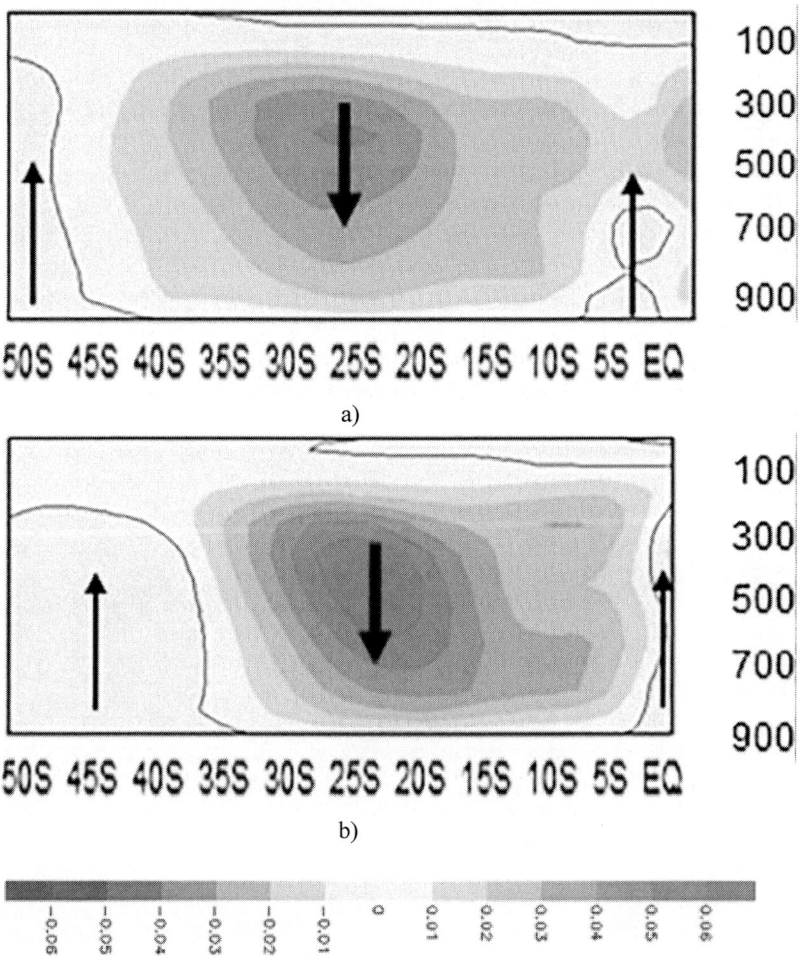

Figure I-2-9 – Latitude-heigh cross section of the vertical velocity (omega, Pa s^{-1}) in summer (a) and winter (b) according to the global climate model GFDL-ESM2M. *The arrows show the air lifting and sinking.*
Source: adapted fragment from Reboita et al., 2019

The vertical structure of SASH is presented in the latitude-height sections of the vertical velocity (Reboita et al., 2019). SASH is represented by positive values in the vertical velocity (in Pa s^{-1}) or sinking. Comparing summer and winter (Figures 9a and b) shows that in winter the subsidence associated with the Hadley cell is more intense and reaches lower latitudes.

Example: October 9, 2022

Anticyclone – High or SASH located at sea level above the Atlantic centered at 25°S, 18°W and has a maximum pressure of 1021 hPa (Figure I-2-10). The influence of this SASH has reached the continent of South America. Near the coastal region of the BNE, the 1014 hPa and 1012 hPa isobars change their anticyclonic curvature, and along the coast, the map shows a trough (*bold solid line* in the figure I-2-10) over the ocean and a ridge (*bold dashed line* in the same figure) over the continent. This trough and ridge along the coast can only be seen at low levels, and the map at 850 hPa shows the influence of SASH on the BNE.

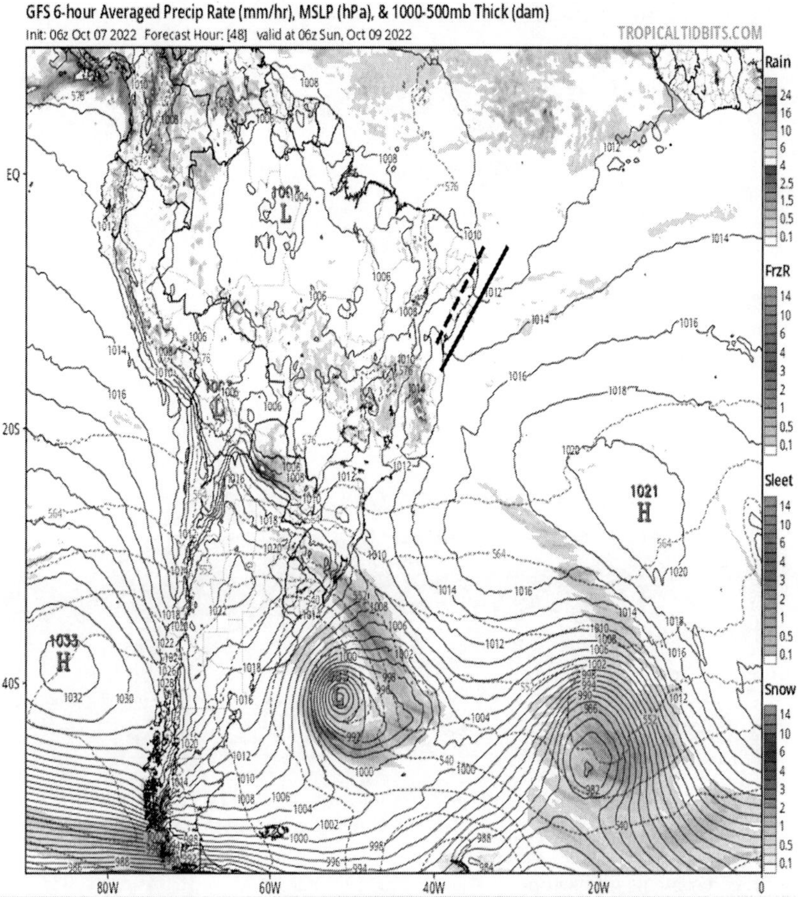

Figure I-2-10 – SASH on MSLP (hPa, solid line), thickness 500-1000 hPa (dam, dashed line) and average precipitation rate (mm/h) on 9 October 2022, 06 UTC. *Bold solid and bold dashed lines near the BNE – trough and ridge, respectively.* Source: TropicalTidbits.com

The same SASH can be seen on the 850 hPa geopotential height map centered on 22° S, 22° W with isoline 155 dam (Figure I-2-11). This SASH has influence over the entire BNE area at this level. This can be seen from the isoline of 150 dam and the wind direction.

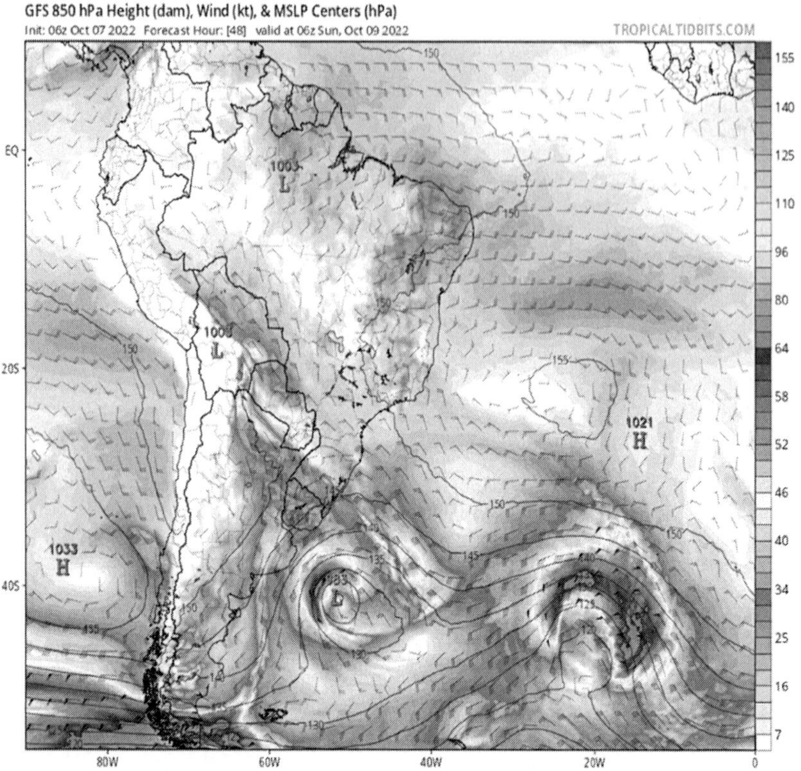

Figure I-2-11 – SASH at 850 hPa level on 9 October 2022, 06 UTC.
850 hPa height (dam; solid line, and colors), wind direction and velocity (wind barbs) and MSLP centers (hPa)
Source: TropicalTidbits.com

At high levels (the map shows 250 hPa level), the SASH center was located near 20°S, 20°W (Figure I-2-12). The location of the centers on the maps at low and high levels shows the inclination of the anticyclone axis to the NE from low to high levels. It is important to note that this SASH was located only over the ocean and did not reach the BNE. An Upper Tropospheric Cyclonic Vortex (UTCV, Chapter I-4) has been identified above the BNE.

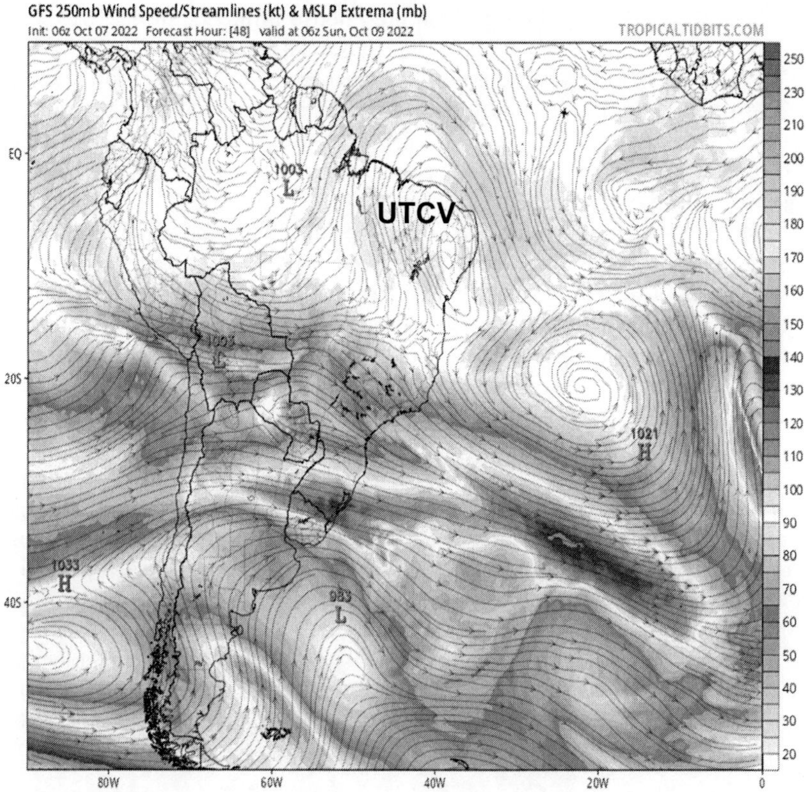

Figure I-2-12 – SASH at 250 hPa level on 9 October 2022, 06 UTC
Streamlines (solid lines with arrows) and wind speed (knots) at 250 hPa and MSLP centers (hPa.)
UTCV – Upper Tropospheric Cyclonic Vortex
Source: TropicalTidbits.com

4. Association with synoptic systems

SASH affects the sea surface temperature (SST) and consequently all meteorological systems in that region (Bombardi and Carvalho, 2011; Bombardi et al., 2014; Sun et al., 2017).

Being at the root of the trade winds (as shown in Chapters I-3 and I-7), such anticyclones often act as centers of tropical circulation, either blocking the northward advance of cold fronts or enhancing the movement of tropical storms within the trade winds (Hastenrath, 1985; Sun et al., 2017).

It is necessary to point out the more intense influence of the SASH on the formation and development of the "South Atlantic Convergence Zone", which is a stationary or cold front (Chapter I-7). As a result of this tropospheric block, the northward propagation of the frontal zone is prevented, so the frontal zone stops and does not affect BNE weather (Sun et al., 2017). Frontal zone blocking contributes to the trade winds intensification (Chapter I-3).

The vertical structure of the SASH periphery can be modified due to approximation of another synoptic system. This can occur at different levels. The above example on 9 October 2022 shows the effect of a baroclinic cyclone with a frontal zone at low levels on the SSW periphery of the SASH (Figure I-2-10). This can be identified by the trough formation on the periphery of this anticyclone. This frontal zone location is confirmed by: 1) the cyclonic curvature of the 1016 hPa isobar and 2) a weak warm wave in the forefront of this trough axis and a weak cold wave in the rear side on the thickness 500-1000 hPa map. Detailed information on frontal zone identification is presented in Chapter I-7. As a result, the average precipitation rate map (Figure I-2-10) shows the rain forecasted along this frontal zone at the SASH periphery. In addition, the map of the geopotential height of 850 hPa and the average precipitation rate show the continuation of the frontal zone with precipitation on the western SASH periphery.

The same example on 9 October 2022 shows the influence of the Upper Tropospheric Cyclonic Vortex (UTCV, Chapter I-4) at high levels on the northwestern periphery of SASH (Figure I-2-12). This UTCV was located only at a high level, above the periphery of the anticyclone.

5. SASH formation

The South Atlantic Subtropical High (SASH) is the dominant circulation feature over the South Atlantic Ocean and part of the subtropical high-pressure belt (Palmén & Newton, 1969). It is characterized by an anticyclone in the lower-tropospheric flow, the origin of which is associated with such factors as: 1) downward movements of the global circulation pattern, 2) intrusion of anticyclonic vorticity from extratropical anticyclones, and 3) heat and humidity exchange with the surface (Hastenrath, 1985).

The influence of global circulation on the SASH formation is associated with the Hadley cell. The global circulation shows the Hadley cell between the equatorial trough and the subtropical high at about 30°S. "*Solar heating heats the atmosphere directly on the equator. This air rises and flows toward the poles. Because of the Coriolis force the air begins deflecting right and converges in subtropical latitude. This forces air to sink towards the surface. The result is a band of high pressure at approximately 30° latitude, called the subtropical high*" (Vasquez, 1994). This means that the root cause of SASH formation is the "*solar heating heats on the equator*". A similar explanation of the Subtropical High formation is described in many books (for example, Djuric, 1994; Palmén & Newton, 1969). Djuric found direct thermal circulation in the Hadley cell: "*Warm air lifts near the equator and cold air sinks in the sub-tropical anticyclones*".

A similar description, namely that the polar branch of the Hadley cell is one of the main drivers of SASH formation, is given by many authors, for example: Rodwell and Hoskins, 2001; Dima and Wallace, 2003; Seager et al., 2003. Therefore, it is important to conclude that changes in this cell affect the SASH formation.

At the same time, according to climatological data (Chapter I-1), the belt of solar heating and warm temperature is wide, between 20°N and 20°S, approximately (Palmén & Newton, 1969; Lockwood, 1974; Fedorova, 2001). Therefore, it cannot by itself explain the subsidence of the polar branch of the Hadley cell by 20°S.

The second factor, such as the intrusion of extratropical anticyclones, plays a very important role in the SASH formation (Pedelaborde, 1958; Fedorova, 2001). The Rossby wave, or family of baroclinic cyclones and anticyclones, is a baroclinic anticyclone intrusion mechanism. This mechanism can be described in different ways: 1) the last anticyclone of the Rossby wave connects/couples with a subtropical anticyclone; 2) the baroclinic anticyclone "feeds" the barotropic subtropical anticyclone or 3) SASH is regenerated by the entry of the baroclinic anticyclone.

The climatological data, described in section 3 above show the location of the *steering* or *main current* on the south side of the SASH. It is known that the last stage in the development of a baroclinic anticyclone is a barotropic anticyclone with a warm core, when a *steering current* is observed on the High periphery (Peteresen, 1956; Fedorova, 2001). Therefore, the approach of the *steering current* to the SASH confirms that the Rossby wave (or a family of baroclinic and migrating cyclones and anticyclones) plays an important role in the SASH formation.

The entry of the final anticyclone of the Rossby wave intensified the High in the eastern regions of the Atlantic Ocean, which is confirmed by the SASH climatological location (Pedelaborde, 1958). The formation of SASH was primarily associated not with the sinking of air masses, but with the inflow of polar air and, consequently, with the dynamic subsidence into the equatorial side of the jet stream (Pedelaborde, 1958, who quoted Rossby). SASH is associated with the Rossby waves; the stable location of SASH exists only on climatological maps; daily maps show its location along with the influx of cold air. The final anticyclone of the Rossby wave developed and was observed at approximately the same place for several days (on average, one week). The trade winds are unstable in the equatorial regions due to meridional air supply, and not due to surface radiation heating (Pedelaborde, 1958).

Examples of the meridional entry of the Rossby wave will be presented in Chapters I-7 and I-9. In addition, one example of SASH regeneration due to baroclinic Rossby wave entry was shown in Fedorova, 2008b using IR satellite imagery and low-level streamlines. Other examples are presented below.

Example 1: from 1 to 15 November 2022

The first example shows the intrusion of the baroclinic anticyclone *Hma* into the barotropic cyclone *H* in the period from 1 to 15 November 2022 (Figure I-2-13). At the initial moment of 1 November, the center of the baroclinic anticyclone *Hma* was in the south of Brazil and Uruguay (Figure I-2-13a). This migrating anticyclone was located west of baroclinic cyclone *Lma*, between *Hma* and barotropic anticyclone *H* over the South Atlantic (or SASH). In 15 days, the *Lma* displaced in an east-southeast direction and weakened. During the same period, *Hma* was connected and coupled with *H*. Finally, one anticyclonic circulation with two centers was observed over the South Atlantic.

This process can be seen on the METEOSAT-11 IR satellite image (Figure I-2-14). The Rossby wave was visible as a cloudiness band on satellite imagery at the start of the process on 1 November. This band reached Antarctica south of Africa, crossed the South Atlantic south of the BNE. This band passed between *H* and *Hma*. Only the remnants of this band were observed at the end of the process 15 November between the same Highs. The remnants of these clouds were associated on the pressure map (Figure I-2-13b) with a weak trough in the southeast sector of *Hm*. During these days, the entire region of the BNE, except for the extreme south, was under the influence of the SASH.

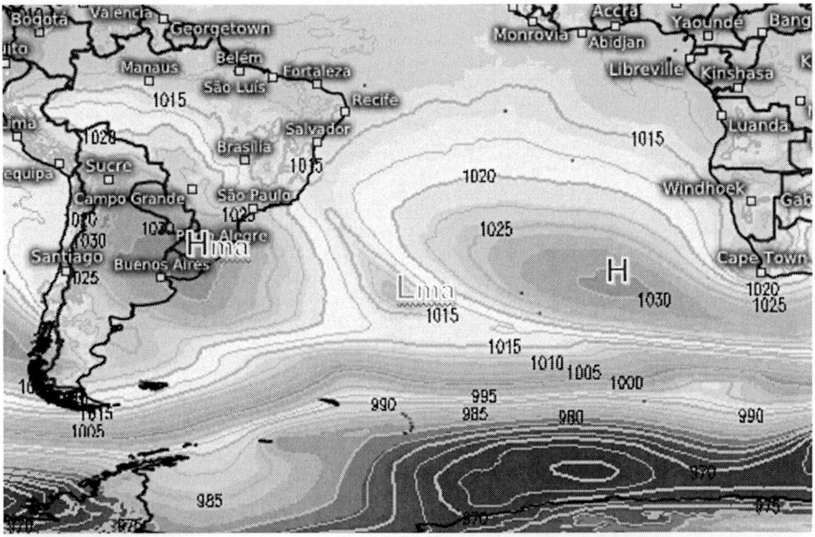

a)

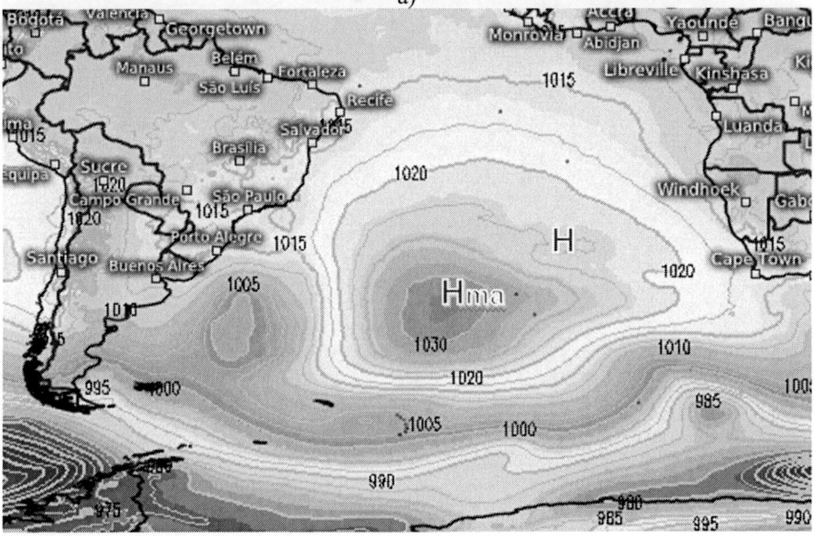

b)

Figure I-2-13 – MSLP map (hPa) on a) 1 November 2022, 21 UTC and b) 15 November 2022, 21 UTC.
Source: global model of ECMWF IFS HRES; meteologix.com; GIScience Research Group, Heidelberg University

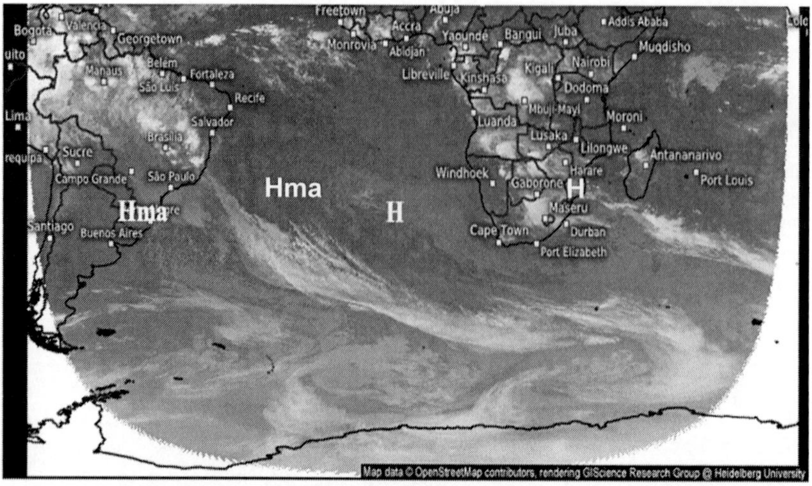

a)

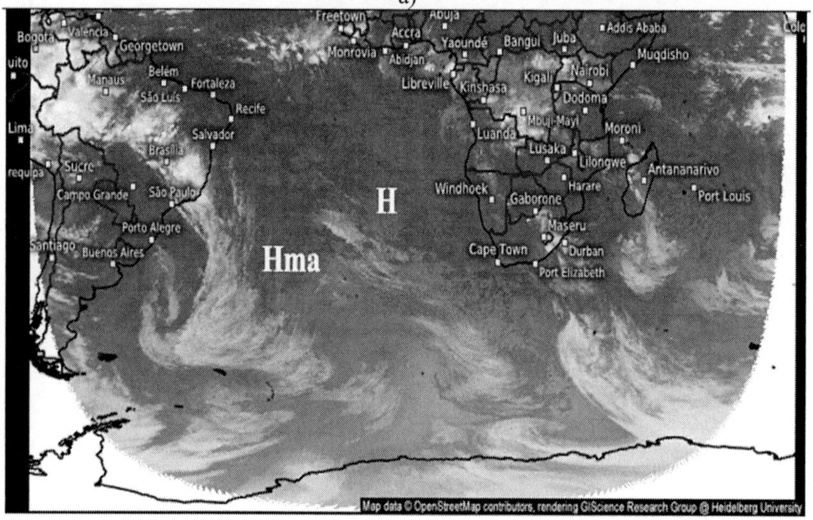

b)

Figure I-2-14 – Infrared satellite imagery (IR channel) from Meteosat-11, on a) 1 November 2022, 21 UTC and b) 15 November 2022, 21 UTC.
Source: EUMETSAT; meteologix.com
GIScience Research Group, Heidelberg University.

Example 2: from 2 to 7 June 2022

The intrusion of *Hm4* into the *H* circulation is shown in Figure I-2-15. On June 2, the anticiclonic circulation *H* was in a typical climatological position in the South Atlantic (Figure I-2-15a). The baroclinic High (*Hm4)* was observed that day over Uruguay and southern Brazil. The baroclinic Low (*Lm4)* was between these Highs. The trajectory of this vortex (*Lm4*) was directed to the southeast. Thus, the two anticyclonic circulations *Hm4* and *H* were combined into one anticyclonic circulation (denoted *Hm4* in Figure I-2-15b).

The June 2 Meteosat -11 IR satellite image shows a typical baroclinic cyclone cloud vortex (*Lm4)* in the South Atlantic (Figure 16a). This cyclone separated the baroclinic migratory High (*Hm4*) and SASH (*H* in the same figure). After that, this cyclonic vortex developed and displeased to the southeast. The vortex cloudiness is more curved (*Lm4* in the Figure I-2-16b). This vortex displacement allowed *Hm4* to be introduced into the *H* circulation. Finally, a single anticyclonic circulation (*Hm4*) can be seen in a typical SASH location.

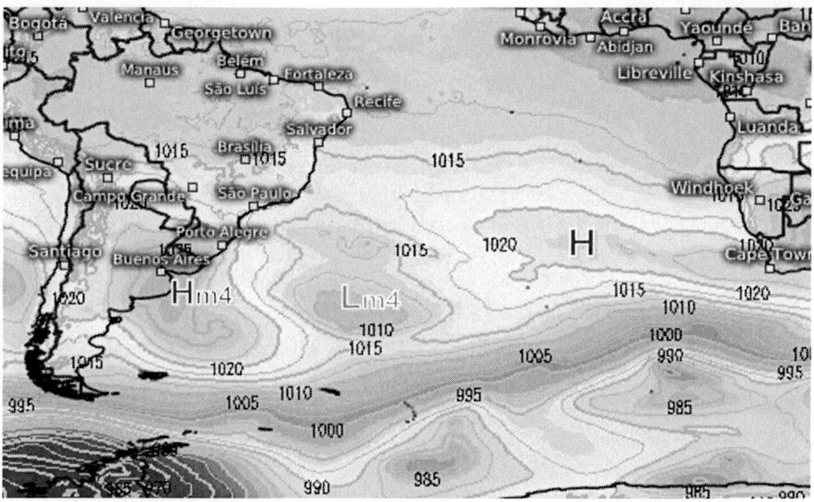

a)

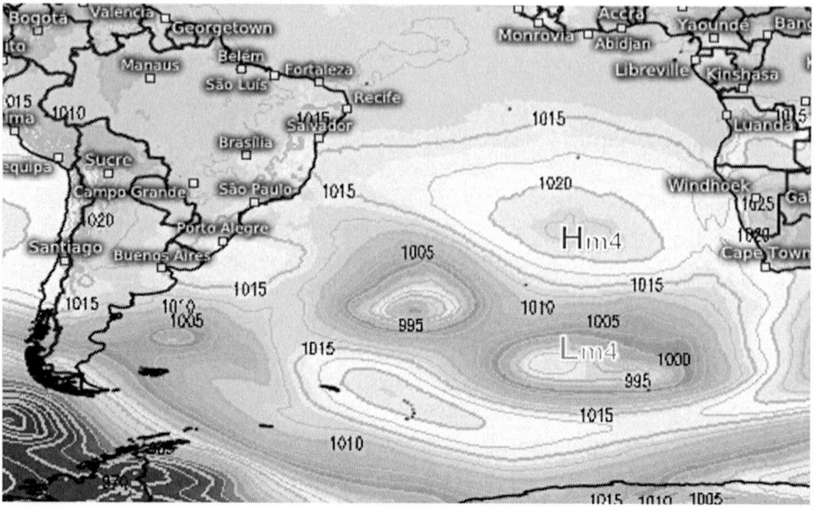

b)

Figure I-2-15 – MSLP maps (hPa) on a) 2 June 2022, 21 UTC, b) 7 June 2022, 21 UTC

Source: global model of ECMWF IFS HRES; meteologix.com; GIScience Research Group, Heidelberg University

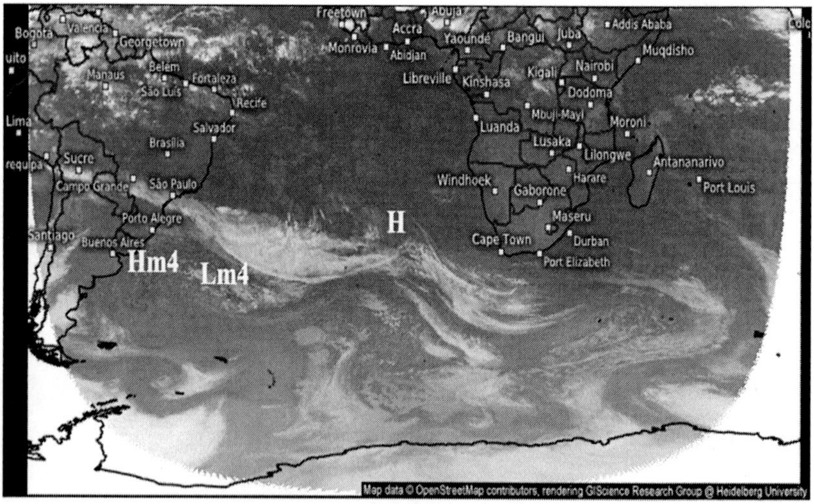

a) 2 June

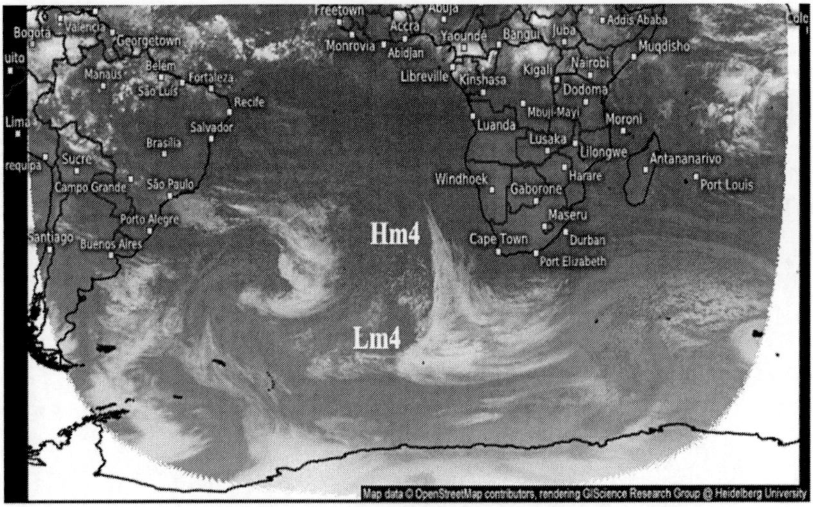

b)7 June

Figure I-2-16 – Infrared satellite imagery (IR channel) from Meteosat-11, on a) 2 June 2022, 21 UTC, b) 7 June 2022, 21 UTC
Source: EUMETSAT; meteologix.com GIScience Research Group, Heidelberg University

6. Meteorological phenomena

Wind

Observing variations in SASH is essential to understanding the regional climate in the BNE because a change in its intensity or position will alter the wind patterns in that region. Winds on the northeast coast blow from the east/northeast in summer and from the southeast in winter (Servain & Lukas, 1990). Thus, the winds are perpendicular to the coast and, together with land breeze, contribute to the formation of the nighttime convergence zone. But also, the sea breeze can combine with winds and interact with the terrain (valley breeze), intensifying the air lifting and displacing the coastal convergence zone further inland during the daytime.

Precipitation

Typically, SASH weather is associated with stable atmosphere phenomena such as stratus clouds; fog formation is common (Vasquez, T.,1994; Fedorova 2001, 2008b). Vertical developed clouds are rarely observed. For example, on the of average precipitation rate (mm/h) chart on 9 October

2022, 06 UTC, there was no precipitation in the SASH region (Figure I-2-10).

In this system, closed cells of stratiform clouds are formed, which are visible on satellite images (Figure I-2-17). However, a few cumulonimbus clouds and a TW convergence line were observed over the ocean. Precipitation in the coastal zone is confirmed by the precipitation map (Figure I-2-18).

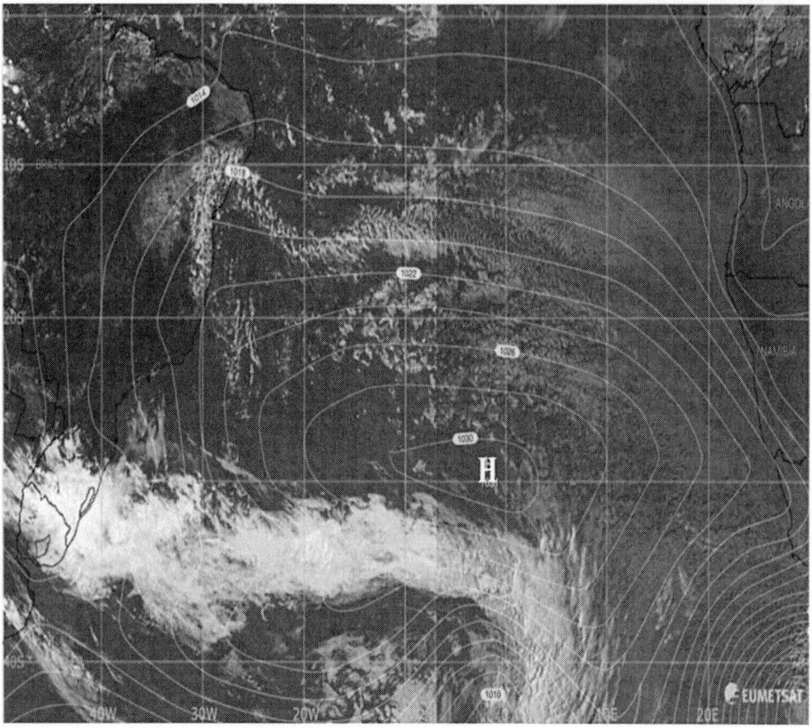

Figure I-2-17 – Typical clouds in the SASH on 13 September 2020, 17:10 UTC. GOES and METEOSAT visible satellite image composite.
Source: Windy/GOES/EUMETSAT (2020).

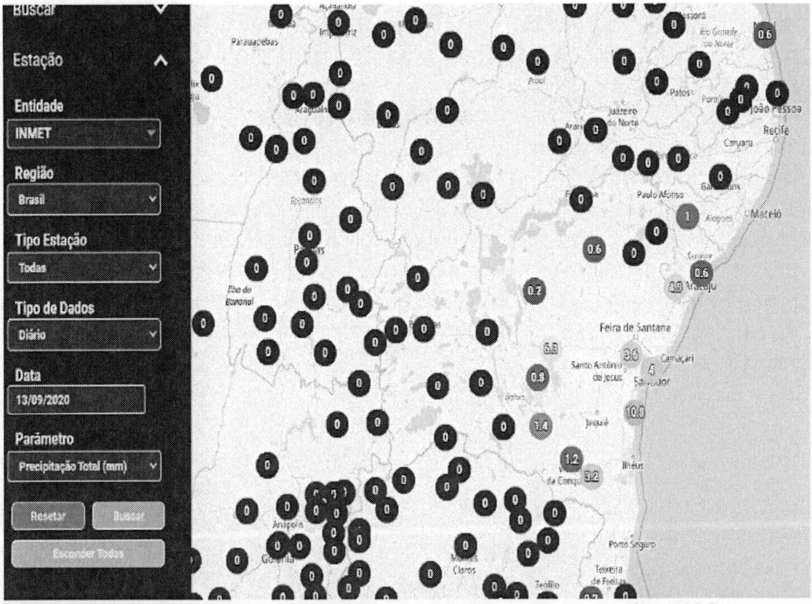

Figure I-2-18 – Precipitation (mm/24h) in the BNE on 13 September 2020.
Source: INMET

SASH favors or disadvantages precipitation over the BNE coast due to its position and the intensification of the trade winds in its northern branch (Lima, 1991). When approaching the continent, high pressure prevents precipitation, while its eastern position supports the contribution of the moisture flow from the ocean trough the trade winds.

The maximum amount of precipitation in the east of the BNE falls from May to July, probably due to the large convergence of the trade winds with the sea breeze. The breeze is stronger in autumn and winter, when the contrast between sea and land temperatures is higher (Kousky, 1980). Thus, the convergence of the trade winds with the sea breeze is stronger during this period.

The Sea Surface Temperature (SST) over the South Atlantic influences the amount of precipitation in the BNE. This type of analysis is usually developed using South Atlantic Dipole (SAD) data, which is characterized by a dipole in SST anomalies centered over the tropical and extratropical South Atlantic. The persistent SST anomalies can be identified for an ENSO duration of 5-6 months. A significant correlation between SAD and ENSO showed SAD leading ENSO by 5-12 months (Bombardi and Carvalho, 2011) or 7-8 months (Nnamchi et al., 2011). El Niño plays an important role in SST anomalies over the tropical and extra-tropical South Atlantic. The response of Atlantic SST anomalies to weak and short-term El Niño events is opposite to the response to strong and prolonged El Niño events (Rodrigues et al., 2011). During positive SAD events, positive SST anomalies were observed over the north of the BNE and an increase in precipitation over the north-northeast of Brazil (Figure I-2-19; Bombardi et al., 2014). During the negative SAD events, the increase in precipitation ended in the southern region of the BNE. The difference between SAD events composites (negative minus positive) shows a high amount of precipitation in the southern region of the BNE. At the same time, similar differences for the SST recorded a negative anomaly over the ocean near the BNE.

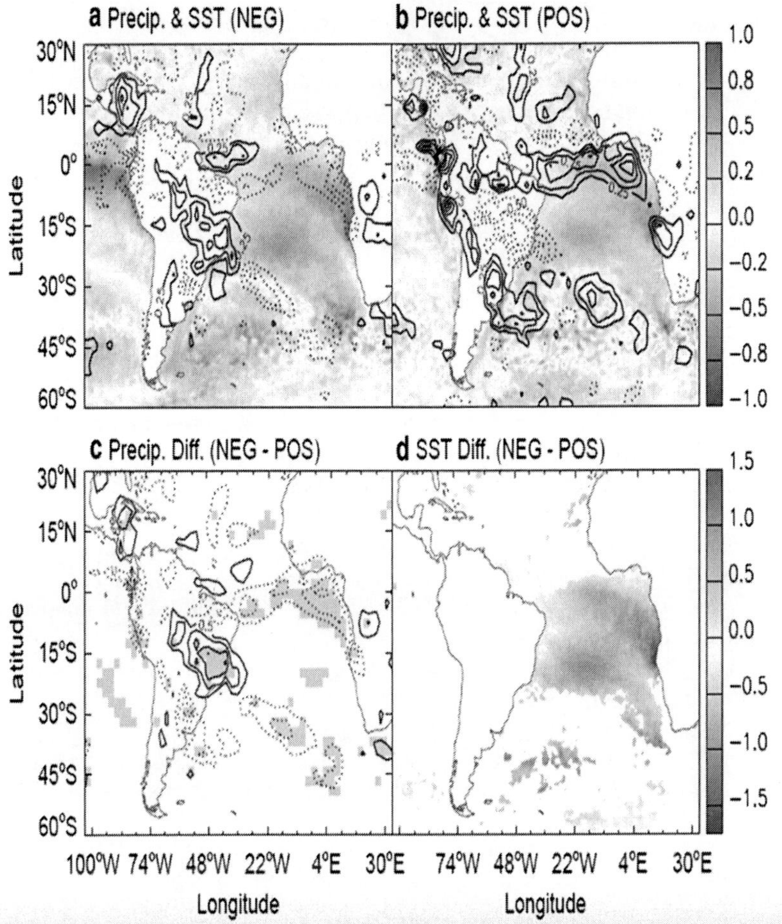

Figure I-2-19 – Composites of SST anomalies (*shade*) and precipitation (*contour*) during:
a) negative SAD events (NEG); b) positive SAD events (POS);
Difference between composites of SAD events (negative minus positive) for: c) precipitation and d) SST
Shading in c and d indicate regions where the difference is statistically significant at 5 % level according to the z-test of the difference of means.
Contour interval equal 0.25 mm /day in (a) and (b) and equal 0.5 mm/day in (c)
Source: Bombardi et al., 2014

Pollution

SASH is responsible for the absence of rainfall and therefore increased air pollution during episodes (periods) of low relative humidity; it depends on the SASH position relative to the continent (Padilha & Satyamurty, 2004; Ferreira et al., 2006).

7. References

Bryson, R. A., & Lahey, J. F. *The march of the seasons. Final Report Contract*, University of Wiscounsin, Department of Meteorology, Madison, 1958.

Bombardi RJ, Carvalho LMV "*The South Atlantic dipole and variations in the characteristics of the South American Monsoon in the WCRP-CMIP3 multi-model simulations.*"

Climate Dynamics 36: 2091–2102, (2011). doi:10.1007/s00382-010-0836-9.

Bombardi RJ, Carvalho LMV, Jones C& Reboita MS "*Precipitation over eastern South America and the South Atlantic Sea surface temperature during neutral ENSO periods.*" Climate Dynamics 42, 1553-1568, (2014), doi: 10.1007/s00382-013-1832-7.

Degola, T. S. D. *Impacts and Variability of the South Atlantic Subtropical Anticyclone on Brazil in the Present Climate and in Future Scenarios.* MSc thesis, University of São Paulo, São Paulo, 2013.

De Lima, M.C. *Variabilidade da precipitação no litoral Leste da Região Nordeste do Brasil.* MSc thesis, Nacional Institute of Space Science, São José dos Campos, 1991.

Dima, I. M., Wallace, J. M. "*On the seasonality of the Hadley cell.*" J. Atmos. Sci. 60, 1522–1527, (2003). doi: 10.1175/1520-0469(2003)060< 1522: OTSOTH> 2.0.CO;2.

Djuric Dusan, *Weather Analysis*. New Jersey: Prentice Hall, 304p, ISBN: 0-13-501149-3, 1994.

Fedorova Natalia. *Meteorologia Sinótica Volume 2.* Pelotas: Ed. Universitária/UFPel, 242p, ISBN: 85-7192-145-7, 2001.

Fedorova Natalia. *Sinótica III: Frentes, correntes de jato, ciclones e anticiclones.* Maceió: Ed. Universitária/ UFAL, 192 p, ISBN:978-85-7177-421-6, 2008a.

Fedorova Natalia. *Sinótica IV: Sistemas e processos sinóticos atuantes na américa do sul.* Maceió: Ed. Universitária/ UFAL, 192 p. ISBN: 978-85-7177-422-3; CDU: 551.5, 2008b.

Ferreira, EA, Melo, ML, Almeida, JMV, Bezerra, WA, Reinke, CK, Seabra, MS. "*Estudo de Caso: ConFiguração da atmosfera na ocorrência de baixo índice de umidade relativa do ar no Centro-Oeste do Brasil.*" In: XIV Congresso Brasileiro De Meteorologia, Florianópolis, (2006).

Hastenrath, S., Lamb, P. *"Climatic Atlas of the Tropical Atlantic and Eastern Pacific Oceans."* University of Wisconsin Press, 113 p, 1977.

Hastenrath, S."*Climate and circulation of the tropics.*" Springer Netherlands, (1985).

Kousky, V.E. "*Diurnal rainfall variation in Northeast Brazil.*" Mon. Weather Rev 108, 488-498, (1980).

Lockwood J. G. *World climatology. An environmental approach.* (Lecturer in Geography, University of Leeds). Now York: St Martin's Press. 330p., 1974.

Loon H. *"Pressure in the Southern Hemisphere. Chapter 4.*" In meteorology of the Southern Hemisphere. Edited by Newton C.W. 59-86, 1972.

Nnamchi H.C, Li J., Anyadike R.N.C. "*Does a dipole mode really exist in the South Atlantic Ocean?*" J Geophys Res 116, D15104, 2011, doi:10.1029/2010JD015579

Padilha, CK, Satyamurty, P. *"Um Caso de Estagnação de Ar Quente e Seco sobre a Região Central do Brasil.*" In: XIII Congresso Brasileiro De Meteorologia, Fortaleza, (2004).

Palmén, E., Newton C.W. *Atmospheric circulation systems: their structure and physical interpretation.* Academic press, New York. 603p. 1969.

Petterssen, Sverre. *Weather analysis and Forecasting.* v. 1, 2 New York, Toronto, London: McGraw-Hill, 428, 266p, 1956.

Reboita MS, Ambrizzi T, Silva BA, Pinheiro RF, Rocha RP. "*The South Atlantic Subtropical Anticyclone: Present and Future Climate.*" Front. Earth Sci. 7: 8. 2019. doi: 10.3389/feart.2019.00008

Rodrigues RR, Haarsma RJ, Campos EJD, Ambrizzi T. *"The impacts of inter–El Nino variability on the tropical Atlantic and Northeast Brazil Climate."* J Climate 24, 3402–3422, (2011). doi: 10.1175/2011JCLI3983.1

Rodwell, M. J., Hoskins, B. J. *"Subtropical anticyclones and summer monsoons."* J. Clim. 14, 3192–3211, (2001). doi: 10.1038/srep21346

Sun, X., Cook, K. H., Vizy, E. K. *"The South Atlantic subtropical high: climatology and interannual variability".* Journal of Climate 30, 3279–3296, (2017). doi: 10.1175/ JCLI- D- 16- 0705.1

Servain J., Lukas S., *"Climatic Atlas of the tropical Atlantic wind stress and sea surface temperature 1985-1989.*" Oceans tropicaux/Atmosphere global. IFREMER. ORSTOM. UH-NOAA Joint Institute for Marine and Atmospheric research. 133p, 1990.

Seager, R., Murtugudde, R., Naik, N., Clement, A., Gordon, N.,Miller, J. *"Air–sea interaction and the seasonal cycle of the subtropical anticyclones." J. Clim.* 16, 1948–1966, (2003). doi: 10.1175/1520-0442(2003)016<1948: AIATSC> 2.0.CO;2.

Vasquez, Tim, *Weather forecasting handbook.* Weather graphics technologies. Garland, Texas, 75p. 1994.

U.S. Navy, Naval Weather Service Command, U.S. *"Navy Marine Climatic Atlas of the World. vol. 1",* North Atlantic Ocean. NAVAIR 50-IC-528, U.S. Government Printing Office, Washington, D.C., 20402; 371 p., 1974.

U.S. Navy, Naval Oceanography and Meteorology, U.S. *"Navy Marine Climatic Atlas of the World. vol. 2, North Pacific Ocean"*, NAVAIR 50-1C-529, U.S. Government Printing Office, Washington, D.C., 20402; 388 p., 1977.

U.S. Navy, Naval Weather Service Command, U.S. *"Navy Marine Climatic Atlas of the World. vol. 3, Indian Ocean."* NAVAIR 50-IC-530, U.S. Government Printing Office, Washington, D.C., 20402; 348 p., 1976.

U.S. Navy, Naval Oceanography and Meteorology, U.S. *"Navy Marine Climatic Atlas of the World. vol. 4, South Atlantic Ocean."* NAVAIR 50-IC-531, U.S. Government Printing Office, Washington, D.C., 20402; 325 p, 1978.

U.S. Navy, Naval Oceanography Command, U.S. *"Navy Marine Ctimatic Atlas of the World, vol. 5, South Pacific Ocean."* NAVAIR 50-IC-532, U.S. Government Printing Office, Washington, D.C., 20402; 350 p, 1979.

CHAPTER I.3

DISTURBANCES IN THE TRADE WINDS

BRUCE FRANCISCO PONTES DA SILVA AND NATALIA FEDOROVA

1. Introduction

Trade winds (TW) can be defined as equatorward air currents from subtropical Highs that prevail at tropical latitudes, typically centered at 30°N and 30°S over the oceans (Vasquez 2000). TW act as a steering current for tropical disturbances, including tropical cyclones, transporting African dust westward across the Tropical Atlantic Ocean (Riehl 1954; Varejão-Silva 2000).

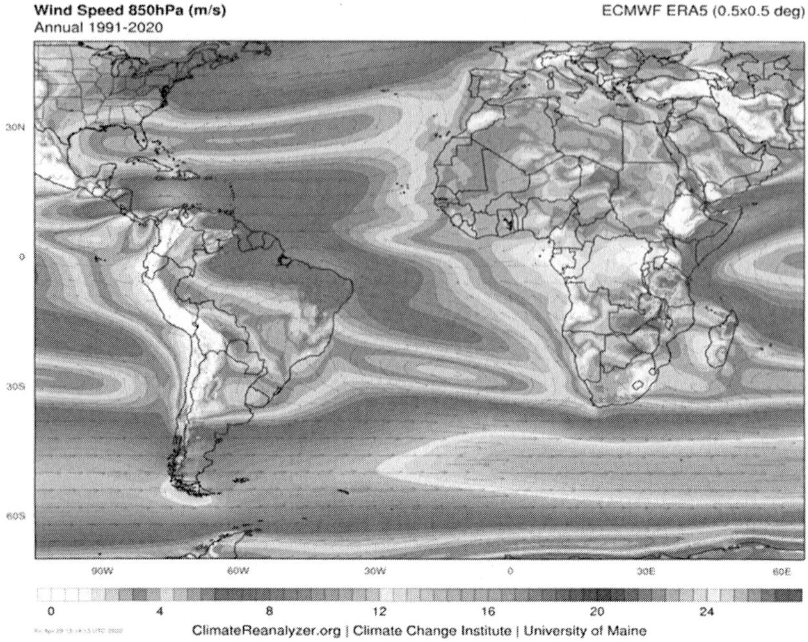

Figure I-3-1 – Wind speed (black/white scale) and direction (vectors) annual average (1991-2020) at 850 hPa.
Source: University of Maine with data from ECMWF Reanalysis version 5 (ERA5).

TW represent currents from the northeast (NE) in the North Atlantic and from the southeast (SE) in the South Atlantic (Figure I-3-1).

TW represent the maximum speed in winter and when the Arctic Oscillation (Northern Hemisphere) or Antarctic Oscillation (Southern Hemisphere) are in the warm phase. The associated ocean currents helped early African and European ships sailing to the Americas and ships from the Americas to reach Asian regions. To this day, ships still use these normally calm winds.

The TW from SE over the Tropical South Atlantic (TSA) blow to Northeast Brazil (NEB – Figure I-3-2) throughout the year (Fedorova 2008). When disturbances forming in the TW reach the BNE, they can cause intense precipitation in the Eastern BNE (EBNE – Figure I-3-2), especially during the rainy season from April to July. Often, heavy rains associated with disturbances in the TW damage local agriculture, tourism, and other important economic activities, and can account for more than half of all EBNE rainy season precipitation (Pontes da Silva 2011, 2020; Gomes et al., 2019). However, in areas about 100-250 km inland from the coast, heavy rainfall is infrequent, even during the rainy season in the BNE (Molion & Bernando 2002).

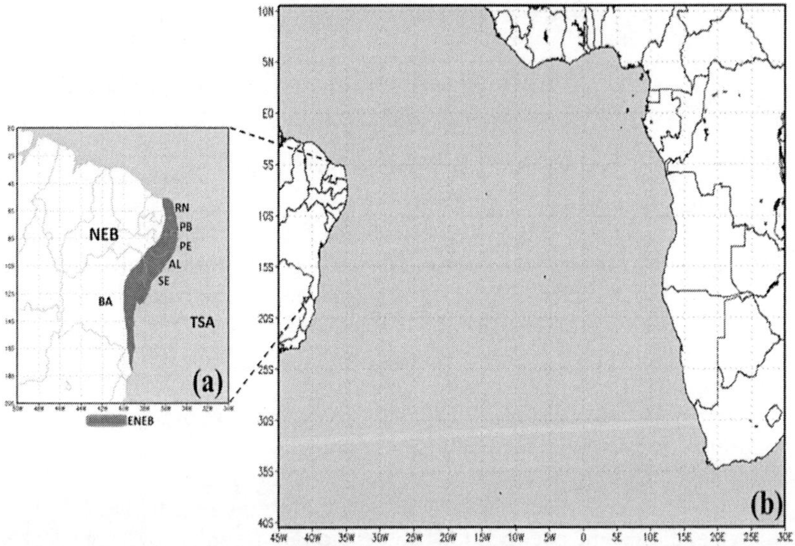

Figure I-3-2 – Location of Northeast Brazil (BNE), Eastern BNE (EBNE, grey shading area) (a) and part of the tropical South Atlantic Ocean (TSA) (b).
Names of the EBNE states: Rio Grande do Norte (RN), Paraiba (PB), Pernambuco (PE), Alagoas (AL), Sergipe (SE) and Bahia (BA).
Source: adapted from Pontes da Silva (2020).

Other synoptic systems, such as convergence zones associated with cold fronts or cold fronts themselves, are also associated with extreme precipitation over the BNE (Kousky 1979, 1980; Kouadio et al. 2012). All these weather systems frequently interact with sea-to-land and land-to-sea breezes to increase precipitation (Kousky 1980; Alestalo & Savijärvi 1985; Roeloffzen et al. 1986).

The main purpose of this chapter is to provide a brief overview of the literature on TW disturbances, focusing on disturbances in the South Atlantic (SA) and their identification.

This chapter is organized as follows: Section 2 provides an overview and discussion about the disturbances in TW according to the literature; Section 3 proposes to discuss the disturbance in TW and the methodology for their identification; and Section 4 summarizes the results.

2. Overview of disturbances in the trade winds

This Section contains a brief overview and discussion of some literature about disturbances in the TW.

2.1 North Atlantic, Pacific, and Indian oceans trades winds disturbances

Easterly Waves (EW) are one of the most important weather systems, generating precipitation in many tropical regions of the Northern Hemisphere (Asnani 1993). They can be defined as synoptic disturbances associated with troughs and warm sea surface temperatures (Riehl 1945; Rosenthal 1960; Yanai 1963, 1968; Krishnamurti & Baumhefner 1966; Williams 1970; Keshavamurty 1971; Reed and Recker 1971; Burpee 1974; Chang & Miller 1977; Tai and Ogura 1987; Asnani 1993; Céron and Guérémy 1999).

According to Vasquez (2000), EWs are migratory wave-like disturbances in a tropical easterly air current with a cycle of about 3 - 4 days and a wavelength of about 2000 - 2500 km, maximum intensity in the lower and middle troposphere, convection developing parallel to the low-level current and, in some events, they can develop into tropical cyclones.

One of the main reasons for the formation of Tropical North Atlantic EWs was appointed the intense heating of the surface over the Sahara Desert, which creates a semi-permanent warm-core barotropic low (Vasquez 2000).

The main causes of EW genesis, according to some researchers, are:

1) EW can appear due to the trade winds convergence (between the northeast and southeast trade winds or convergence along the north or south trade winds); at the same time, it should be recalled that the TW convergence is the Intertropical Convergence Zone – ITCZ (Chapter I-1). This is consistent with the review and conclusions of Asnani (1993);
2) EWs can develop due to the intensification of the Upper Tropospheric Cyclonic Vortex to a middle or low level (Asnani 1993; Vaz 2011), which propagates to the west. However, in these cases, EW should also be considered as part of the UTCV circulation and not as another system. Operational practice shows that UTCV did not create any disturbance if previously there were no disturbances at low levels.
3) EWs can develop due to a mid-latitude trough displacement towards the equator (frontal troughs that will separate from their fronts,

heading west along with the trade winds) or due to an equatorial trough displacement to higher latitudes (may be associated to the equatorial trough) (Yamazaki and Rao 1977; Asnani 1993; Espinoza 1996; Fedorova 2008). Some of these events of separation from frontal zone are described in Chapter I-7.

According to Asnani (1993), the greatest difficulties in EW research in various regions of the globe are:

a) relatively low intensity: EWs are short compared to troughs and stationary ridges over tropical latitudes and cannot be identified on conventional synoptic maps;
b) lack of a homogeneous structure: the EWs are in a zonal current, and this current has different characteristics in each Earth location; if the eastward flow is shallow (only at low levels), shallow EWs are expected; however, in a deeper eastward current (up to middle or high levels), deeper EWs are expected;
c) little data is available for the tropics: most of the tropical areas are covered by oceans, where it is difficult to install and operate meteorological stations.

Even over continents, the density of meteorological stations is limited. In the meantime, satellites have provided data for remote areas such as deserts and oceans. Even with this problem, a lot of EW research has been carried out. Many authors have analyzed EWs through case studies using synoptic charts (Riehl 1945; Yanai 1963, 1968), using composite methods (Williams 1970; Reed and Recker 1971; Burpee 1974; Mota and Gandu 1996, 1998), using spectral analysis (Rosenthal 1960; Keshavururty 1971; Chan 1990), using orthogonal empirical functions (Espinoza 1996; Céron and Guérémy 1999) and through dynamic simulations (Krishnamurti and Baumhefner 1966).

With regard to cloudiness in particular, Merritt (1964) found five different cloud distributions associated with tropical disturbances over the Caribbean Sea, showing that not all EWs have synoptic scale convective cloud bands. Often these systems contain only mesoscale shallow clouds in one period, and convective clouds develop in another period. According to Merrit (1964), some EWs do not have significant cloudiness at any stage of their life cycle, but this does not mean that they are necessarily weak disturbances in synoptic fields. However, EWs usually have a cloudiness characteristic that is different from other systems. This cloudiness may be predominantly stratiform or cumuliform, depending on the environment in which the waves propagate. According to Asnani (1993), cloudiness and precipitation associated with EWs commonly mix with ITCZ cloudiness and rainfall.

Possible mechanisms for the formation and maintenance of African EWs were discussed by Burpee (1972, 1974), Rennick (1976); Karyampudi and Carlson (1988), among others.

Reed and Recker (1971), Keshavamurty (1971), Chang and Miller (1977) and Tai and Ogura (1987) investigated EWs in the tropical Pacific Ocean. Overall, as Asnani (1993) concluded, EWs in the Northern Hemisphere have a wavelength of ~3000 km, a period of 4-5 days, and a phase velocity of 5 to 7° day^{-1} (~7ms^{-1}) using research results available up to the earlier 1990's.

In summary, the methodologies used are somehow confusing and none of these studies have indicated practical or universal criteria for identifying and predicting trade winds disturbances on a regular operational manner.

2.2 South Atlantic trade winds disturbances

Almost all precipitation over the BNE during the rainy season (from April to July) is caused by trade winds disturbances, convergence zones associated with stationery fronts, or cold fronts themselves (Kousky 1979, 1980; Molion & Bernardo 2002; Pontes da Silva 2011, 2020; Kouadio et al. 2012; Gomes et al. 2019). These weather systems frequently interact with sea-to-land and land-to-sea breezes. Heavy rainfall is not common even during the BNE rainy season in areas 100-250 km inland from the coast (Molion & Bernardo 2002).

The terminology most commonly used in Brazil to describe TW disturbances is "eastern wave disturbances" (free translation from the Brazilian Portuguese "*Distúrbios Ondulatórios de Leste*"). However, not all disturbances that move from east to west are associated with the trade winds, so we have chosen to use the term Wave Disturbances on Trade Winds (WDTW) in this chapter to avoid confusion.

According to some studies, WDTW events occur at any time of the year, often affecting BNE (Chan 1990; Espinoza 1996). Its maximum frequency occurs in the austral autumn/winter during the rainy season in the BNE (Yamazaki, Rao 1977). Most of the time, these systems can be observed on satellite images as shallow clouds that spread from the east into the ocean and intensify near the coast, forming convective clouds. Sometimes WDTWs spread to the BNE already with convective activity, weakening as they enter the continent (Molion, Bernardo 2002).

A theoretical study of the barotropic instability in the eastern zonal current over the Atlantic and BNE, has shown that it is directly related to the occurrence and maintenance of these disturbances Yamazaki (1975). As a result of the study of barotropic instability, a wavelength of ~6000 km was obtained. In the cited paper, the results of this theory were tested by observing cloudiness from satellite data over the southern tropical Atlantic. It was concluded that disturbances propagate only in the austral winter with a wavelength of 4000 km, which corresponds to the theoretical result. Yamazaki noted a westward spread of well-defined cloudiness between 5° and 10°S, from 10°E to about 40°W.

Chan (1990) showed that in the autumn and winter of 1979 WDTW had a wavelength of 6200 km.

The results of a number of EW studies using the southern wind component in the equatorial zone of the globe for 10 years (1980-1989) were summarized in Espinoza (1996). This author noted that WDTW-like systems spread throughout the year with different wavelengths: in the austral summer, the wavelength ranged from 6000 to 7000 km, and in the austral autumn it was from 5000 to 6000 km. In the austral winter, the waves were shorter, their length varied from 3500 to 4000 km.

Dislocation of some WDTWs towards the BNE during the rainy seasons of 2005 and 2006 was studied (Torres 2008). The author used satellite data and Hovmöller-type diagrams. Twenty-six (26) WDTWs with a wavelength of 4000 km were detected. The WDTW amplitude gradually increased from April to July (BNE rainy season), which indicates the possible existence of a well-defined seasonal cycle for this system, i.e., seasonal progressive increase in the number of WDTW.

The WDTW displacement in the BNE during the rainy season (from April to July) of 2006 and 2007 was analyzed (Gomes et al. 2015). This study was based on cases identified by Pontes da Silva (2011, 2020). Composite fields of 700 hPa showed a wavelength of 4500 km. These results are consistent with Pontes da Silva (2011), who found an average wavelength of 4300 km in the same region and same season during 2006 and 2010. These WDTWs contributed at least 70% of the total precipitation observed over the BNE during the same period (Pontes da Silva, 2011, 2020). WDTWs lasted 4 days and these systems were associated with the heavy precipitation observed in the austral winter (Yamazaki 1975, Yamazaki and Rao 1977).

WDTWs moved to the Amazon basin with a period of 3 to 5 days and maximum humidity close to the WDTW trough (Kayano, 1979). Different authors indicate the similar periods of WDTW reaching the BNE (5°S): 1) an average period of 5- 6 days in 1979 (Chan,1990), 2) a 5-day period, 2005 -2006, (Torres 2008), 3) a period of 5.3 days (Pontes da Silva 2011) and 4) a period of 5.5 days (Gomes et al. 2015).

As for the WDTW phase speed, the results of studies by different authors also turned out to be similar: 1) 10 $m.s^{-1}$ (Yamazaki 1975); 2) 10-14 $m.s^{-1}$

in the austral summer and 10-13 $m.s^{-1}$ in the austral autumn and winter (Espinoza 1996); 3) 12 ms^{-1} in autumn and winter 1979 (Chan 1990); 4) 10 $m.s^{-1}$ (Torres 2008); 5) 9.5 $m.s^{-1}$ (Pontes da Silva 2011, 2020; Gomes et al. 2019).

Summarizing the kinetic characteristics of WDTW, the waves have an averaged wavelength of 4000 to 6000 km, a period of 5 days and a phase speed of about 10 $m.s^{-1}$.

Due to the differences between the EW and WDTW systems, some researchers suggest that WDTW cannot be classified as Easterly Waves (Merritt 1964; Riehl 1979; Diedhiou & Machado, 2010).

According to Molion and Bernardo (2002), the WDTW number will increase since the cumulonimbus clusters cause sinking and gusty winds, generating disturbances that will propagate outside the ITCZ.

In addition, the cold front extremities reaching tropical latitudes could theoretically contribute to the WDTW formation. Nevertheless, convection in WDTW will rarely intensify over the ocean, where the trade winds inversion (humidity and temperature inversion) is stronger. Although when WDTW reaches the BNE continent, convection intensifies by moisture convergence. However, Molion and Bernardo (2002) did not provide a single example to clarify exactly what they mean by disturbances. It is very likely that any undulating disturbance over the currents of the south-easterly trade winds was considered.

In order not to discuss the presence or absence of "classic" EW over the South Atlantic, Pontes da Silva (2011, 2020) and Gomes et al. (2015, 2019) called the wave disturbances they detected over the ocean and reaching the BNE during the rainy season, "Easterly Wave Disturbances" (EWD). However, the term can also be confusing if the characteristics described by the authors are not properly followed. Nevertheless, the authors have made it very clear what type of system, referred to here as WDTW, they are referring to.

The main results of the Pontes da Silva (2011, 2020) studies show the following. The most observed synoptic process was the propagation of a trough in the TW at 850 hPa and/or 700 hPa (nearly 90% from 116 cases). 28% of WDTWs were associated with cold fronts. It was concluded that low levels (trade winds levels) are key to identifying and monitoring these disturbances.

In addition, Gomes et al. (2019) studied the synoptic systems associated with disturbances and concluded that cold fronts play an extremely important role in the formation of most disturbances in trade winds, as 374 (~72%) of 518 cases were associated with cold fronts over or near the BNE.

Table I-3-1 presents the main characteristics of WDTWs identified using satellite images and ERA-Interim reanalysis (Pontes da Silva, 2011). WDTWs are more frequent over the BNE in June. On average, nearly 60% of the WDTWs occurring in each rainy season were associated with convective clouds, at least in interaction with sea/land breeze and coastal convergence near the BNE coastline (Alestalo & Savijärvi 1985).

The mean lifetime of a WDTW is 5 days (from its first detection to dissipation, at a standard deviation of ±2 days), with each rainy event averaging 2.6 days over the BNE (standard deviation of ±1.5). Using composite fields at 700 hPa, the average lifetime and wavelength of WDTW are 5.3 days and 4300 km, respectively. The average phase velocity of WDTW, equal to 9.5 $m.s^{-1}$, are close according to the results of Pontes da Silva (2011) and Mota (1997). According to Pontes da Silva (2011), convective clusters associated with the WDTW contributed approximately 50% of the precipitation observed over the BNE in 2006-2019 (from April to July; BNE rainy season). It needs to be clarified that this percentage was higher than 60-70% near the BNE coast (Figure I-3-3).

Table I-3-1 – Mean WDTW characteristics identified during April, May, June, and July (AMJJ) from 2006 to 2010.
Source: adapted from Pontes da Silva (2020).

Reference	2006	2007	2008	2009	2010	Total	%	Average
Total of EWs (AMJJ)	26	22	24	22	22	116		23.2
Convection detected (at least moderate)	11	12	17	16	12	68	58.6	13.6
Number of events in April	6	5	5	5	6	27	23.3	5.4
Number of events in May	5	6	4	4	5	24	20.7	4.8
Number of events in June	7	5	9	7	5	33	28.5	6.6
Number of events in July	8	6	6	6	6	32	27.6	6.4
Mean duration (days)	4.8	5.0	5.4	6.4	4.5			5.2
Wavelength (km)	4773	4662	3996	3996	4107			4307
Phase Velocity ($m\ s^{-1}$)	11.8	9.8	8.9	8.4	8.6			9.5
Period (days)	4.7	5.5	5.2	5.5	5.5			5.3
Number of rainy days in EBNE due EWs	2.4	2.5	2.9	3.2	2.2			2.6

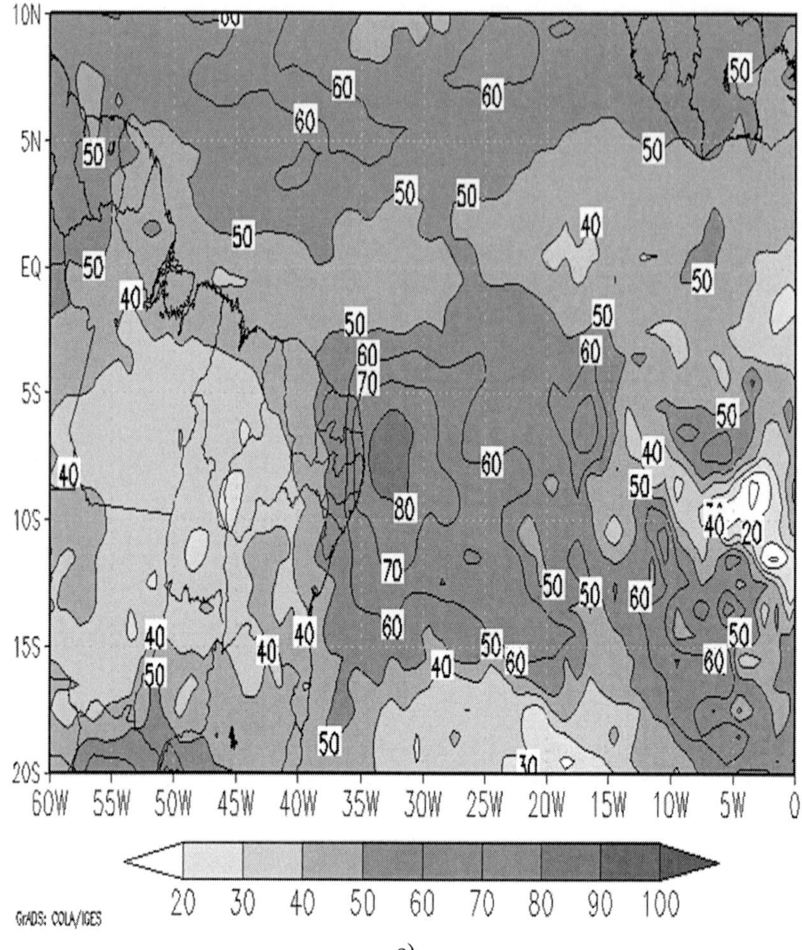
10N
5N
EQ
5S
10S
15S
20S
60W 55W 50W 45W 40W 35W 30W 25W 20W 15W 10W 5W 0
20 30 40 50 60 70 80 90 100
GrADS: COLA/IGES

a)

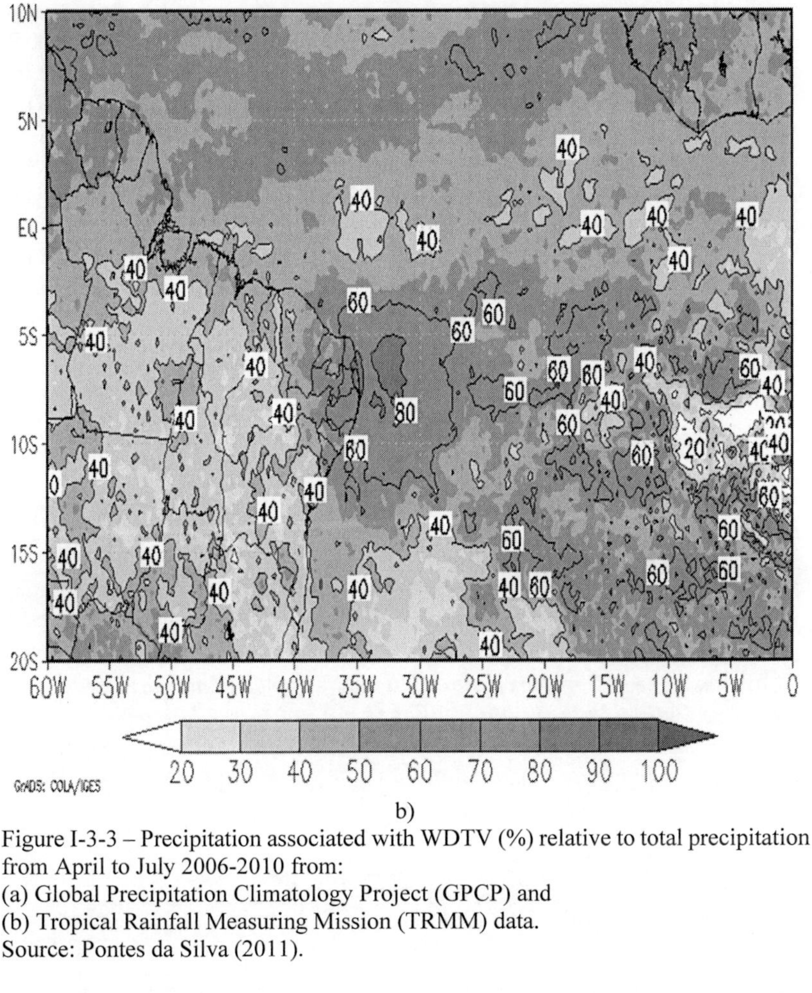

b)

Figure I-3-3 – Precipitation associated with WDTV (%) relative to total precipitation from April to July 2006-2010 from:
(a) Global Precipitation Climatology Project (GPCP) and
(b) Tropical Rainfall Measuring Mission (TRMM) data.
Source: Pontes da Silva (2011).

Similar results were obtained by Gomes et al. (2015, 2019), using the same methodology as Pontes da Silva (2011): at least 60% of the total precipitation observed in the BNE fell during WDTW in 1989-2009 (from April to August).

2.3 Overview of average basic characteristics

The main average characteristics of the trade wind disturbances over the globe, according to the literature data, are presented in Table I-3-2.

Table I-3-2 shows that the most frequent meteorological system used to describe or identify the term EW is trough, but at different levels. Also, EW were identified from satellite images, but without any information about the average tropospheric vertical profile in some studies (Yamazaki 1975; Hall 1989; Chan 1990; Torres and Ferreira 2011).

The horizontal dimensions of the systems studied by all authors ranged from 500 km to 6000 km, approximately. This difference is very large and certainly corresponds to different systems.

Table I-3-2 – The main characteristics of the trade wind disturbances, identified from the data of several studies.
N - Number of events, V - Variables, HD - Horizontal Dimension, L - Analyzed vertical levels, ML - Main Level/
maximum intensity level, SS - Synoptic System or process,
SLP - surface level pressure (hPa), Stl - streamlines; Q - omega (Pas^{-1}),
Pw - precipitable water (mm), T - temperature (°C),
v and u - meridional and zonal components of the wind, respectively (m.s^{-1})
Mdiv - moisture flux divergence (10^{-5}gkg^{-1}s^{-1}), OLR - Outgoing Long Wave Radiation (Wm^{2}), SH - Specific humidity (gkg^{-1}), SST - sea surface temperature (°C)
IR - infrared satellite imagery, WV - water vapor satellite imagery, VIS - visible image;
Conv - convergence of air current; Div - divergence of air current;
CRV - cyclonic relative vorticity; Wave - cyclonic curvature at streamlines;
RH - relative humidity, P - precipitation.

Reference	N/ Study period	V	HD/ region	L (hPa)	ML (hPa)	SS
			Textbook / manual / guide			
Varejao& Silva 2006	-	Stl, SLP	2000-4000 km	-	-	Trough Low SLP
Vasquez 1994	-	*Stl, Low, middle, high*	2000-2500 km	*Low, Middle, High levels*	Middle levels	Trough in low and middle levels
Riehl 1945		SPL, Stl, T	15° long. (2000 km)	1000, 700, 500	700, 500,	Barotropic wave, Trough, TW

Research article						
Torres 2008	26 events 2005 2006 rainy seasons	IR, v, soundings	4000 km	850, 700		Trough
Torres & Ferreira2011	2 events May June 2006	Sti, Q, Pw, Mdiv	8° long 10° lat (800x 1000 km)	700	700	trough
Vaz 2011	1999-2009	SLP, OLR, IR, WV, VIS, SST	20-35° W 0-20°S *Region II*	500, 700	200-500	UTCV 200, Trough 500, 700
Vaz 2011	1999-2009	SLP, OLR, IR, WV, VIS, SST	30-35°W 0-20°S *Region V*			Frontal zones; Trough low levels
Chan 1990	12/ 1978-11/ 1979	OLR, v 850	4800-6200 km	850		Trough/ ridge
Coutinho1999	1988-1997	v SH		850, 775, 700, 600, 500		Trough/ ridge
Espinoza 1996	1980-1989	v	6000-7000 km	1000, 850, 700, 500	850	Trough/ ridge
Hall 1989	198419851986 3events 1 event per year	u, v				Trough
Yamazaki 1975		IR	4000 km			Trough Barotropic wave
Molion Bernando 2002	experience					Wave/ trough on frontal zone, TW;

						ZCIT, MCC
Barbosa 1997	1984-1998		Northern Coast of Brazil			convection
Coutinho Fisch 2007	64 events 96 events 100 events			850-700 700-500 850-500		trough
Espinoza1996		u, v anomaly		Low, middle levels		
Gomes et al, 2019	1989-2009	Stl, Q, OLR, IR, WV, VIS, div CRV, RH, P v, u	4500 km	All standard levels	1000 conv 700 div *CRV*	Wave IR CRV, div
Pontes da Silva, 2011, 2020	2006-2010 April to July	Stl, Q, OLR, IV, WV, VIS, div, CRV, RH, P, v, u	4307 km	All standard levels	1000 (*CRV, conv* anomalies)	Wave IR CRV, div

Molion and Bernardo (2002) present a summary of some interesting empirical data, but without a precise definition of WDTW.

In addition, the analyzed levels or layers vary greatly. Troughs in different layers (64 events in the 700-850 hPa layer; 96 events in the layer of 500 -700 hPa, 100 events in the 500-850 hPa layer) were identified as EW without cloudiness analysis (Coutinho & Fisch, 2007). A trough in the 500-700 hPa layer (20-35°W and 0-20°S) associated with UTCV is also included in the EW study (Vaz, 2011). Low level troughs in another region (30-35°W, 0-20°S) very similar to WDTW (Vaz, 2011*)* confirm the existence of these systems.

A more complete WDTW analysis was done by Pontes da Silva (2011) using satellite images, streamlines (average and anomaly), relative vorticity (anomaly) and divergence (anomaly) at all standard tropospheric levels. Pontes da Silva (2011) also used cross sections of relative humidity (anomaly), relative vorticity (anomaly), omega (anomaly) and wind vector (anomaly) in the range of 200 to 1000 hPa, and also, tested anomalies in precipitation and outgoing longwave radiation, associated with EW events in the period 2006-2010. This method was applied to WDTW climatology by Gomes et al. (2019).

According to Pontes da Silva (2011) and Gomes et al. (2015), a higher frequency of WDTW is observed between April and July. However, Gomes et al. 2019 found that there was also a high incidence of WDTW cases in August. This conclusion confirmed that WDTW is an important weather system for the BNE rainy season (429 cases, April-August) compared to other months (89 cases, dry season). Gomes et al. (2019) also concluded that there is a large interannual variability in the WDTW frequency, especially with ENSO episodes, when it tends to be lower (higher) in El Niño (La Niña) years, especially in years with more strong or longer ENSO episodes.

3. Identification

3.1 Trade winds

The subtropical High (pressure at the center 1023 hPa according to the Global Forecast System – GFS model) is an intense ridge from the H center to the NWW and creates TW with wind direction from E, ESE, SE over the South Atlantic (Figure I-3-4). The same Figure shows weak disturbances (troughs at the low levels). These disturbances in TW formed extensive areas of shallow clouds (cumulus and stratocumulus) over the South Atlantic Ocean, as seen in the satellite image in Figure I-3-5.

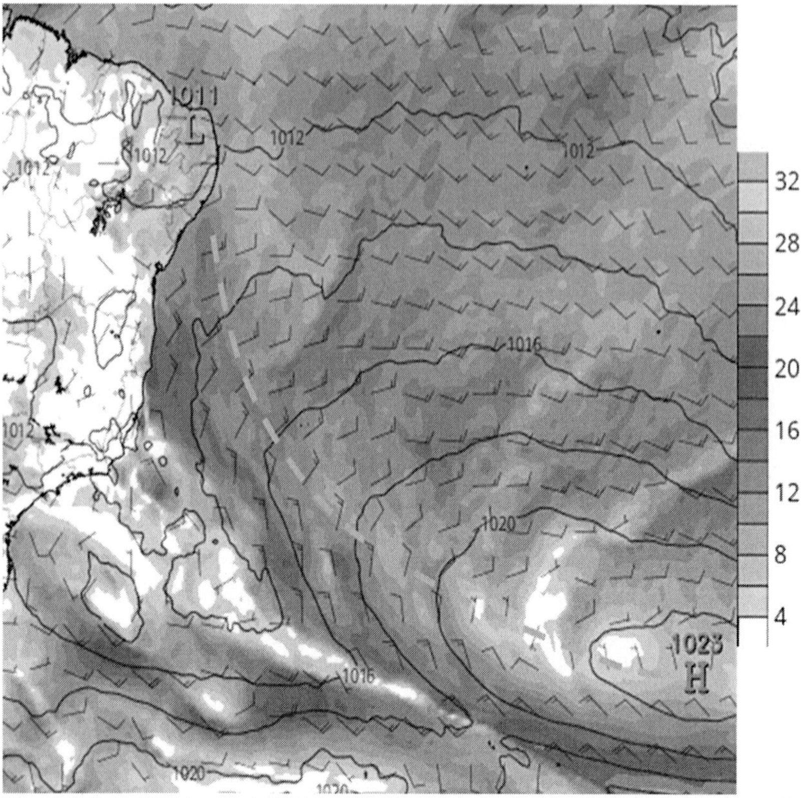

Figure I-3-4 – Subtropical High and trade winds in the South Atlantic between the equator and ~39 °S and between 0 and ~39 °W, 08 February 2021, 12 UTC.
Elaborated by Global Forecast System (GFS) model, 10m wind, wind magnitude ($m.s^{-1}$), direction (barbs) and mean sea level pressure (hPa). The dotted curve shows the axis of the ridge.
Source: adapted from Tropical Tidbits, with NOAA data.

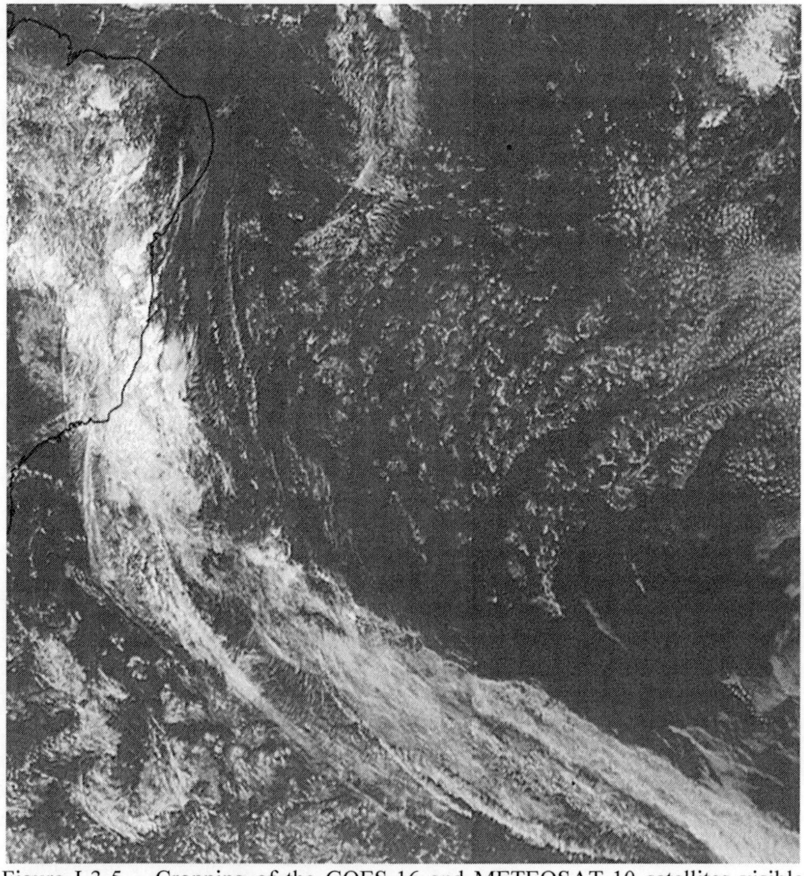

Figure I-3-5 – Cropping of the GOES-16 and METEOSAT-10 satellites visible channel, 08 February 2021, 12 UTC.
Source: adapted from Windy, with NOAA and EUMETSAT data.

In the Southern Hemisphere, the trade winds have a more meridional flow than in the Northern Hemisphere, from the southeast to the northwest (Figure I-3-1). The height of the base of low cloudiness depends on the height of the temperature inversion, since below this inversion level, clouds usually form (Figure I-3-6). The cloudiness has a maximum vertical development above the base of the inversion; usually along the western edges of subtropical ridges (Figure I-3-6).

The height of the base of the trade wind inversion is usually higher near the BNE coast (1500 - 2000 m; the inversion is weaker; Figure I-3-7), which provides a better cloudiness vertical development and is not observed on average over the ITCZ region (Palmen & Newton 1969, adapted from Ficker 1936).

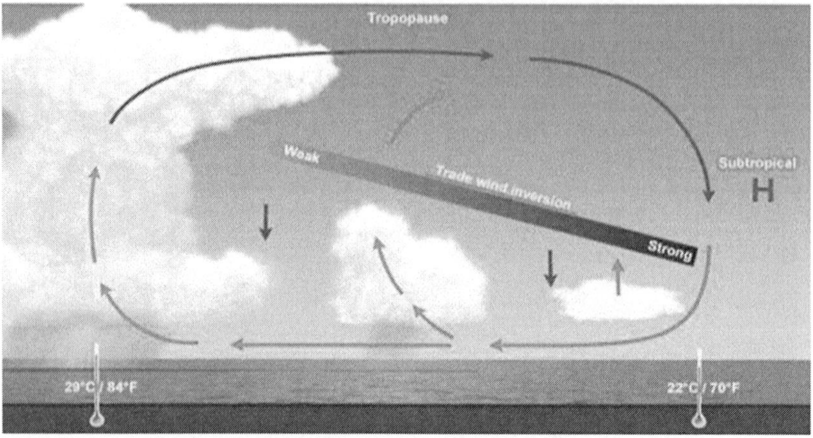

Figure I-3-6 – Conceptual model of TW inversion vertical profile.
Source: The COMET Program (2007).

The vertical profiles show an example of a typical inversion or stable layer in TW on the BNE coast (Figure I-3-8a) and over the South Atlantic (Figure I-3-8b), according to the GFS analysis on February 8, 2021, at 12UTC. Even when averaging data using a large area, as in Figure I-3-8b, one can clearly see the temperature inversion/stable level in TW between ~800 and 900 hPa. The vertical distribution of temperature and humidity shows the subsidence process with inversion or the formation of a stable layer at this level. Below 900 hPa, absolute instability is observed, and above 800 hPa, conditional instability with vertical air movement. However, there is too little humidity over the BNE (Figure I-3-8a).

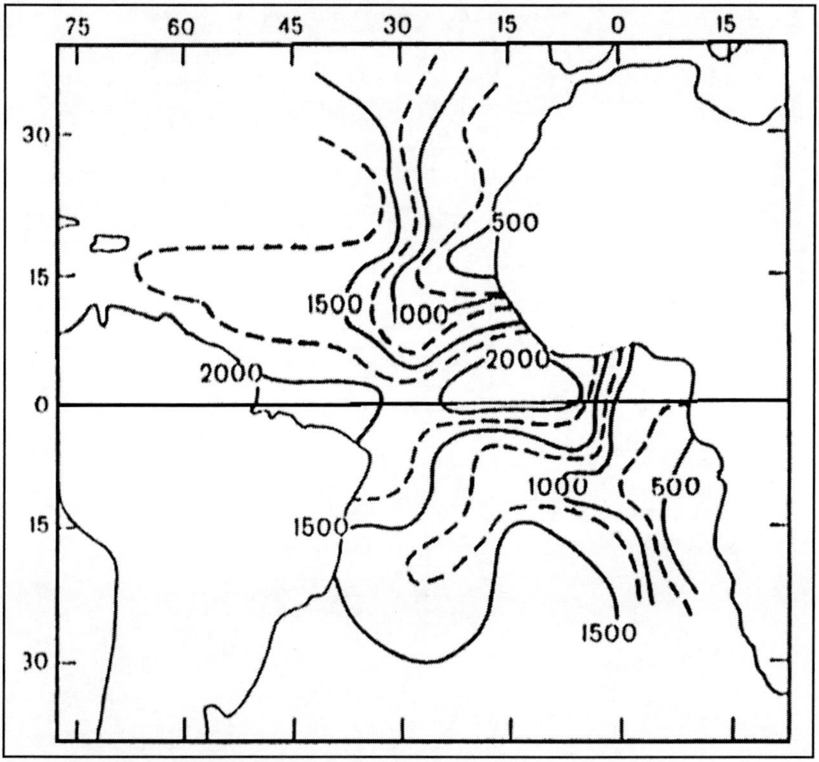

Figure I-3-7 – Base height (m) of the inversion layer in the trade winds over the Atlantic Ocean.
Source: Palmen & Newton 1969, adapted from Ficker 1936.

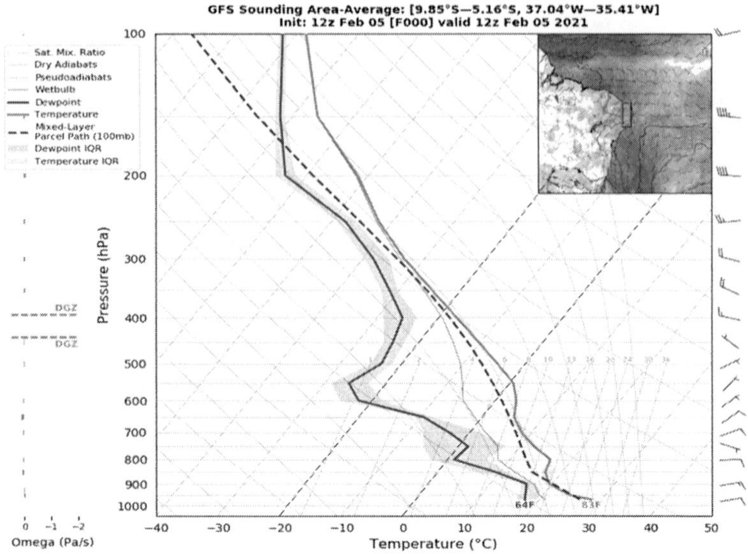

(a)

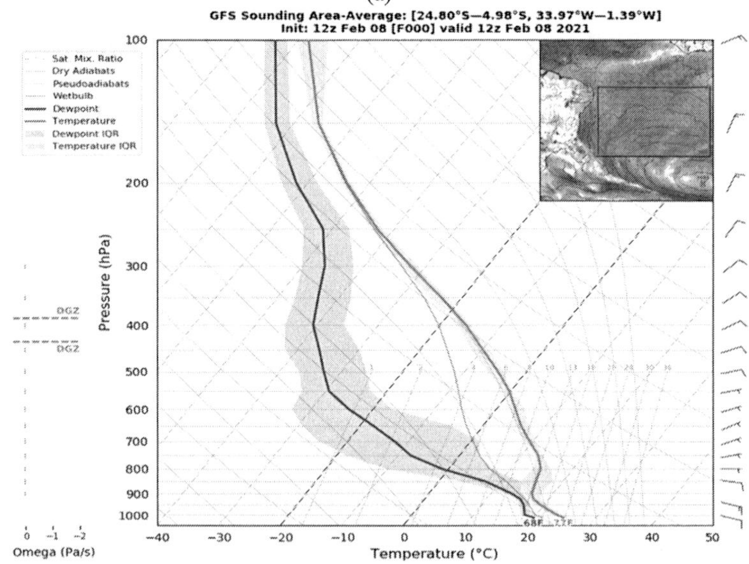

(b)

Figure I-3-8 – Averaged area of the GFS vertical profile over part of the BNE (a) and a larger area in the South Atlantic Ocean (b) on 08 February 2021, 12 UTC. Source: adapted from Tropical Tidbits, with NOAA data.

3.2 Wave Disturbances in the Trade Winds – Guidance and Example

Based on the literature presented and for operational purposes, the following list is proposed with some criteria (guidance) for identifying, monitoring, and forecasting Wave Disturbances in the Trade Winds.

Look for cloudiness on satellite images, spreading along the trade winds (east flow) over the South Atlantic; it may or may not have deep convection at some point in its cycle (cumulonimbus clusters are found in 63.7% according to Gomes et al. 2019).

Frequently WDTW is shallow/weak convection (cumulus congestus and stratocumulus clusters with little cumulonimbus activity). Convection is most noticeable near the coastal BNE.

- Look for a synoptic or meso-α trough at low levels in the streamlines (not necessarily all levels at the same time) extending westward (propagating westward).

  Usually, a trough is found at 850 and/or 700 hPa, but it is not uncommon to see a trough at the surface (1000 hPa) or 925 hPa. The trough may be stationary for some reason, so further study may be required.
- It may also help to detect cyclonic relative vorticity minima and/or convergence minima along the eastern trough or at least a weak low level cyclonic circulation.

To better explain this guide, below is a WDTW example (represented by Pontes da Silva, 2011). Figures 9 and 10 show two successive analyzed disturbances.

The METEOSAT-7 satellite image (infrared channel) dated 22 May 2006, 00 UTC (Figure I-3-9a) shows convective cloudiness (arrow) over the Atlantic Ocean near the BNE. The convection intensifies, reaching the BNE in the following hours (Figure I-3-9b). This was the first (W1) of two WDTW events. On 24 May 2006, W1 crossed the BNE and reached its northern part, the cloudiness of which was associated with the ITCZ cloudiness (Figure I-3-10b). The satellite image from May 22, 2006 already shows the lower and middle clouds (in addition to small convective cores) of the second WDTW (W2) centered at ~7°S. / 18°W in TSA (Figure I-3-10 – ellipses).

At 1000 hPa, the ECMWF Reanalysis (ERA-Interim) streamlines show a cyclonic flow east of the BNE (trough with north-south orientation), with a confluence of winds converging between the PE and RN coast, suggesting convergence at this location. At 850 hPa, the relative vorticity (RV) field shows an area of negative relative vorticity in the region of the trough that extends meridionally, between 1°N and 6°S, east of the BNE, which was the strongest signal from W1 (Figure I-3-9c).

At 700 hPa, the vorticity core and trough are still visible in the same field, but are less intense and displaced to the north (Figure I-3-9d). At a level of 500 hPa (figure not shown), the streamlines showed a cyclonic circulation in the region between the two ridges, where, according to the satellite image of that time, the WDTW passed. Streamlines at 200 hPa (figure not shown) also showed a trough but moving eastward with winds converging at its forefront edge.

a)

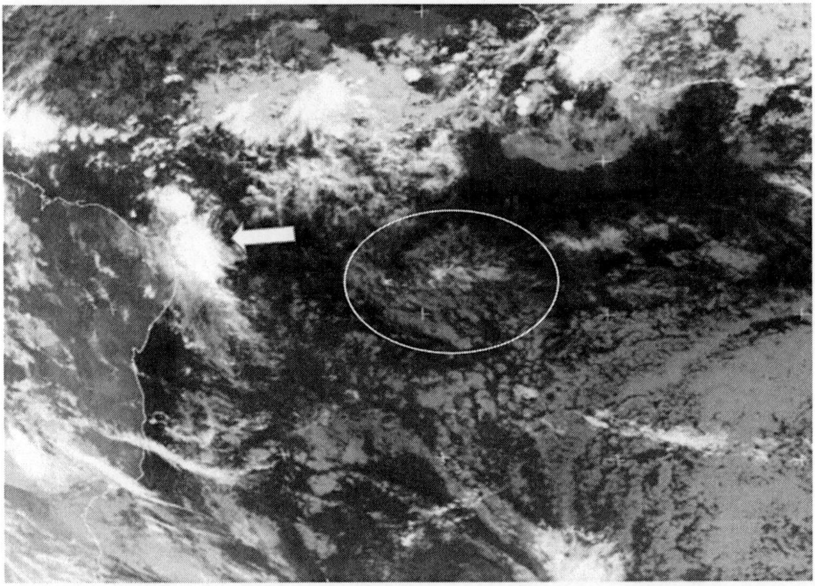

b)

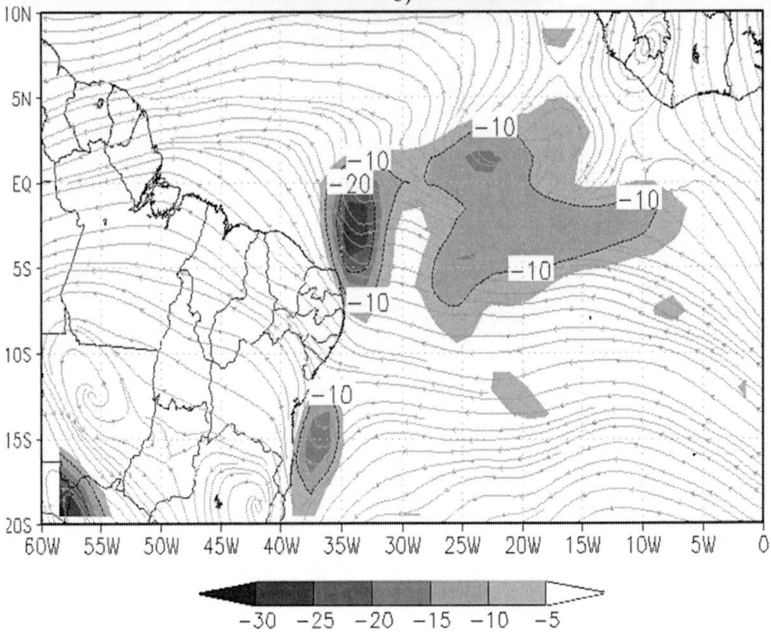
10N
5N
EQ
5S
10S
15S
20S
60W
55W
50W
45W
40W
35W
30W
25W
20W
15W
10W
5W
0
-10
-20
-10
-10
-10
-10
-10
-10
-30 -25 -20 -15 -10 -5

(c)

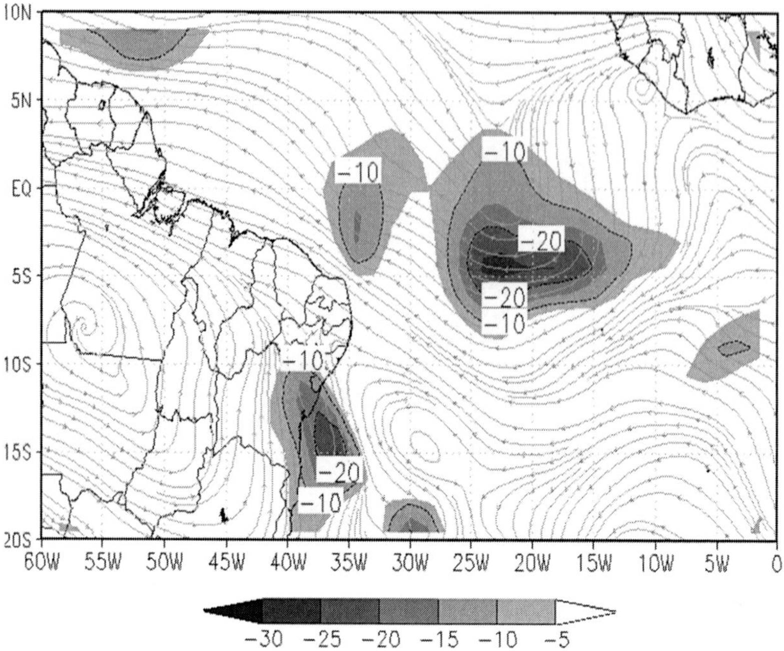
10N
5N
EQ
5S
10S
15S
20S
60W
55W
50W
45W
40W
35W
30W
25W
20W
15W
10W
5W
0
−10
−10
−20
−20
−10
−10
−20
−10
−30 −25 −20 −15 −10 −5

(d)

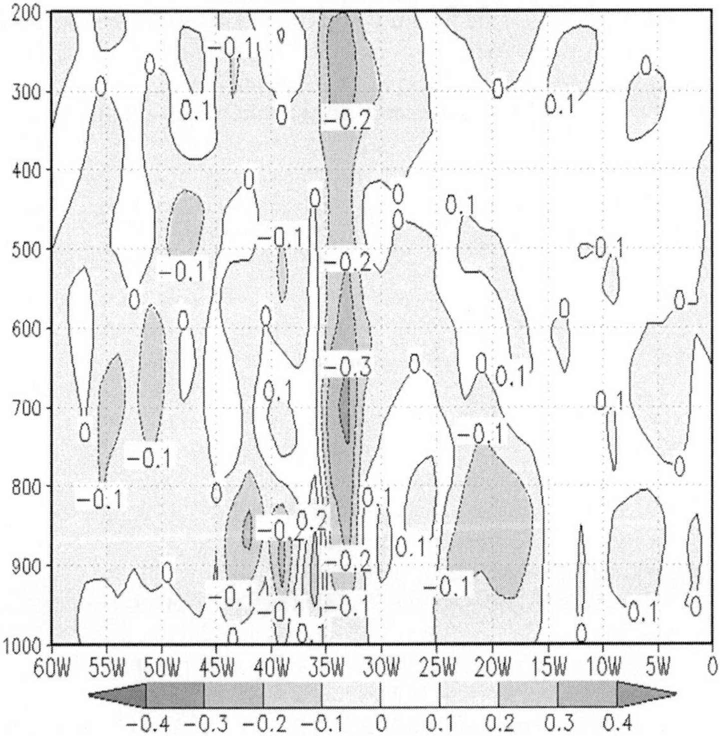

(e)

Figure I-3-9 – WDTW example on 22 May 2006, 00 and/or 06 UTC:
(a) and (b) cuts of the IR channel of METEOSAT-7 satellite images, 00 and 06 UTC, respectively.
(c) and (d) show the ERA-Interim reanalysis streamlines and cyclonic relative vorticity at 850 and 700 hPa, respectively, 06 UTC.
(e) omega vertical cross-section at 6 °S from the ERA-Interim reanalysis, 06 UTC. Arrows and ellipses show the cloudiness of two WDTW.
Source: adapted from Pontes da Silva (2011), Dee et al., 2011).

This may have contributed to the de-intensification and reduction of the wide distribution of W1 towards the continent, as this may have caused subsidence, thereby weakening this system.

At low and middle levels, disturbances associated with W2 are also detected, such as the trough and wind confluence at its rear at 700 hPa (Figure I-3-9d). W2 was not yet been intense in Figure I-3-9, probably due to convergence at high levels, which may have inhibited intense convective activity.

A few days earlier, the frontal system passed in the South Atlantic, and its extremity (FE-frontal extremity) reached the BNE between 15 and 20 May. This process disturbed the TW and FE, eventually merged with the southern winds of the anticyclone and thereby increased the transfer of moisture from the ocean to the BNE. These factors may contribute to the formation and intensification of W1, which advanced across the continent, causing rainfall in the North and Northeast. It is important to note that in this case the W1 cloudiness was already coupled with the ITCZ cloudiness.

Omega section along 6°S (Figure I-3-9e) on May, 22, 2006, 06 UTC, showed a predominant lifting between 35 and 31°W, in the area with the highest convective activity on W1. The most intense omega nuclei were observed in the 600-800 hPa and 200-350 hPa layers, which means that low and high levels indicate greater W1 activity. There is also a negative omega core around 20°W, between 800 and 900 hPa, associated with W2. Intense subsidence can be detected between 5 and 10°W, in the area closest to the South Atlantic Subtropical Ridge (SASR).

The METEOSAT-7 satellite image from May 23, 2006, 06 UTC (Figure I-3-10a), shows the W2 intensification in the east of the BNE, which is visible though horizontal magnification and clouds with weak vertical development.

W1 continues to approach the ITCZ, affecting weather in the north of the Northeast (Figure I-3-10; arrows). W2 reaches the BNE on May 24, 2006 at 00 UTC (Figure I-3-10b). Unlike W1, W2 did not penetrate the continent, dissipating on the night of 24-25 May.

Looking at the circulation on May 24, 2006, 00 UTC, a trough was noted at 1000 hPa over the BNE (figure not shown); the southeastern TWs were more intense and, therefore, contributed to the transport of moisture to the coastal region. At 850 hPa, in terms of cyclonic Relative Vorticity (RV), W2 was more intense than at 1000 hPa and W2 increased from 23 May. The RV core associated with W2 was visible even before the wave could be identified from satellite images. At 700 hPa, a trough and this cyclonic RV core can also be identified, only less intense and located to the north of its position at 850 hPa, as happened in W1 (Figure I-3-10d).

At middle levels (500 hPa), the confluence of the wind over W2 and a wide ridge to the south of it indicate lifting conditions at these levels (figure not shown). At 200 hPa, the circulation was anticyclonic and divergent above W2, which may contribute to lifting. The dissipation of W2 occurred mainly due to the weakening of the troughs at low levels (figure not shown).

Omega vertical section along 8°S on May 24, 2006, 00 UTC shows lifting over the W2 (between 35 and 30°W) (Figure I-3-10e). The most intense negative omega core was located in the 700-900 hPa layer, which indicates that the most important processes for W2 intensification occurred at low levels.

a)

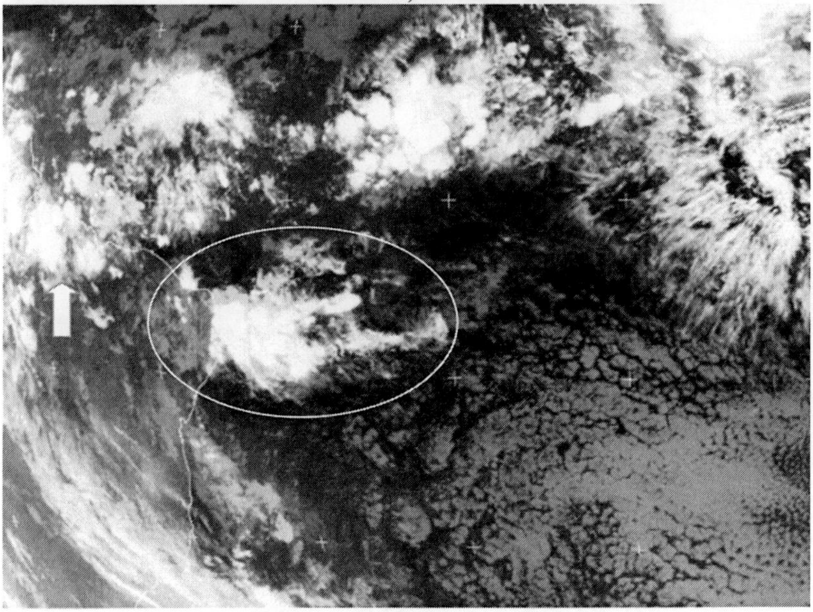

b)

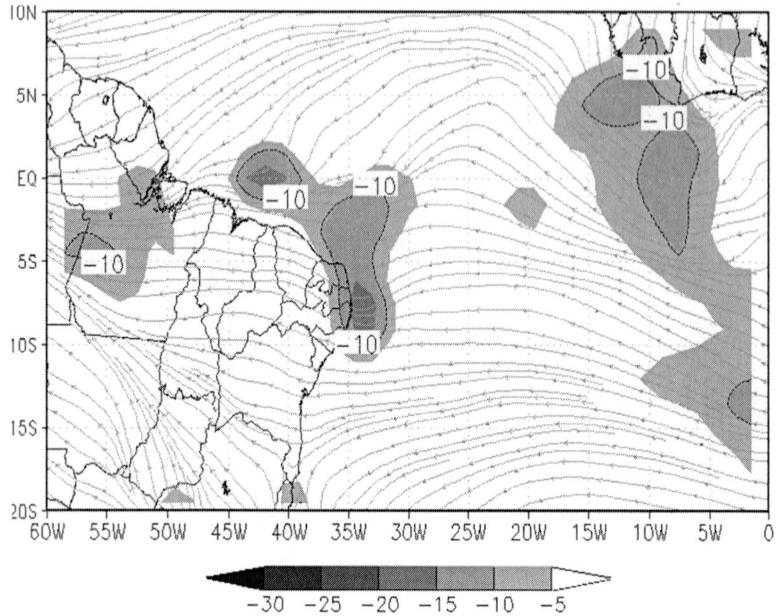
10N
5N
EQ
5S
10S
15S
20S
60W 55W 50W 45W 40W 35W 30W 25W 20W 15W 10W 5W 0
−10
−30 −25 −20 −15 −10 −5

c)

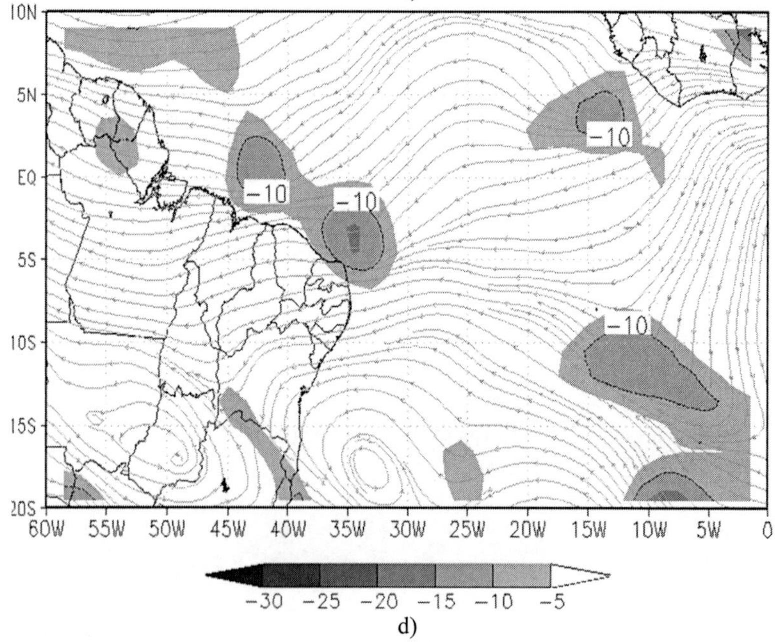
10N
5N
EQ
5S
10S
15S
20S
60W 55W 50W 45W 40W 35W 30W 25W 20W 15W 10W 5W 0
−10
−30 −25 −20 −15 −10 −5

d)

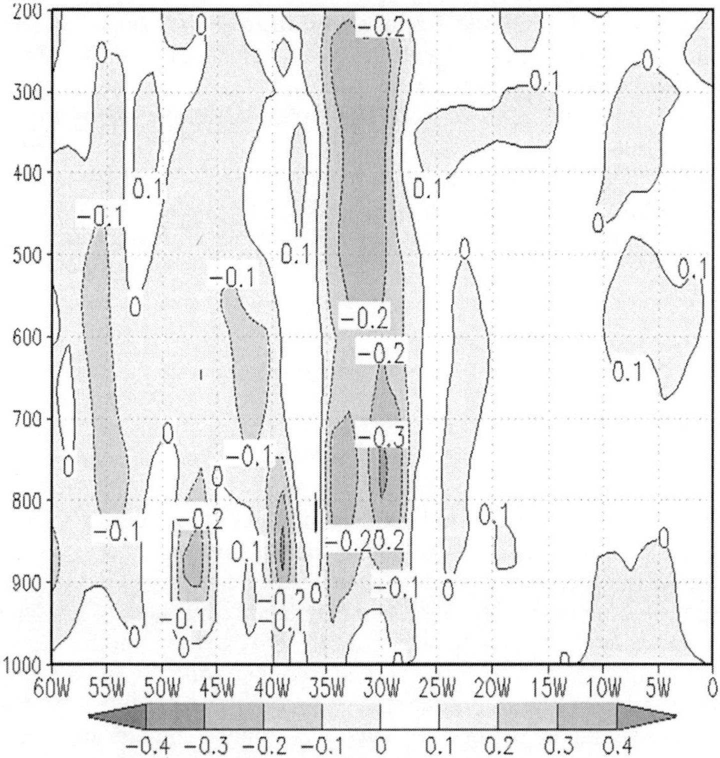

e)

Figure I-3-10 – WDTW example on 23 and 24 May 2006, 06 and 00 UTC:
(a) and (b) cuts of the IR channel of METEOSAT-7 satellite images for 23 May 2006, 06 UTC and 24 May 2006, 00 UTC, respectively
(c) and (d) show the ERA-Interim reanalysis streamlines and cyclonic relative vorticity at 850 and 700 hPa, respectively on 24 May 2006, 00 UTC
(e) omega vertical cross-section along 8 °S by the ERA-Interim reanalysis on 24 May 2006, 00 UTC.
Arrows and ellipses showing two WDTW.
Source: adapted from Pontes da Silva (2011), Dee et al., (2011).

Synoptic analysis showed that in the two WDTWs studied, the main process was detected at low levels, especially between 700 and 850 hPa. The W1 event was associated with the frontal zone passage 7 days before the WDTW formation in the BNE, which disturbed the southeast trade winds. The W1 crossed north of the BNE, joining the ITCZ and convection over the continent. The second event (W2) was already observed in the Atlantic on May 20, 2006.

This WDTW was displeased westward, reaching the BNE four days later. The trough associated with this wave was more intense at 700 hPa and moved to lower levels as it approached the BNE.

It is very important to note that multiple small troughs in the TW (usually smaller than meso-α, Figure I-3-4 e I-3-5) can be identified throughout the year, but mostly in autumn and winter. These small waves can form low clouds that can produce light precipitation (at times light to moderate), especially when they interact with the coastal convergence near the BNE. These multiple meso-β disturbances can be mistaken for WDTWs, however, they are not the same systems, but much less intense.

4. Conclusion

A brief review of wave disturbances in the trade winds is carried out, focusing on systems that affect the weather conditions over the BNE.

Most authors suggest that WDTW occur at low levels of the troposphere, such as troughs, cyclonic and convergent vorticity centers with lifting and westward movement. Climatological data show that these systems are common in the Southern Hemisphere from autumn to winter and are responsible for most of the annual precipitation over the BNE.

Thanks to the main characteristics of WDTW found in the literature, it was possible to create a practical guide to identifying this synoptic system. The criteria are mainly to identify and monitor of westward propagating cloudiness (convective or not), low level troughs and centers of cyclonic vorticity and divergence.

Further study of multiple small troughs in TW (usually smaller than meso-α) is recommended as these perturbations can be mistaken for WDTW.

5. References

Alestalo M, Savijärvi H. *"Mesoscale circulations in a hydrostatic model: coastal convergence and orographic lifting."* Tellus A: Dynamic Meteorology and Oceanography, 37:2, 156-162, DOI: 10.3402/tellusa.v37i2.11662, (1985).

Asnani G. C. *Tropical Meteorology.* Indian Institute of Tropical Meteorology, 1202p, ISBN: 8190040006, 1993

Barbosa, R. L. *Interação das perturbações convectivas iniciadas na costa Norte do Brasil com Distúrbios Ondulatórios de Leste.* São José dos Campos. 81 p. MSc thesis, Instituto Nacional de Pesquisas Espaciais – INPE, 1997.

Burpee RW. *"The origin and structure of easterly waves in the lower troposphere of north African."* J. Atmos Sci 29, 77-90, (1972).
Burpee R. W. "*Characteristics of North African easterly waves during the summers of 1968 and 1969"*. Journal of the Atmospheric Sciences 31(6), 1556-1570, (1974).
Céron JP, Guérémy JF *"Validation of the Space–Time Variability of African Easterly Waves Simulated By The CNRM GCM."* J. Climate 12, 2831–2855, (1999).
Chang C P, Miller C R. "*Comparison Of Easterly Waves In The Tropical Pacific During Two Contrasting Periods Of Sea-Surface Temperature Anomalies.*" J. Atmos. Sci. 34, 615- 28, (1977).
Chan CS *"Análise de distúrbios ondulatórios de leste sobre o Oceano Atlântico Equatorial Sul."* MSc thesis, Instituto Nacional de Pesquisas Espaciais, São José dos Campos, 104 p, 1990.
Coutinho, E.C.; Fisch, G. *"Easterly wave disturbances over Centro de Lanxcamento de Alcantara –MA.", Revista Brasileira de Meteorologia* 22, n.2., 193-203, 2007.
Dee, D.P., Uppala, S.M., Simmons, A.J., Berrisford, P., Poli, P., Kobayashi, S., Andrae, U., Balmaseda, M.A., Balsamo, G., Bauer, P., Bechtold, P., Beljaars, A.C.M., van de Berg, L., Bidlot, J., Bormann, N., Delsol, C., Dragani, R., Fuentes, M., Geer, A.J., Haimberger, L., Healy, S.B., Hersbach, H., Hólm, E.V., Isaksen, L., Kållberg, P., Köhler, M., Matricardi, M., McNally, A.P., Monge-Sanz, B.M., Morcrette, J.-.-J., Park, B.-.-K., Peubey, C., de Rosnay, P., Tavolato, C., Thépaut, J.-.-N. and Vitart, F. *"The Era-Interim Reanalysis: Configuration And Performance Of The Data Assimilation System.*" Quarterly Journal of The Royal Meteorological Society 137, 553-597, https://doi.org/10.1002/qj.828 (2011).
Diedhiou AS, Machado LAT. *"Laurent H Mean kinematic characteristics of synoptic easterly disturbances over the Atlantic."* Adv Atmos Sci 27(3),1-17, (2010).
Espinoza, E. S. *"Distúrbios nos ventos de leste no Atlântico Tropical."* MSc thesis., Instituto Nacional de Pesquisas Espaciais, São Jose dos Campos, 123p, 1996.
Fedorova Natalia. *Sinótica IV: Sistemas e processos sinóticos atuantes na américa do sul.* Material didático: sinopses, figuras, equações Maceió: Ed. Universitária/ UFAL, 192 p. ISBN: 978-85-7177-422-3; CDU: 551.5, 2008.
Ficker H., "*Die Passatinversion", Veroeffentl. Meteorol. Inst. Univ. Berlin 1, n4,* 1936.

Gomes HB, Ambrizzi T, Herdies DL, Hodges K, Pontes da Silva BF *"Easterly wave disturbances over northeast Brazil: an observational analysis."* Adv Meteorol. (2015) https://doi.org/10.1155/2015/176238

Gomes, H. B., Ambrizzi, T., Pontes da Silva, B. F., Hodges, K., Silva Dias, P. L. da, Herdies, D. L. *"Climatology of easterly wave disturbances over the tropical South Atlantic." Clim Dyn* 53 (3-4), 1393–1411 https://doi.org/10.1007/s00382-019-04667-7, (2019).

Hall, B.A. *"Westeward moving disturbances in the South Atlantic coinciding with hevy rainfall events at Ascension Island."* The Meteorological Magazine 118, n. 1405, p. 175–181, (1989).

Karyampudi VM, Carlson TN *"Analysis and numerical simulations of the Saharan air layer and its effect on easterly wave disturbances"*. J Atmos Sci 45, 3102–3136, (1988).

Kayano, M.T. *Um estudo climatológico e sinótico utilizando dados e* radissondagem (1968 - 1976) de Manaus e Belém. São José dos Campos, (INPE 1559 - TDL/013), 1979.

Kouadio YK, Servain J, Machado LAT, Lentini CAD *"Heavy Rainfall Episodes In The Eastern Northeast Brazil Linked To Large-Scale Ocean-Atmosphere Conditions In The Tropical Atlantic."* Advances In Meteorology, Article Id 369567, 16 Pages, (2012). Doi:10.1155/2012/369567.

Kousky VE *"Diurnal Rainfall Variation in Northeast Brazil."* Mon. Weather Rev. 108, 488-498, (1980).

Kousky VE *"Frontal Influences on Northeast Brazil."* Mon. Wea. Rev. 107, 1142-1153, (1979).

Keshavamurty RN *"Vertical Coupling In The Indian Summer Monsoon."* Nature Physical Science 232, 169-170, (1971).

Krishnamurti TN, Baumhefner D *"Structure Of A Tropical Disturbance Based On Solutions Of A Multilevel Baroclinic Model."* Journal of Applied Meteorology 5, Issue 4, 396-406, (1966).

Merrit, E. S. *"Easterly waves and perturbations: a reappraisal."* Journal of Applied Meteorology 3, 367-382, (1964).

Molion, L. C. B.; Bernardo, S. O. *"Uma Revisão da Dinâmica das Chuvas no Nordeste Brasileiro."* Revista Brasileira de Meteorologia 17(1), 1-10, (2002).

Mota GV *"Estudo Observacional De Distúrbios Ondulatórios De Leste No Nordeste Brasileiro."* MSc thesis, University of Sao Paulo, (1997)

Mota GV, Gandu AW *"Análise de Distúrbios Ondulatórios de Leste que Afetam o Nordeste Brasileiro: Um Estudo De Caso." In:* IX Congresso Brasileiro De Meteorologia, (1996).

Mota GV, Gandu AW *"Análise Dos Padrões Ondulatórios De Leste No Nordeste Brasileiro Durante O Inverno De 1994.*" In: X Congresso Brasileiro De Meteorologia, Brasília, (1998)

Palmén, E., Newton C.W. *Atmospheric circulation systems: their structure and physical interpretation.* Academic press, New York. 603p, 1969.

Pontes da Silva BF, Fedorova N, Levit V, Peresetsky A, Brito BM *"Sistemas Sinóticos Associados Às Precipitações Intensas No Estado De Alagoas."* Revista Brasileira De Meteorologia 26, n.3, 323 – 338, (2011).

Pontes da Silva BF, Rocha RP, Gomes HB *"Easterly Wave Disturbances Activity Over the Eastern Northeast Brazil During 2006-2010 Rainy Seasons."* Revista Científica Foz, 3(2), 30, (2020).

Reed JR, Recker EE *"Structure And Properties Of Synoptic-Scale Wave Disturbances In The Equatorial Western Pacific."* J. Atmos. Sci. 28, 1117-1133, (1971).

Rennick, Mary Alice. *"The Generation of African Waves."* Journal of the Atmospheric Sciences 33, 1955-1969, (1976).

Riehl, H. *Waves in the easterlies and polar front in the tropics.* Chicago: Chicago University, Department of Meteorology, 79p, 1945.

Riehl, H. *"Waves in the easterlies"*. In: Tropical Meteorology. Departament of Meteorology, University of Chicago, McGraw-Hill, New York, 210-234, 1954.

Riehl, H *"Climate and Weather In The Tropics.* London: Academic Press, 611 p., 1979.

Rosenthal SL "*A Simplified Linear Theory Of Equatorial Easterly Waves."* J. Meteor. 17, 484–488, (1960).

Roeloffzen, J. C., Van Den Berg, W. D., Oerlemans, J. *"Frictional convergence at coastlines."* Tellus 38A, 397-411, (1986).

Tai KS, Ogura Y "*An Observational Study Of Easterly Waves Over The Eastern Pacific In The Northern Summer Using FGEE Data."* Journal of The Atmospheric Sciences, 44(2), 339-361, (1987).

Torres Roger Rodrigues. *"Analise e Modelagem dos Distúrbios Ondulatórios de Leste que atuam na Costa Leste do Nordeste do Brasil."* MSc thesis, Instituto Nacional de Pesquisas Espaciais, São José dos Campos, (INPE-15459-TDI/1421), 2008.

Torres, RR, Ferreira, NJ. "*Case Studies of Easterly Wave Disturbances over Northeast Brazil Using the Eta Model",* Weather and Forecasting 26, 225-235, (2011).

Varejão-Silva, M. A. *Meteorologia e Climatologia*, Brasiilia: INMET, Stilo, 515 p., 2000.

Vaz JCM. *"Análise das Ondas de Leste sobre a costa leste do Nordeste do Brasil para o período entre 1999-2009."* MSc thesis, UFRJ/IGEO/CCMN, (2011).

Vasquez, T. *Weather forecasting handbook*. Weather graphics technologies. Garland, Texas, 75p, 2000.

Williams KT "*Characteristics Of The Wind, Thermal, And Moisture Fields Surroundings The Satellite-Observed Mesoscale Trade Wind Cloud Clusters In The Western North Pacific."* In: Symp. Tropical Meteorology, Honolulu, Amer. Meteor. Soc., D Iv-1 A D Iv-6, (1970).

Yamazaki, Y. "*Estudos teóricos e sinóticos dos distúrbios tropicais."* MSc thesis, São José dos Campos. Instituto Nacional de Pesquisas Espaciais, (INPE-624-LAFE), 46p,1975.

Yamazaki, Y.; Rao, V. B. *"Tropical Cloudiness over the South Atlantic Ocean."* Journal of the Meteorology Society of Japan, 55, n. 2, 204-207, (1977).

Yanai M *"A Preliminary Survey of Large-Scale Disturbances Over The Tropical Pacific Region."* Geofis. Int. 3, 73–84, (1963).

Yanai M. *"Evolution of a Tropical Disturbance in the Caribbean Sea Region."* Journal of the Meteorology Society of Japan, 46, 86-109, (1968).

CHAPTER I.4

UPPER TROPOSPHERIC CYCLONIC VORTEX

BRUNO MIRANDA DE BRITO
AND NATALIA FEDOROVA

1. Introduction

This chapter presents typical and atypical characteristics of the Upper Tropospheric Cyclonic Vortex (UTCV). UTCV is a high troposphere low pressure system operating over the Brazilian Northeast (BNE) region and the adjacent South Atlantic Ocean. These vortices are known as Palmer type UTCVs and are tropical in origin. The UTCV patterns represent a cyclonic circulation at high levels, which are sometimes also deeper and reach middle levels. Cold air sinks at the vortex center. At the UTCV edges, warm air rises with the formation of cloudiness and rain.

UTCV is one of the meteorological systems that cause rain, thunderstorms and hail in the BNE (Gan, 1982; Brito et al., 2011; Glickman, 2000; Kulivoc & Rudnev, 1980; Knight & Knight, 2021).

These meteorological phenomena are detrimental to local communities. Examples of havoc can be seen in Figure I-4-1, and its formation will be explained in paragraph 6.

It is important to know that similar systems have names such as "slack lows", "low Palmer", "upper troposphere cyclonic vortices", "high-level tropical cyclones" and "kona cyclones" (Gan, 1982). The first UTCV studies for South America were carried out by Dean (1971) and Aragao (1975).

Figure I-4-1 – Images of events that occurred in the east of the BNE during UTCV between 2 and 5 November 2013.
Source: G1.

2. UTCV identification

2.1 UTCV identification using streamlines

UTVC can be identified using streamlines and geopotential height isohypes. The streamlines represent a cyclonic circulation (clockwise in the Southern Hemisphere) at high levels (200-300 hPa).

An example of UTCV is shown in Figure I-4-2. UTCV was registered between 2 and 4 April 2022 in the northeast of the Bahia state. UTCV changed the location of the center in a short time. This vortex system was restricted at high levels of the atmosphere.

This UTCV was not observed at middle levels of 500 and 700 hPa (Figures I-4-2c and I-4-2d). Also, there was almost no cloudiness in the UTCV circulation on the satellite image, and only a slight cloudiness was observed in the center of the Bahia State (figure I-4-2b).

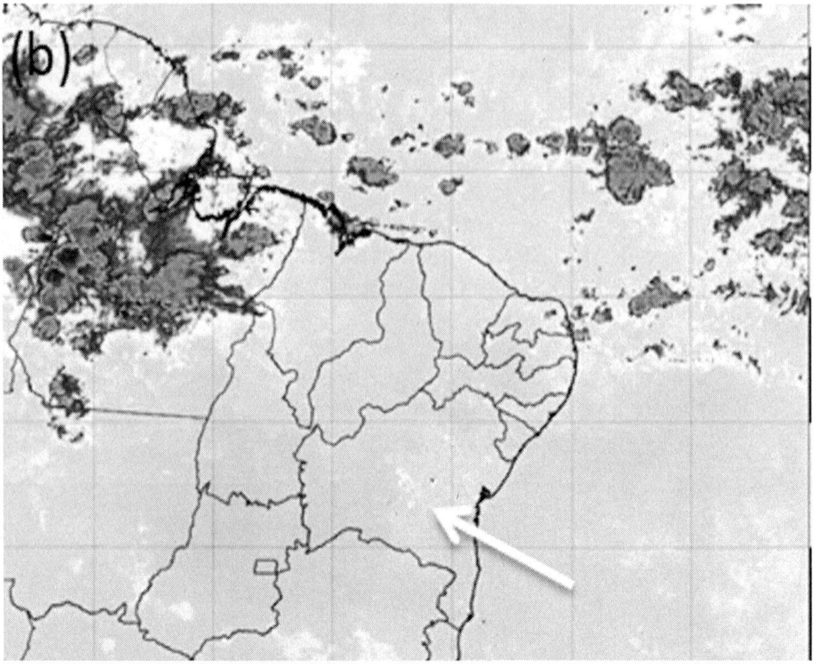
(b)

(c)
INPE-DIPTC
03-04-2022 12Z

Figure I-4-2 – Synoptic analysis at the levels of 250 hPa (a), 500 hPa (c), 700 hPa (d) and IR satellite image (b) on 3 April 2022.
White arrows indicate the center of the UTCV.
Light gray with arrows shows streamlines and white lines geopotential height.
Source: INPE.

High levels are the main location of UTCV, but this system is not limited to this height. There are episodes when the system expands and deepens to other levels. UTCVs in some cases reached the middle level (500-700 hPa) and rarely deepened to 800 hPa (Morais et al., 2020).

2.2 UTCV identification using satellite images

UTCV can be identified from satellite images in the water vapor (WV) and infra-red (IR) channels. The WV channel presents humidity above 600 hPa and the new GOES-16 data shows Upper-Level Tropospheric Water Vapor in Band 8 (6.2 μm). The black and white scale shows the decrease in humidity from white as high humidity to black at no humidity.

UTCV is identified by the dry part in the center (black image color indicated by gray dotted arrow) (Figure I-4-3). The white arrow marks high humidity with intense convection at the vortex periphery.

Not all UTCVs can be identified on satellite images. When UTCVs are deeper, they tend to have more humidity and cloudiness. Satellite images identify the clouds that reveal the UTCV, especially in summer and, as a result, UTCVs are visible on satellite images. Dry UTCVs are not visible on satellite images (Gan, 1982). Generally, dry cases occur in winter season.

The UTCV of a classic form with a typical structure can be confirmed by satellite data. If the typical structure is not visible on satellite images, the absence of UTCV cannot be confirmed. Final confirmation can only be given by the streamlines at high and middle levels.

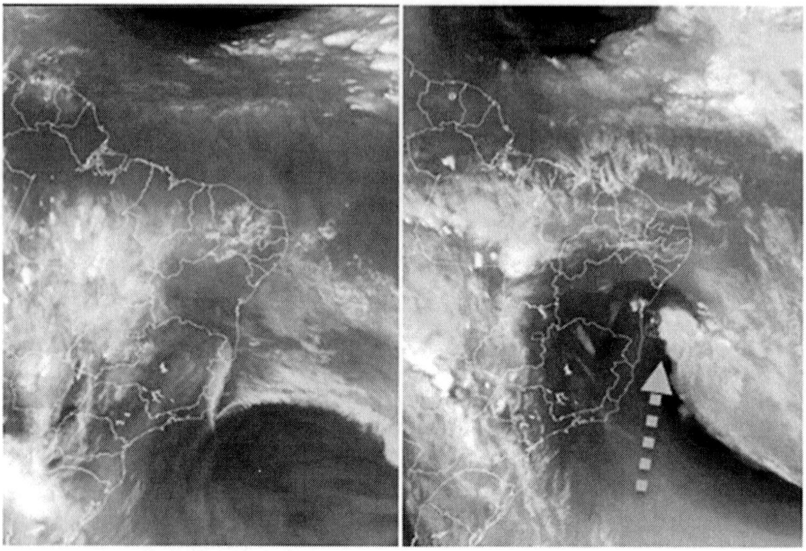

(a) (b)

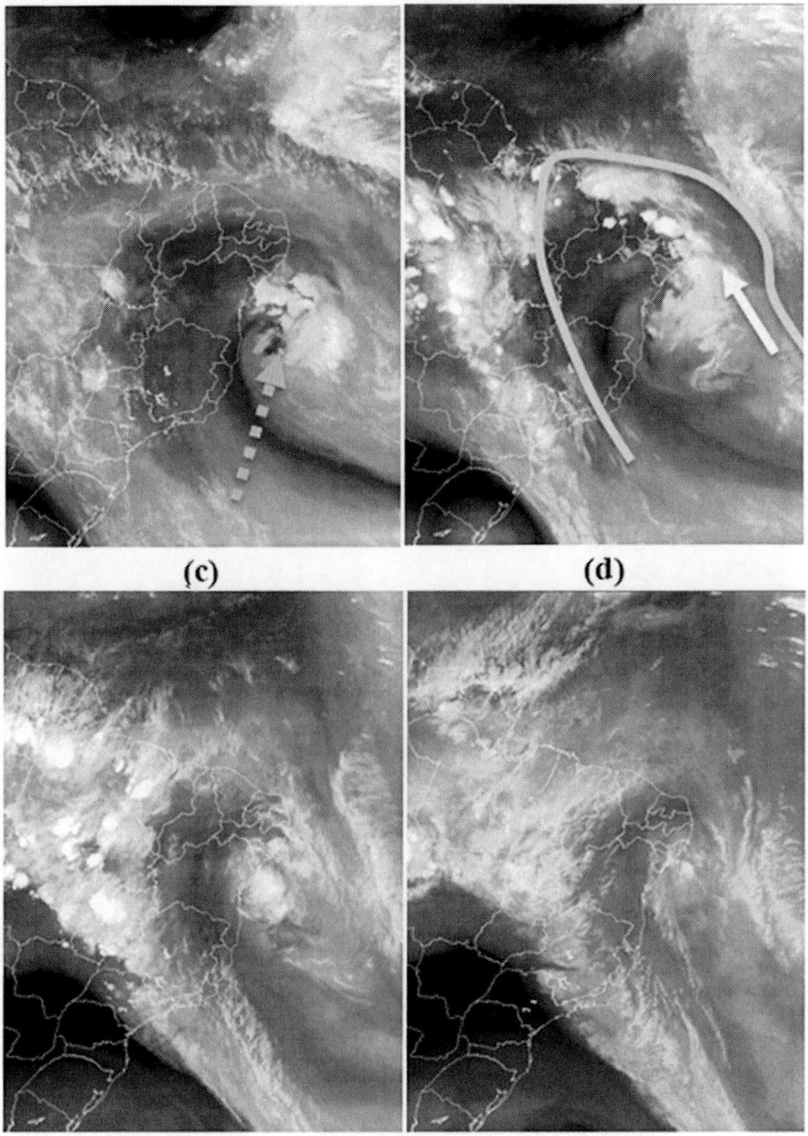

Figure I-4-3 – Fragments of VW GOES -13 satellite images, 2-5 November 2013: a) 2 November 00 UTC, b) 3 November 00 UTC, c) 3 November 12 UTC, d) 4 November 00 UTC, e) 5 November 00 UTC, f) 5 November 12 UTC.

The dark gray dotted arrow points the UTCV center.
The white arrow shows intense convection at the UTCV periphery.
The gray line passes through UTCV cloudiness.
Data source: INPE.

The UTCV pattern on satellite images depends on the associated systems. In cases where UTCVs are associated with Frontal Systems (FS), an S-shaped cloud pattern is observed when the FS moves over Brazil (Figure I-4-7D, Figure I-4-4a) (Ferreira et al., 2001). In these cases, usually cloudiness situation persists in the same place for a longer period. In cases where UTCV, FS and BH cloudiness merge, a cloudiness pattern is formed in the form of the letter Y (Figure I-4-4b).

There are also cases in which cloud patterns don't have any of the listed shapes (Fedorova, 2008). These events cannot be identified by cloudiness (Figure I-4-3b). Dry UTCVs did not show any conceptual cloudiness model on satellite imagery. In these cases, the cloudiness did not look like the S or Y pattern (Figure I-4-4).

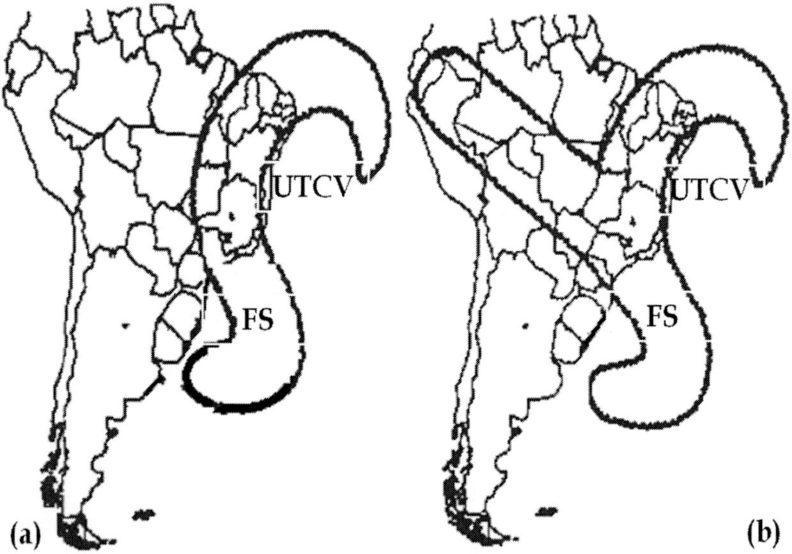

Figure I-4-4 – Conceptual model of cloudiness shape associated with UTCV, in the shape of "S" (a) and "Y" (b).
Source: Adapted from Ferreira et al. (2001).

2.3 Numerical method for UTCV identification

The steps of an objective method for identifying UTCV have been established (Coutinho et al., 2010; Morais et al., 2020). First, a semi-automatic method for identification the location of UTCV was developed (Coutinho et al., 2010). It is based on the determination of the vortex center by the minimum negative relative vorticity (ζ) at 200 hPa and the closed cyclonic circulation around this minimum. The relative vorticity calculation is determined for its minimum threshold less than $-2.5\times10^{-5}s^{-1}$. These minimum values are compared to eight neighboring grid points to find the location of the minimum values. This method is used at ten isobaric levels. In addition, more than one UTCV has been found in some situations. In these cases, the following criterion was adopted: if the difference between latitudes and/or longitudes is found to be more than five degrees, UTCVs are considered different.

A more advanced method (Morais et al., 2020) used automatic determination of the UTCV vertical extension and its trajectory tracking. To determine the UTCV vertical extent, the position of the center at different levels was analyzed. Thus, it was guaranteed that the same UTCV would be at other levels.

The UTCV trajectory is determined at 200 hPa every two consecutive steps at 6-hours intervals. A maximum distance of 9° UTCV center displacement was established between these two steps. If the displacement was less than or equal to 9°, the system was considered one UTCV; at larger values, they were considered two different UTCVs.

The results of the objective method (Morais et al., 2020) were compared with one year of streamline map identification. UTCVs lasting more than 24 hours were analyzed. It was confirmed that 90% of vortexes were the same by the two methods.

3. UTCV frequency, duration, and trajectories

Frequency

The UTCV frequency has been determined in many studies (e.g., Gan, 1982; Costa, 2010). UTCVs occur throughout the year, mainly during the summer and transition seasons (autumn and spring). In summer (Table 1), the largest number of cases is observed, especially in January. This information has been confirmed in several studies that have become generally accepted (Kouskly & Gan 1981; Gan, 1982; Calbete et al., 1996; Ramirez et al. 1999, 2000; Costa, 2010; Silva, 2007).

In winter, UTCVs are rarely recorded (Costa, 2010; operational practice). But a few cases were reported in the winter between 2000 and 2002. The interannual and seasonal frequency of UTCV according to the objective method is presented in Table 1 (Morais et al., 2020).

Duration

UTCVs are usually persistent and last more than a week (Cavalcanti et al., 2009). The duration of UTCV varies from 4 to 11 days (Cavalcanti et al., 2009). The average duration of UTCV events was 6 days, which were analyzed over 9 years between 1988 and 2000 in El Niño, La Niña and Neutral years (Repinaldo et al., 2020). Sometimes UTCV can last less than 24 hours, just a few hours. UTCV can quickly weaken and return to a trough. For example, this chapter presents an episode lasting less than 4 days (items 1 and 4).

The frequency of UTCV is higher for events lasting from 1 to 3 days (Table 1, Morais et al. 2020). Longer UTCVs were recorded at a lower frequency. In winter, UTCV events lasting 7 days or more were not observed. There were also no episodes lasting 10 days or more in the spring in the Southern Hemisphere.

The annual distribution of UTCV events from 1981 to 2015 is shown in Figure I-4-5 (Reis, 2018). The frequency of UTCV has been increasing during these years. The events showed no correlation with El Niño Southern Oscillation (ENSO) events. The occurrence of UTCV did not show significant variability in extreme ENSO years (Repinaldo et al., 2020). There is a slight increase (+1.4%) in the number of cases in El Niño years and a decrease (-6.4%) in La Niña years compared to neutral years.

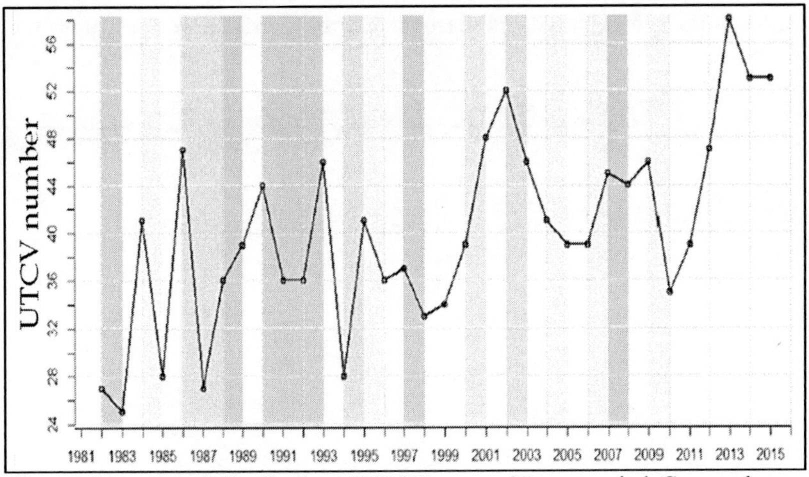

Figure I-4-5 – Annual distribution of UTCVs over a 37-year period. Gray scales are the intensity of the ENSO phenomenon.
Source: Adapted from Reis (2018).

Table I-4-1 – Interannual and seasonal distribution of the number of UTCV, from 1984 to 2013.
Adapted from Morais et al. (2020).

Life cycle (days)	1-3				4-6				7-9				≥10				
Season	SON	DJF	MAM	JJA	SON	DJF	MAM	JJA	SON	DJF	MAM	JJA	SON	DJF	MAM	JJA	Total
1984	13	32	13	2	2	4	2	0	0	0	0	0	0	0	0	0	67
1985	5	12	10	5	2	5	3	0	0	2	0	0	1	0	0	0	45
1986	15	32	18	1	0	4	0	0	0	1	0	0	0	0	0	0	71
1987	7	14	11	2	1	6	2	0	1	1	0	0	0	1	0	0	46
1988	14	21	10	3	0	4	1	0	0	1	0	0	0	2	0	0	56
1989	15	18	10	0	1	3	3	0	0	2	0	0	0	0	0	0	52
1990	11	21	10	5	5	5	2	0	1	1	2	0	0	1	0	0	64
1991	15	22	9	6	1	3	1	1	0	1	0	0	0	2	0	0	61
1992	9	23	11	6	1	4	2	0	0	1	0	0	0	0	0	0	57
1993	12	22	11	3	1	4	2	0	0	2	0	0	0	2	0	0	59
1994	11	15	12	3	1	5	0	0	2	2	1	0	0	1	0	0	53
1995	12	16	11	5	2	2	0	0	0	2	0	0	0	2	0	0	52
1996	10	13	5	4	0	5	1	0	1	0	1	0	0	1	0	0	41
1997	11	19	6	5	2	3	2	0	3	6	0	0	1	1	0	0	59
1998	11	20	11	3	0	7	2	0	0	2	0	0	0	0	0	0	57
1999	7	18	11	0	1	3	2	0	0	1	0	0	0	2	0	0	45
2000	11	24	10	3	1	2	1	0	0	1	0	0	1	2	0	0	56
2001	6	19	12	6	2	2	3	1	0	1	1	0	1	1	0	0	55
2002	13	11	17	5	1	2	4	0	0	3	0	0	1	2	0	0	59
2003	14	27	17	1	0	4	1	0	0	2	0	0	0	0	0	0	66
2004	13	15	13	1	1	2	2	0	1	2	0	0	0	2	0	0	52
2005	9	24	14	3	4	3	2	0	0	2	0	0	0	1	0	0	62
2006	15	18	8	5	1	6	1	0	1	3	0	0	0	0	0	0	58
2007	18	23	19	6	1	8	1	0	0	3	1	0	0	0	0	0	80
2008	21	22	9	4	0	3	0	0	0	1	0	0	0	1	0	0	61
2009	19	17	10	7	0	4	1	0	0	3	0	0	0	1	0	0	62
2010	9	25	9	8	1	2	1	0	0	2	1	0	0	2	0	0	60
2011	11	27	9	3	0	6	0	0	1	1	0	0	0	0	0	0	58
2012	11	27	15	7	1	4	1	0	0	1	1	0	0	1	0	0	70
2013	13	22	13	10	0	4	4	0	0	2	0	0	0	1	0	0	69

Trajectories

UTCVs can either move east or west or remain stationary (Cavalcanti et al., 2009). UTCVs, located only at high levels, have a predominant displacement to the east, and deeper vortexes (up to middle levels) to the west (Fedorova, 2008).

Based on the simulation of the Weather Research and Forecasting (WRF) model, it was confirmed that the UTCV trajectories to the west are similar to the trajectories obtained in the ERA-Interim reanalysis (Reboita et al., 2017).

4. Three-dimensional structure of UTCV

UTCVs are large-scale systems and form at the highest levels of the troposphere. In the BNE this means an altitude above 10 km (Fedorova, 2006). Sometimes UTCV can extend to middle levels, and in rare cases, to low levels. In these rare cases, only a weak (unclosed) cyclonic circulation at low levels is noted. The UTCV axis is vertical at the high and middle levels, so this system is defined as barotropic. UTCVs are formed at high levels and can then extend to other levels.

At the center of the UTCV, air sinking is observed (Kouskly & Gan, 1981) (Figure I-4-6). This indicates that cold and dry air enters from high altitudes into the middle layers of the troposphere. The opposite is observed at the periphery of the UTCV, where the lifting of warm and humid air causes the formation of convective clouds.

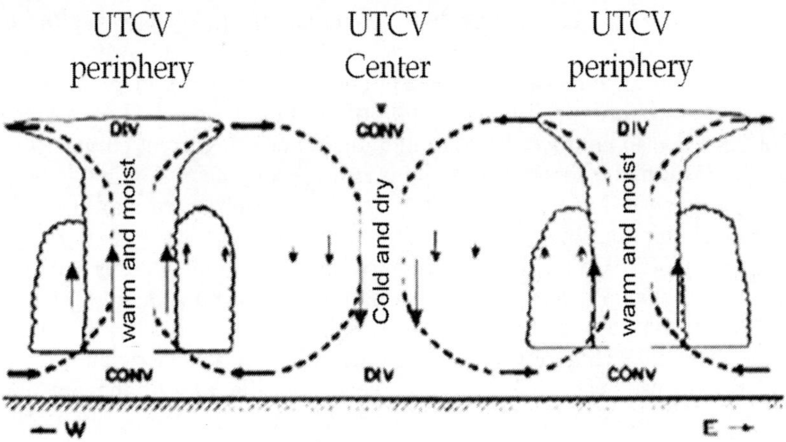

Figure I-4-6 – Scheme of the classical vertical movement in UTCV.
Source: Adapted from Kousky and Gan (1981).

Thus, in the center of UTCV there is a divergence at the middle levels and a convergence at the high levels. At the periphery of the vortex, surface convergence and altitude divergence can be observed. UTCVs have direct thermal circulation, in other words, warm and humid air gets lifted and cold and dry air sinks. Convergence at high levels in the UTCV center causes a sinking. Conservation of mass in the center leads to divergence at the middle levels. Air leaving from the center to the periphery causes convergence and air lifting. On the periphery, at high levels, difluence is observed.

The circulations between UTCV and the Bolivian High (BH) creates difluence at high levels. In these cases, there is low-level air current convergence at the UTCV periphery. In some events, lifting in the UTCV center and the formation of convective clouds or Cbs in this region is possible (Gan, 1982). In these cases, the area of divergence at high levels causes UTCV to weaken.

UTCVs reach low levels or close to the surface in 10% of events (Gan, 1982, adapted from Frank, 1966, 1970). They reach a low level at the stage of the maximum development (Morais et al., 2020). Deep UTCVs are short-lived, usually 1 to 3 days.

Jets with cyclonic curvature contribute to the intensification of cyclonic vorticity below the jet (Ramirez et al., 2000). After that, the jet enters the area of anticyclonic circulation and intensifies the anticyclonic vorticity above the jet. Subsequently, the northern sector of the trough intensifies up to the formation of a vortex.

Potential vorticity is conserved during adiabatic frictionless motion (Holton, 2004). The potential vorticity is measured in terms of absolute vorticity, which determines the effective vorticity depth. In an incompressible fluid, the law of conservation of potential vorticity has a simpler form. The density is constant, and therefore the horizontal area is inversely proportional to the height of the fluid parcel. The absolute vorticity is also conserved during motion. When the current curves to the north or south, the absolute vorticity is still conserved.

The relationship between UTCV and FS was observed and described using the isoline of potential vorticity -1 PVU (1PVU = 10^{-6} m^2 $s^{-1}K$ kg^{-1}) (Ferreira, 2005). Conceptual models showed the deepening of this isoline to the level of 500 hPa in events with strong/intense FS and without FS (Figure I-4-7a and I-4-7c). The depth of this isoline decrease to 200 hPa for events with a weak FS (Figure I-4-7c). The intrusion of stratospheric air through the tropopause was evident in all events. Oscillations of the potential temperature indicated a positive anomaly of the potential temperature at the vortex center with a strong FS and without FS. This oscillation was absent in events with a weak FS.

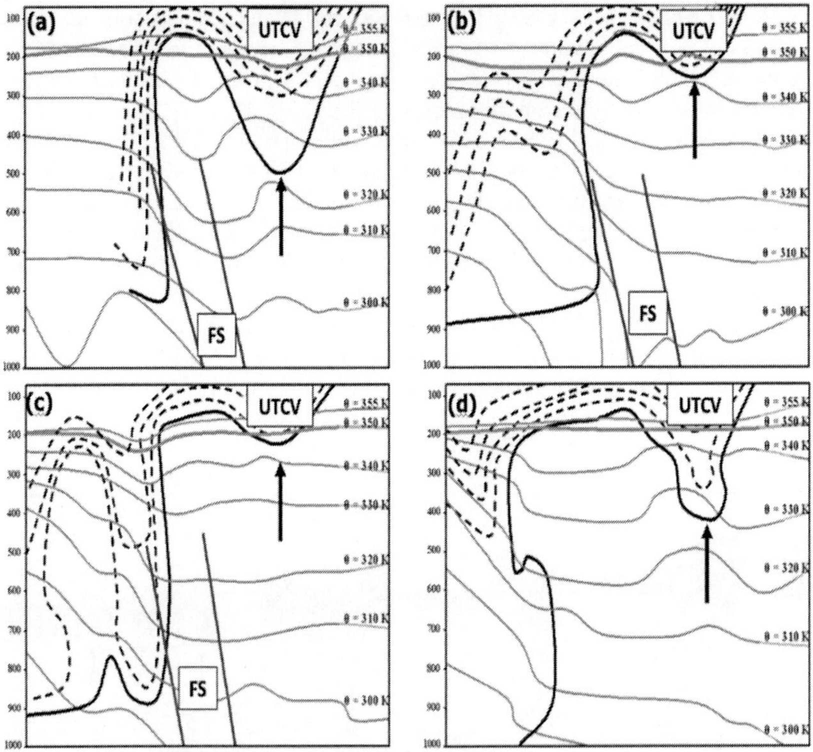

Figure I-4-7 – Relationship between UTCV and FS in the events with a frontal system of different intensity: a) strong, b) middle, c) weak and d) without FS.
Gray lines - potencial temperature.
Dotted lines - potencial vorticity.
Black wide line - isoline of the potencial vorticity -1 PVU.
Source: Ferreira, 2005.

5. Processes of formation and dissipation of UTCV

5.1 Formation of UTCV

The process of UTCV formation in the BNE were discussed by several authors (Rao & Bonatti, 1987; Figueroa, 1997; Mishra et al., 2001; Paixão & Gandu, 2000; Kouskly & Gan 1981; Gan, 1982; Ramírez, Kaiano & Ferreira, 1999, Morais et al. 2020). As a rule, the process of UTCV formation occurs mainly due to the intensification of the ridge associated with the BH. First, such an intensification of BH occurs due to the large amount of latent heat released by convection. Usually, such convection is intensified by the advance of the FS over Brazil. As a cold front advances through Brazil, there is strong advection of warm air on the eastern and equatorial sides of the frontal zone.

The UTCV genesis is not associated with polar air (Fedorova, 2008). In other words, the cold air behind the cold front does not affect the UTCV formation. The warm air at the forefront of the cold front, also called the "warm conveyor belt", plays a very important role in the whole process. This hot air current intensifies the low levels trough of the frontal zone and, subsequently, the ridge from the BH at high levels. The intensification of the BH ridge at high levels provokes trough intensification at the same levels to the east and over the ocean. As a sequence, UTCV formed from this trough. This UTCV formation process is shown in Figure I-4-8. UTCV formation occurs when trough intensification persists long enough to become closed circulation. This is the classic UTCV formation process.

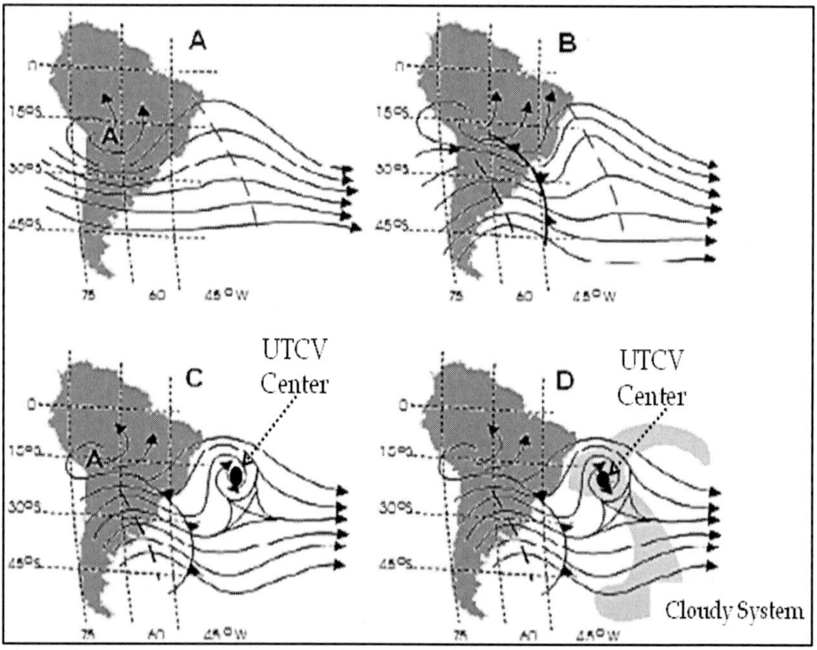

Figure I-4-8 – Scheme of the classic UTCV formation.
Black lines with arrows are streamline at the high level; the grey area shows the cloudiness system.
Source: Adapted from Gan (1982).

UTCVs are enhanced when the available potential energy is converted to kinetic energy because latent heat is released at the UTCV periphery in the BH circulation. The barotropic instability associated with the conversion of the zonal kinetic energy into the kinetic energy of the perturbation also affects the UTCVs (Rao & Bonatti, 1987).

Other mechanisms for the UTCV formation have also been proposed. The first and main one is to provide latent heating due to the approach of the leading edge of the cold front, which amplifies the BH ridge above the frontal trough (Kousky & Gan, 1981; Gan & Kousky, 1986; Figueroa, 1997). The UTCV formation also depends on barotropic instability (Mishra et al., 2001, 2007). Furthermore, a potential vorticity intrusion causes an anomalous cyclonic circulation in the upper troposphere above the BNE (Rao et al., 2007). This leads to convergence in the upper troposphere and subsidence. UTCV is formed by this convergence and sinking.

In addition, there are known cases of UTCVs without cloudiness and the absence of BH influence (Reboita et al. 2010). These events usually occur in winter and are not associated with rain. However, weather changes can occur indirectly as UTCVs affect atmospheric circulation in the tropics.

It is also important to emphasize that four types of UTCVs were cataloged (Paixão & Gandu, 2000): Classic, High, African I and African II (Figure I-4-9). The African heat source influences the formation of the African I and African II types of UTCV. The formation of the UTCV High is preceded by convective activity associated with the South Atlantic Convergence Zone (SACZ). A strong release of latent heat from condensation generates an anticyclonic circulation in the upper troposphere in the Southeast Atlantic. A trough is also formed to the northeast of the anticyclone or below the jet (Paixão & Gandu, 2000; Ramírez et al., 1999). The two types of UTCV formation, i.e. Classic and High, have a direct impact on the BNE weather.

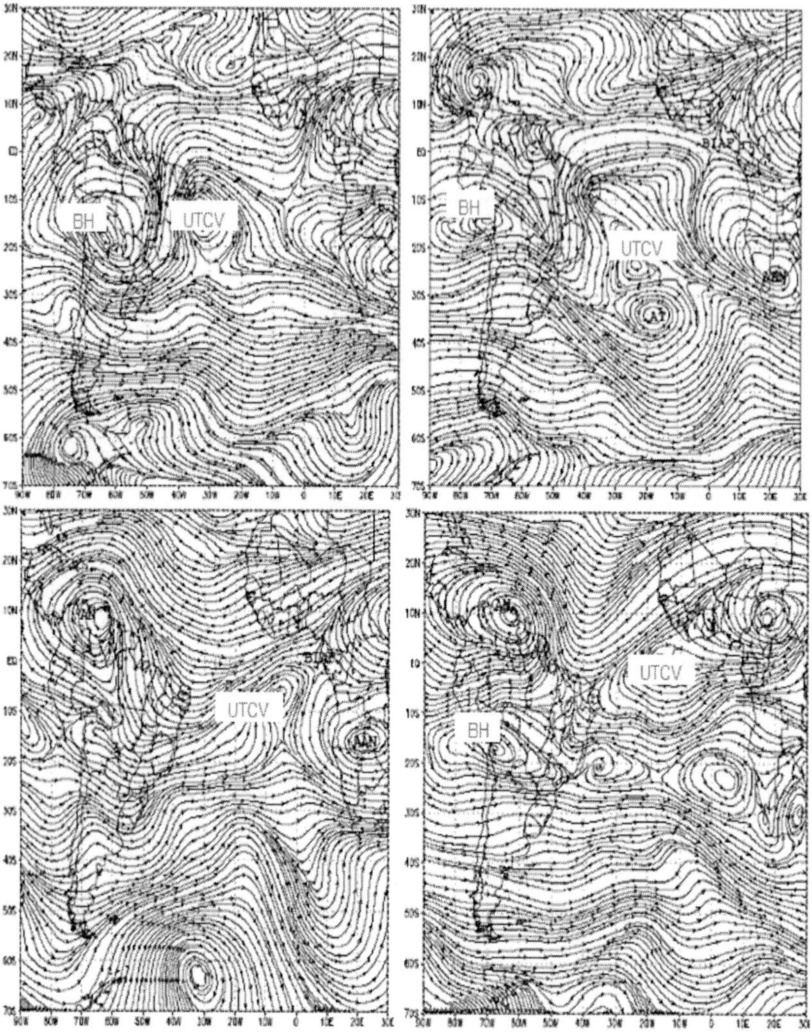

Figure I-4-9 – Examples of four types of UTCV: Classic (a), High (b), African I (c) and African II (d).
Black lines with arrows are streamline at the 200 level.
Source: adapted from Passion and Gandu (2000).

5.2 An example of the UTCV formation and development

The process of UTCV formation and development in November 2013 is shown in Figure I-4-10. The trough intensification in the BNE can be seen on high level streamline and geopotential height maps between 06:00 and 12:00 UTC 2 November 2013. During this period, the narrowing of the trough is observed, which indicates an intensification in cyclonic circulation. The trough intensification process occurred between 12:00 and 18:00 UTC 2 November and the UTCV was formed on 3 November 00 UTC.

The cyclonic circulation of UTCV is shown as an example in Figures *I-4*-10 and *I-4*-11. It can be seen that UTCV started at 18:00 UTC on November 2, 2013. This UTCV was identified from streamlines and geopotential height isolines for three days and twelve hours. These closed isolines of the geopotential height indicate an intense system. The Brazilian Northeast Jet Stream (BNEJS, Chapter 1.6) was associated with this UTCV from its formation to maximum development. BNEJS can have different origins (Chapter 1.6), in this situation in the tropics it is part of the Subtropical Jet Stream of the Southern Hemisphere (STJSSH) and therefore INPE still refers to it as the Subtropical Jet Stream (SJS). This current is shown as a dark grey line (indicated by the white arrows in Figure I-4-10a).

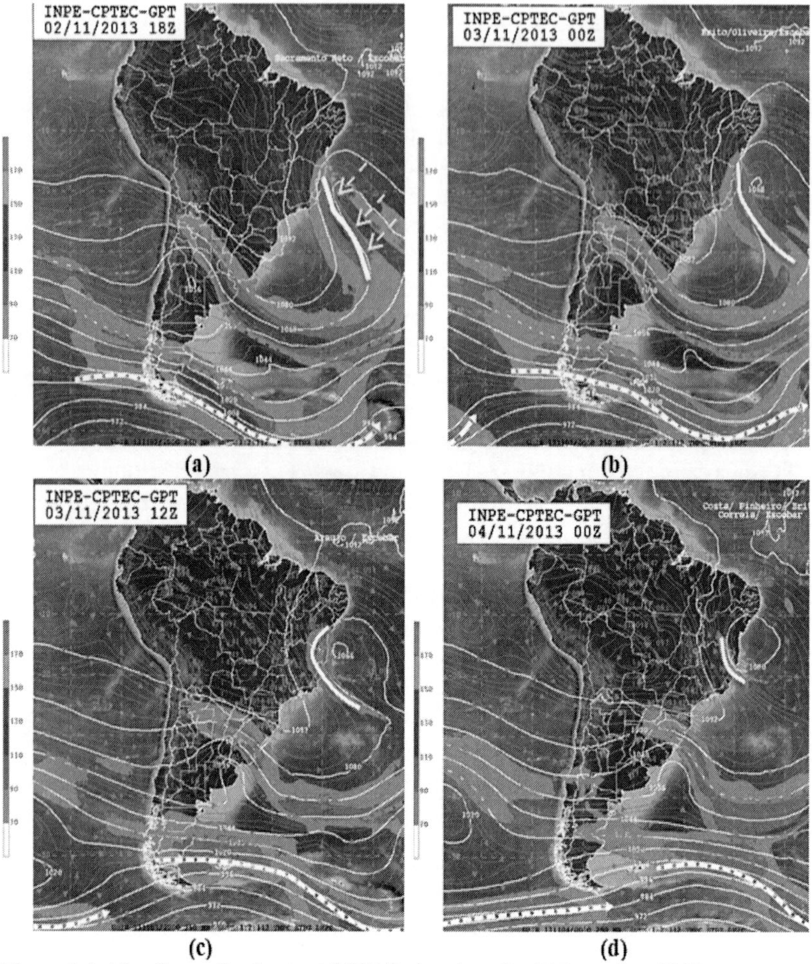

Figure I-4-10 – Synoptic charts at 250 hPa level on 2 - 4 November 2013.
Geopotential height at 250 hPa (a - d) - white lines.
Jetstream lines at 250 hPa - grey lines with arrows.
Polar Jet stream - white dashed line.
Subtropical Jet stream or Brazilian Northeast Jet Stream (BNEJS) - white thick line.
Source: INPE.

This UTCV was very intense and reached middle to low levels. At middle levels (500 hPa), this vortex lasted 3 days, from 2-5 November. This vortex was more intense with two closed geopotential height isolines from 3 November 12 UTC to 4 November 6 UTC (Figure I-4-11). It is also worth noting that due to the presence of this vortex, low temperatures at 500 hPa level were recorded (-15°C over the ocean and -12°C over the continent).

An indication of a strong deepening of this vortex was also observed at the levels of 700 and 850 hPa (Figures I-4-12 and 13).

A cyclonic circulation was observed at 700 hPa (Figure I-4-12). The circulation center at this level (-17°S/-36°W, Figure I-4-12b) has shifted eastward relative to the center at the levels of 500 and 250 hPa (-18°S/-38°W, Figure I-4-11b). This cyclonic circulation was also seen at 700 hPa on 3 November 18 UTC and 4 November 3 UTC.

At the level of 850 hPa, a trough was noted throughout the entire period (Figure I-4-13). In cases of deep UTCV, only trough circulation is observed in the 850 -1000 hPa layer.

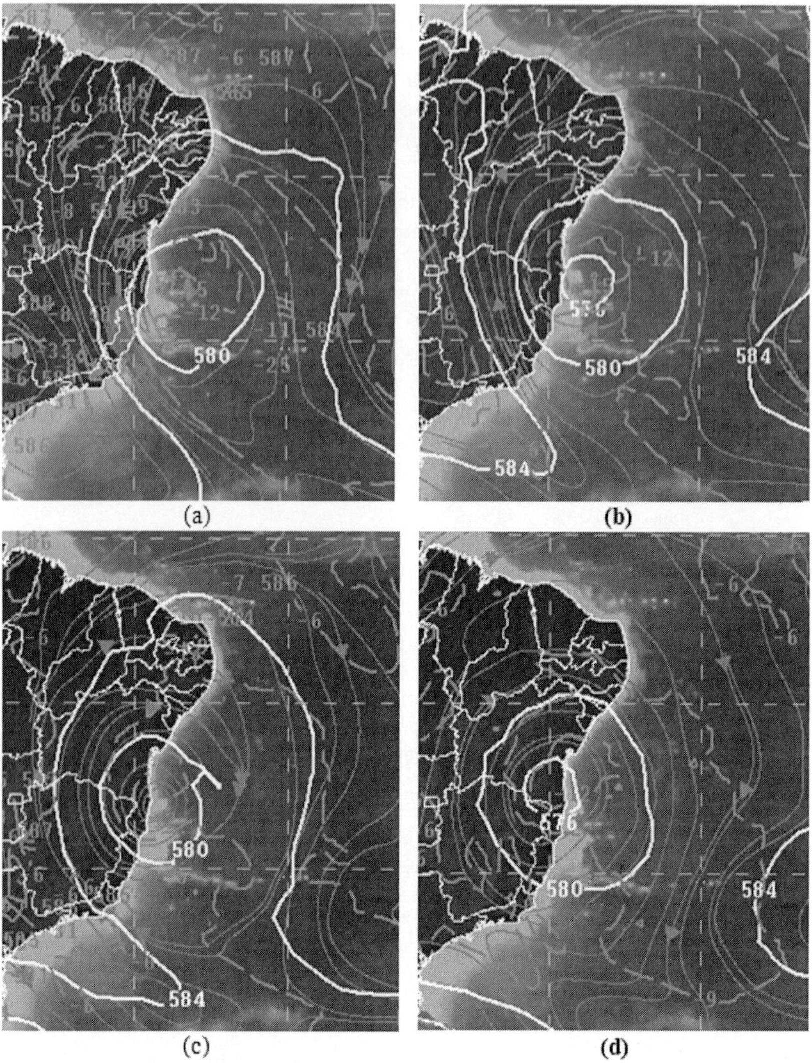

Figure I-4-11 – Synoptic charts at 500 hPa for 3 November 2013,12 UTC – 4 November 2013, 6 UTC.
Geopotential height - white lines. Jetstream lines - grey lines with arrows.
Data source: INPE.

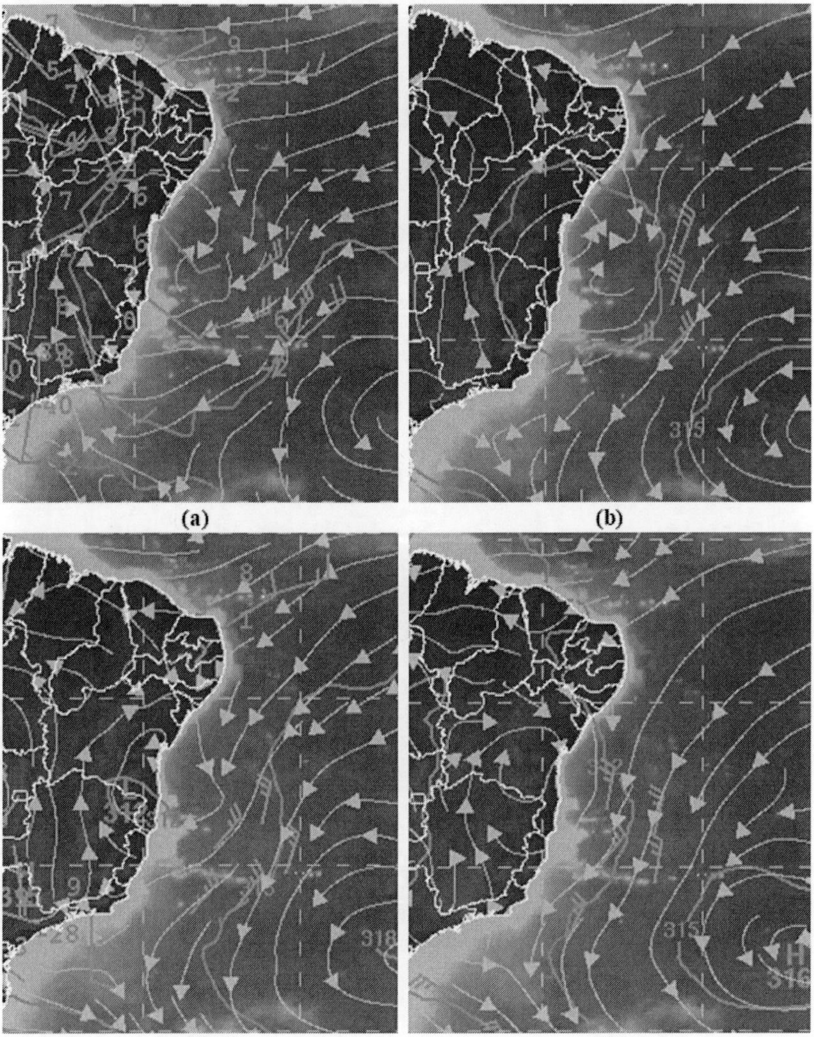

Figure I-4-12 – Synoptic charts at 700 hPa for 3 November 2013,12 UTC – 4 November 2013, 6 UTC.
Jetstream lines - grey lines with arrows.
Source: INPE.

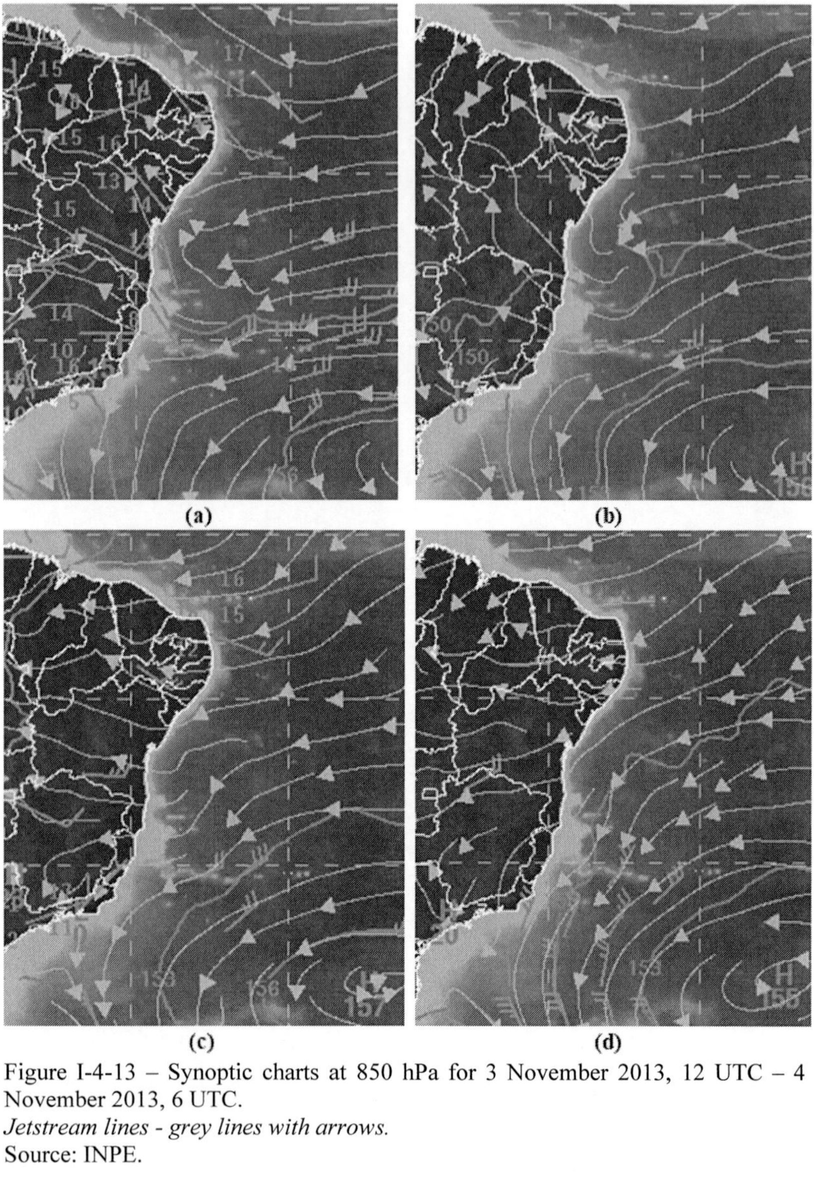

Figure I-4-13 – Synoptic charts at 850 hPa for 3 November 2013, 12 UTC – 4 November 2013, 6 UTC.
Jetstream lines - grey lines with arrows.
Source: INPE.

5.3 UTCV dissipation

UTCV weakens and dissipates when convection develops at or around its center (Morais, 2016, 2020). Convection in the center generates increased pressure at high levels and, consequently, the destruction of the cyclonic circulation.

When UTCVs reach the continent, they begin to dissipate (Gan, 1982; Morais et al., 2020). This is due to surface warming in the tropical region. In this region, temperatures can be very high at the UTCV center in clear weather conditions after UTCV maximum development. Sufficient heating in the surface layer generates convective clouds. These convective clouds release latent heat near the UTCV center. In addition, as UTCV dissipates, the kinetic energy decreases. UTCV dissipation occurred due to an insufficient source of kinetic energy (Carlson, 1967).

During dissipation, UTCVs can extend to low levels (Frank, 1970). As UTCVs extend vertically through the troposphere, they begin to decay. Thus, depth and intensity characteristics are associated with duration (Ramirez et al., 1999).

5.4 UTCV dissipation example

The UTCV dissipation process is shown in Figure I-4-14. A weakening of UTCV was observed between 4 and 5 November, and a dissipation on 6 November, 06:00 UTC. The weakening of UTCV is noticeable by the presence of only one trough. Figure I-4-14 also shows a disappearance of the BNEJS (compare with Figure I-4-10). At this point, BNEJS experienced a drop in the system kinetics. UTCV weakened at lower levels.

INPE-CPTEC-GPT
04/11/2013 12Z
Sacramento Neto / Escobar
170
150
130
110
90
70

(a)

INPE-CPTEC-GPT
05/11/2013 00Z
Costa/ Pinheiro/ Kondraski
Correia/ Escobar
170
150
130
110
90
70

(b)

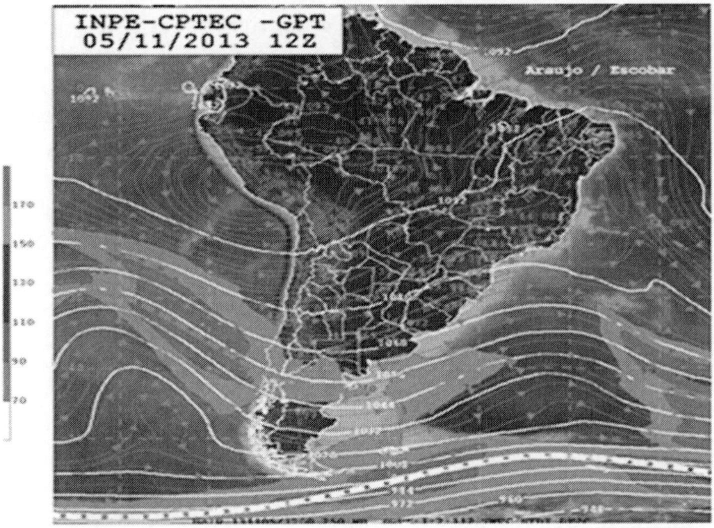

(c)

(d)

Figure I-4-14 – Synoptic charts at 250 hPa on 4 - 6 November 2013.
Geopotential height (a - d) - white lines.
Jetstream lines - grey lines with arrows.
Polar Jet stream - white dashed line.
Source: INPE.

6. Relationships of UTCV with other synoptic and mesoscale systems

UTCVs interact with other meteorological systems due to their main location at high levels of the troposphere. The influence of FS and BH on the UTCVs formation was discussed above (item 4).

6.1 Relationships between UTCV and BNEJS

UTCVs associated with Jet Stream (JS) were observed throughout the years except July and September. BNEJS is more intense from October to November (Costa, 2010). The association of both systems occurs mainly in March. This association of UTCV with BNEJS occurs on average 9 times a year at the BNE center. Just over half of UTCV events were associated with BNEJS (Repinaldo et al., 2020). These jet streams were located near the center of the BNE in 26% of cases.

Cases were identified when JS was born before UTCV (Costa, 2010). The trough coming from the NH helped in the formation of the BNEJS. Subsequently, UTCV is formed, indicating that JS plays an important role in the UTCV formation. Also, a direct connection of the UTCVs with strong winds on the JS periphery was found, which caused a change in its vertical structure.

More intense precipitation (>90 mm/24h in the coastal zone) was associated with the meridional pattern of the BNEJS circulation. Weaker precipitation (50 and 24 mm/24h) were associated with the transversal and zonal patterns, respectively (Costa, 2010).

Most UTCV events are associated with BNEJS (Gomes, 2003; Repinaldo, 2010; Costa, 2010, 2013). The association of these two systems occurs mainly from September to March. In the presence of these two systems, intense precipitation is observed. BNEJS contributes to the formation and intensification of UTCV (Gomes, 2003). Knowing this interaction is important for better forecasting of weather and adverse phenomena.

Due to the interaction of UTCV with BNEJS, the typical structure of vertical movements was changed (Antonio, 2018). For 3 years (2013-2015) there was a significant decrease in the association of BNEJS with UTCV, this relationship was recorded only in 49% of cases. These relationships were not registered in September. The main BNEJS directions (76% of events) in these events were north, east, southeast and northwest. More intense precipitation (about 100 mm/24h) occurred in events where the BNEJS was directed northward.

The UTCV example (Figures I-4-10 and I-4-14) is also associated with BNEJS. As UTCV weakened, BNEJS also weakened and became smaller in area. Strong winds were also observed at middle levels around UTCV, more intense on 3 November 2013, 12 UTC. Thus, the UTCV example presented in this chapter confirms the relationship between UTCV and BNEJS.

6.2 Connection of UTCV and SACZ

High-type UTCVs are influenced by the South Atlantic Convergence Zone (SACZ) (Paixao & Gandu, 2000). The presence of this system causes an intensification in convection between the North, Center-West and part of eastern Brazil, including the eastern North-East. In the area of convection, the latent heat of condensation is released, which subsequently intensifies the trough at low levels and the anticyclonic circulation at high levels.

A similar process was commented on in Section 5. The formation of UTCV in the presence of SACZ or anticyclonic circulation in the Southwest Atlantic and Southeastern Brazil was about 30% of events.

6.3 Connection of UTCV and ITCZ

The Intertropical Convergence Zone (ITCZ) also interacts with tropical UTCVs during vortex displacement. The result of this interaction is increased heating of the atmosphere and condensation of water vapor. High-level circulation patterns in cases of frontal extremity connection with ITCZ were developed by Muanza (2018) and cited in Section 4. In events without this connection, cyclonic circulation was observed at high levels (200 hPa). The example in Figures I-4-15a and I-4-15b shows ITCZ and UTCV respectively. Muanza (2018) presented these figures to confirm the interaction of both systems. Part of the ITCZ associated with the cloudiness of the UTCV periphery.

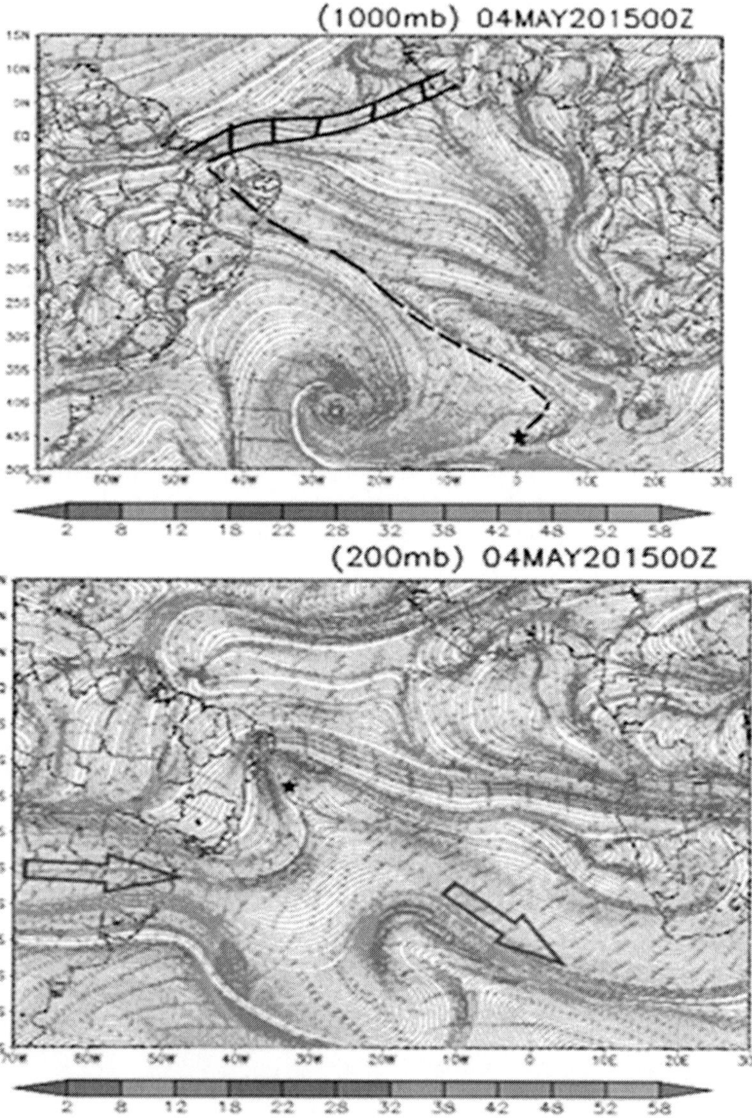

Figure I-4-15 – Connection of UTCV (black star in figure b) and ITCZ (double line in figure a).
Streamlines at 1000 hPa (a) and 200 hPa (b) on May 4, 2015, 00 UTC.
Source: Adapted from Muanza (2018).

6.4 UTCV during El Niño and La Niña years

The changes in the surface temperature of the Equatorial Pacific Ocean influences the UTCV formation. Most often, UTCVs form in La Niña years or in years with neutral temperatures in the Equatorial Pacific Ocean. A lower UTCV frequency was observed during El Niño years.

The vertical depth of UTCV is greater (to middle levels) during El Niño years. As a rule, UTCVs that form during La Niña periods have a smaller vertical extent (Coutinho, 2016).

An analysis of all UTCVs (167 events) over 9 years showed that 37% of them occurred in El Niño years and 29% in La Niña years (Repinaldo, 2010). The remaining UTCV events occurred under neutral condition in the Equatorial Pacific Ocean. There is a slight increase (+1.4%) in the number of cases in El Niño years and a decrease (-6.4%) in La Niña years relative to neutral years. The average duration of UTCV events was 6 days (Repinaldo et al. 2020). Ramirez (1996) obtained and described similar results.

7. Precipitation, thunderstorms and hail associated with UTCV

The main thermodynamic aspects of UTCV were shown using several days (Ferreira & Souza, 2019) and cases observed over 15 years (Cordeiro, 2014). The characteristic troposphere pattern at UTCV tends to be a saturated adiabatic profile (Ferreira & Souza, 2019). That is, the rate of temperature change occurs pseudo-adiabatically in a rising parcel of saturated air. This occurs due to heating (on the surface and in the 300-500 hPa layer) and atmospheric humidification, which is the result of the action of an intense convective system.

Surface heating generates CAPE (Convective Available Potential Energy) above 1000 Jkg^{-1}. Atmospheric instability indices CAPE and CIN (Convective INhibition) in UTCVs get high values (Brito et al., 2011; Cordeiro, 2014).

CAPEs of over 1000 Jkg^{-1} (about 1700 to 3500) were reported for UTCV in convective clouds accompanied by electrical discharges (Brito et al., 2011; Ferreira & Souza, 2019; Cordeiro, 2014). CAPE energy was moderate to high (about 550 to 3000 Jkg^{-1}) in regions energetically prone to convection (Ferreira, 2017), at the UTCV periphery CIN has high values (maximum reached about 280 Jkg^{-1}), which indicates well-dry UTCV center.

Indices K (Instability Index K) and TT (Total Totals Index) did not perform well in thunderstorms associated with UTCVs in the BNE (Brito, 2011; Cordeiro, 2014).

A lifting index (LI) below -2.5 for UTCV events also indicates precipitation and the possibility of thunderstorms (Brito, 2011). This LI value information was determined by the University Corporation for Atmospheric Research (UCAR).

The average values of the instability indexes in thunderstorms with the presence of UTCV were: 1) CAPE, 1319.6 JKg^{-1} (range from 336 to 2034) and 2) LI, -4.5 (range from -8 to -1) (Cordeiro, 2014).

The UTCV example (November 2013) presented in this chapter shows similar characteristics to previous studies. The structure of the troposphere in different parts of the UTCV will be shown below.

An increase in humidity in the layer between the surface and 700 hPa occurred from 3 to 4 November (Figures I-4-16a, I-4-16b and I-4-16c). To the north of UTCV and close to the surface, the layer was unstable, and in most of the troposphere, conditionally unstable (Figures I-4-16d and 16e). Relative humidity reached 100% in the 800 and 950 hPa layer on 3 November 12 UTC (Figure I-4-3d). The K and TT indices at that time were very high. But these indices generally do not give a good result in cases of thunderstorms associated with UTCV in the BNE (Brito et al., 2011). The LI index turned negative, reaching -5.34, indicating strong instability and thunderstorms. Positive CAPE also reached high values above 2000 Jkg^{-1}, indicating intense convection.

To the southwest of the UTCV center, the troposphere was conditionally unstable (Figures I-4-16f and I-4-16g). However, the 750-1000 hPa layer was stable on 3 November 12 UTC. In addition, very thin layers of thermal inversion were confirmed at approximately 500 and 850 hPa (Figure I-4-16f) and between 950 and 900 hPa (Figure I-4-16g). Very high relative humidity (around 100%) was observed between the surface and 600 hPa on 3 November 12 UTC (Figure I-4-16f). The obtained values of instability indices assume strong convection, intense precipitation and thunderstorms. The values of the K and TT indices were high. The LI index was negative, and the CAPE was high (above 1000 Jkg^{-1}) on 4 November 12 UTC.

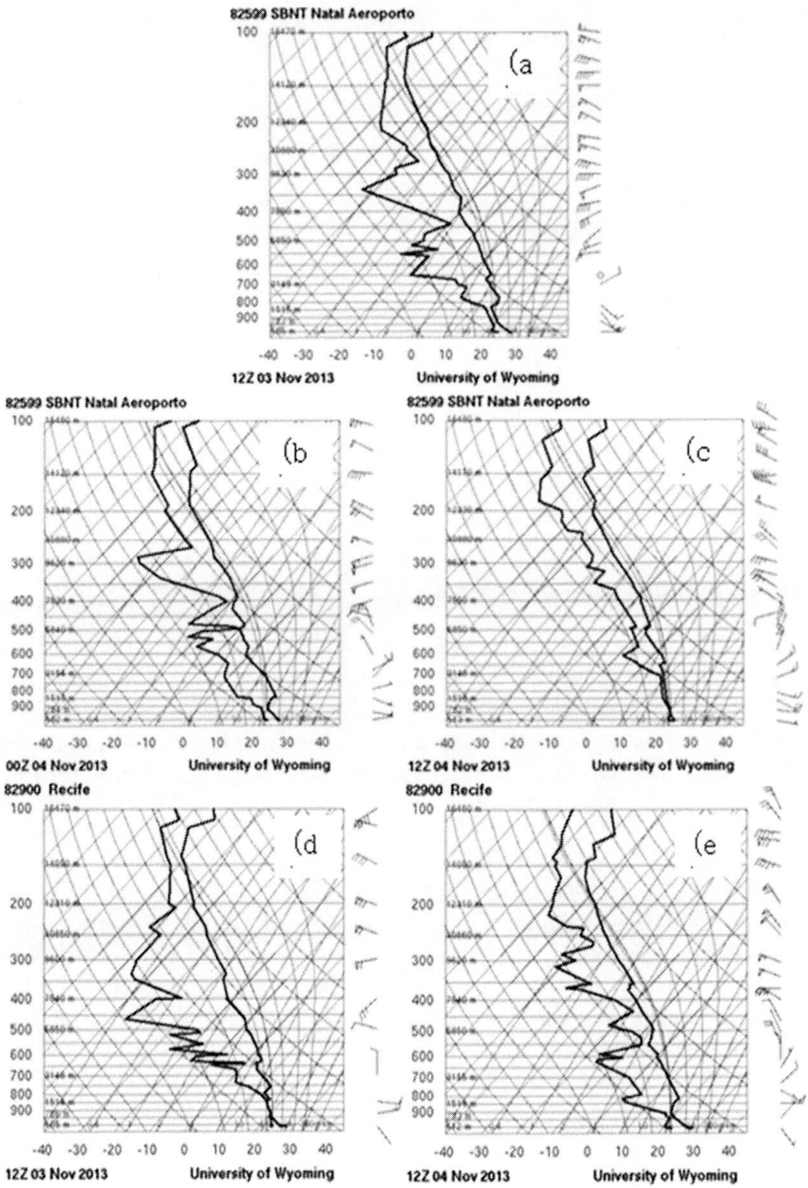
82599 SBNT Natal Aeroporto
(a
100
200
300
400
500
600
700
800
900
-40 -30 -20 -10 0 10 20 30 40
12Z 03 Nov 2013
University of Wyoming
82599 SBNT Natal Aeroporto
(b
00Z 04 Nov 2013
University of Wyoming
82599 SBNT Natal Aeroporto
(c
12Z 04 Nov 2013
University of Wyoming
82900 Recife
(d
12Z 03 Nov 2013
University of Wyoming
82900 Recife
(e
12Z 04 Nov 2013
University of Wyoming

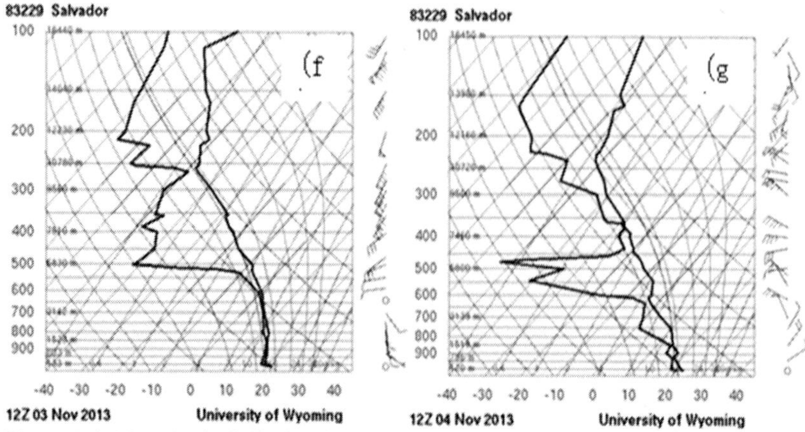

Figure I-4-16 – Vertical distribution of temperature and dew point from radiosonde data in 3 cities in the BNE on 3 and 4 November 2013:
Northwest of the UTCV center in Natal (a, b and c).
North of the UTCV center in Recife (d, e).
Southwest of the UTCV center in Salvador (f, g).
Data source: University of Wyoming.

In the northern part of the UTCV, high relative humidity (90%) is observed in the layer from 500 to 850 hPa (Figures I-4-17a, I-4-4b and I-4-4c). These figures also showed that LI had reached -4. This means a moderate potential for severe weather (according to UCAR) and a potential for thunderstorms (LI < -2.5) (Brito et al., 2011). The Sweat (Severe Weather Threat Index) and K indices obtained in the SE UTCV sector show values of about 270 and 35, respectively (these values correspond to the formation of sparse Cb according to UCAR) (Figure I-4-17D, I-4-4E and I-4-4F). VT (Vertical Totals index) values between 25 and 29 and TT above 45 were observed (Figure I-4-17G, I-4-4H and I-4-4I), indicative of a thunderstorm possibility (UCAR). CAPE received values above 2000 jKg^{-1} (Figures I-4-17J, I-4-4L and I-4-4M), indicating a moderately unstable atmosphere and the possibility of thunderstorms (>1500 JKg^{-1} according to UCAR) (Brito et al., 2011).

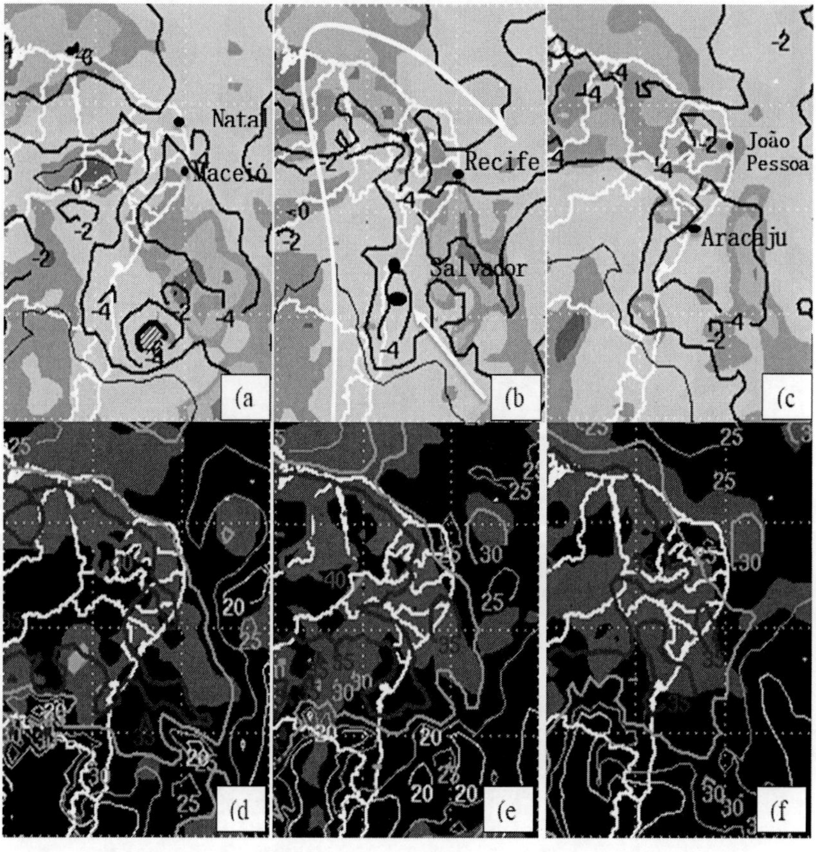
Natal
Maceió
(a
Recife
Salvador
(b
João
Pessoa
Aracaju
(c
(d
(e
(f

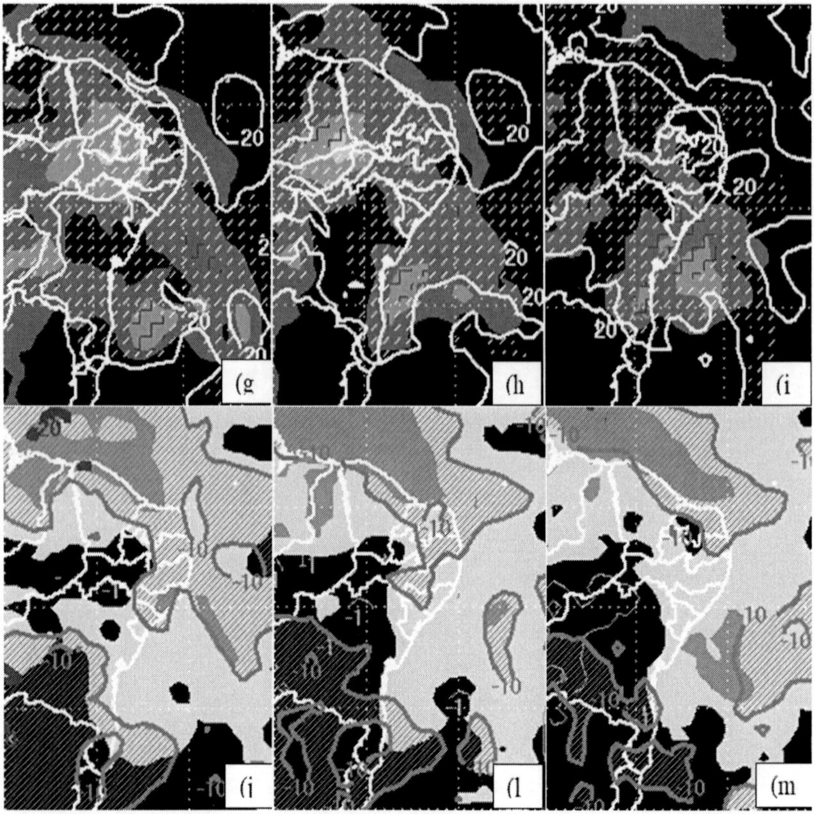

Figure I-4-17 – Indices of atmospheric instability on 3-4 November 2013. Column 1: 3 November 18 UTC; Column 2: 4 November 00 UTC; Column 3: 4 November 06 UTC. a), b), c): Relative Humidity between 500 and 850 hPa (shaded) and Lifted (contours); d), e), f): Sweat (shaded) and K (contours); g), h, i): VT (shaded), CT (Cross Totals, contours) and TT (>45 - white);
j), l), m): CAPE (shaded) and CIN (grey outline)
White arrow in figure (b) indicates the UTCV the center.
White line in figure (b) indicates UTCV periphery.
Source: INPE.

7.1 UTCVs and precipitation

UTCVs are one of the main systems causing precipitation in the BNE. Intense rainfall in the BNE was shown to be associated with UTCVs (Dean, 1971; Aragão, 1975). The cloudiness associated with UTCV may vary depending on its intensity and depth (Frank, 1970). UTCVs are classified as either wet or dry, indicating the cloudiness intensity. Deep vortices (down to low levels) are humid with a lot of cloudiness. Dry vortices are located only in the middle and upper troposphere.

UTCVs are very important in determining the precipitation patterns in these regions (Ramirez, 1997). Although UTCVs occur mainly during the hot season (with a maximum in January), these vortices are responsible for the pre-rainy season precipitation in the BNE (Coutinho et al., 2010; Palharini & Vila, 2017). A five-year study in the state of Alagoas (Pontes da Silva, 2008, 2011) showed different results. UTCVs were associated with a small proportion of intense precipitation.

UTCVs have a strong effect on precipitation, especially when they are located over a continent. Cloudiness is concentrated on the periphery of UTCV and has asymmetric characteristics (Frank, 1970). At the periphery of the UTCV, there is a lifting of hot and humid air, followed by cloudiness and intense precipitation. A clear sky is formed at the vortex center due to the sinking of cold and dry air. It is important to note that this pattern is a typical characteristic. But significant variation is observed in the horizontal distribution of rainfall. Intense convection is formed in small areas in the vortex center (Gan, 1982). Despite the typical location of cloudiness on the vortex periphery, the sinking air destroys it and cloudlessness is recorded.

The cloudiness associated with the vortex is usually larger in the direction in which the UTCV is moving (Kousky & Gan, 1981). In wide UTCVs (with distance from the center to the edge of more than 2000 km), the edge effect is reduced (Silva, 2005). In smaller UTCVs (with a distance from the center to the edge of 1000-2000 km), moderate and heavy rains prevail. Also, when the western edge of the UTCV is close to or above the interior of the BNE, precipitation increases.

7.2 UTCV and hail

UTCVs are also responsible for hail formation. The concept of hail is described as the precipitation of ice, often 5 mm in diameter, spherical or irregular shape (Nunes & Mariano, 2016; Glickman, 2000). The occurrence of hail is not common occurrence in the BNE. The hail phenomenon occurs mainly when there is a strong temperature difference between the surface and middle levels (Fedorova, 2008). This temperature difference creates a strong lifting. Then convective clouds form, having a large vertical extent, mainly in tropical or BNE regions. Clouds acquire temperatures below 0°C and therefore form ice. The sector of the cloud in which this process takes place is called the cold layer, in which temperatures are even lower than minus 40°C. In convective clouds, there are also sinking, and with them, ice. Thus, both the lifting and sinking increase the size of the pieces of ice due to collisions between the ice particles (Nunes & Mariano, 2016; Kulivoc & Rudney; Knight & Knight, 2001). When these pieces of ice grow and reach the size necessary for precipitation, hail occurs.

UTCV is one of the meteorological systems that cause hail because it creates an environment conducive to its formation (Fedorova, 2008). At the vortex periphery, Mesoscale Convective Complexes (MCC) responsible for hail formation are often observed (Gan, 1982; Fedorova, 2008).

The UTCV described above (Figures I-4-10 and I-4-11, Section 4) shows signs of the vortex deepening and reaching middle levels. The isolines of the geopotential height at the level of 500 hPa are closed; this means that it is cold at this level. Cold at 500 hPa contributes to a greater temperature difference between the surface and these levels. This is a synoptic situation with the possibility of hail formation. Low temperatures were observed in this UTCV at middle levels, which contributed to the formation of a strong lifting. The result was the hail formation and damage to society (Figure I-4-1).

Five cases of hail were reported in the BNE from 2014 to 2019 (Melo et al., 2022). The cases occurred during the summer in the Southern Hemisphere. The vertical profile of the atmosphere was very wet at the surface and showed conditional instability above the surface. The variation in the wind with height indicated a local advection of cold air, which contributed to the formation of an instability that generated convective clouds. Hot advection was also observed near the surface.

7.3 UTCV, thunderstorms and lightning

On thundery days, UTCV prevails over the region (in 77% of thunderstorms) (Brito et al., 2011; Cordeiro, 2014). The high-level trough was the second system associated with thunderstorms (Cordeiro, 2014). Thunderstorm events associated with a vortex (over 10 years) ranged from 5 to 15 events, in 1998 and 2007, respectively (Figure I-4-18).

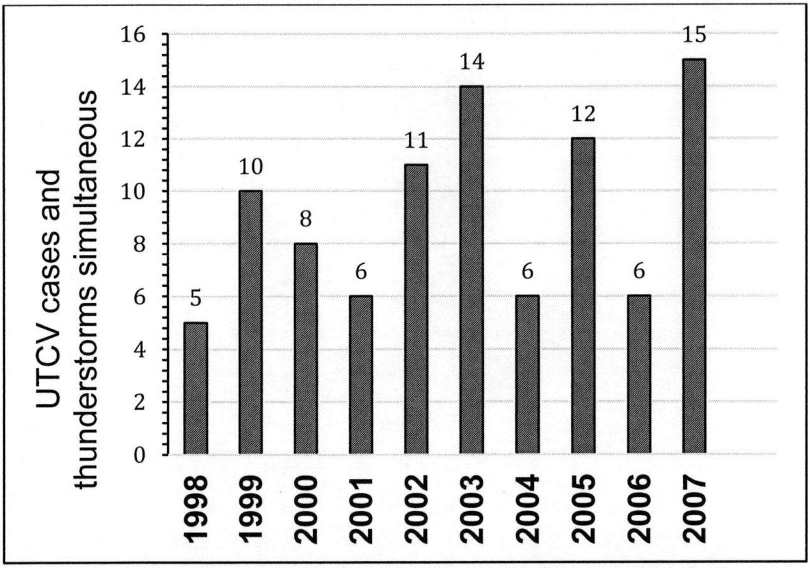

Figure I-4-18 – Numbers of thunderstorms in UTCV in Alagoas State over 10 years (1988-2007).
Source: Brito et al. (2011).

The UTCV center was located over the Atlantic Ocean on most thunderstorm days. Other UTCV locations were rare. There were also thunderstorms in the UTCV center on two occasions in 10 years (Brito et al., 2011). Usually in the vortex center it was cloudless and without thunderstorms. When UTCV weakens over the continent, convection phenomena can occur (Gan, 1982). Lightning and thunderstorms also follow the formation of convection in the UTCV center.

The location of thunderstorms relative to the UTCV center was studied by Brito (2011) (Figure I-4-19a) and Cordeiro (2014) (Figure I-4-19b) for 15 years. Thunderstorms were most often observed in the northwestern and southwestern sectors of the UTCV. The location of thunderstorms relative to the UTCV center was observed to the northwest (NO) in 41 cases and to the southwest (SO) in 13 cases (Brito et al., 2011) (Figure I-4-19a). According to Cordeiro (2014), the NW, N and NE sectors of UTCV were the preferred thunderstorm location (Figure I-4-19b).

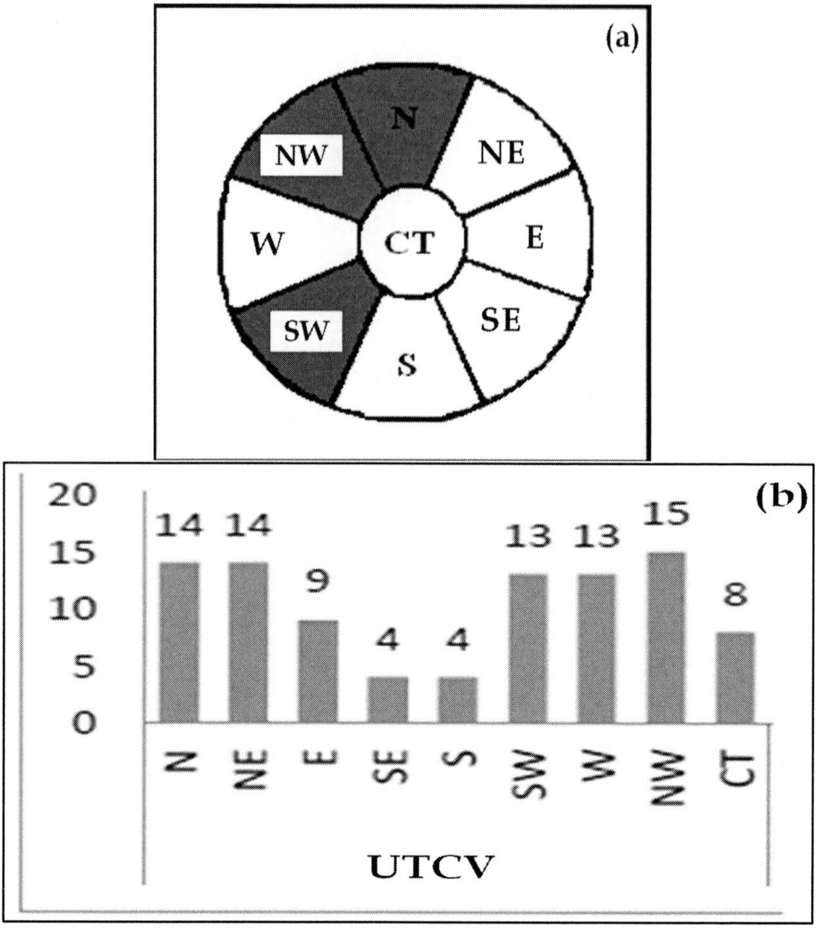

Figure I-4-19 – Thunderstorm's location in UTCV region. a) dark gray indicates a higher thunderstorms frequency; b) numbers show thunderstorm frequency in UTCV sections
Source: (a) Brito et al (2011); and (b) Cordeiro (2014).

In 2018 and 2019, thunderstorms were more frequent in January, February, and March (Silva et al., 2022). The thunderstorm peaked in March. UTCVs were also associated with thunderstorms. The places most affected by lightning are the places with the highest relief elevation.

8. References

Antonio Manuel Agostinho Victor. *"Jet stream connection in the NEB with High Level Cyclonic Vortex."* Thesis., Universidade Federal de Alagoas, Maceio, *2018.*

Aragão, J. O. R. *"A study on the structure of synoptic perturbations in Northeast Brazil."* M. Sc. thesis, National Institute for Space Research, (INPE-789-TPT/017), 1975.

Brito, B.M.; Levit, V; Fedorava, N; Molion, L.C.B; Tenorio, R.S; Rodrigues, R.N; Silva, B. F. P. *"Analysis of Thunderstorm Behavior in the State of Alagoas and Short-Term Forecast."* Brazilian Journal of Meteorology 26, 243-256, (2011).

Calbete, N.O, Gan, M.A, Satyamurty, P. *"Chapter 21. Upper troposphere cyclonic vortices acting over northeastern Brazil.*" In: Climanalysis Special - 10 Years Commemorative Edition. Mct/Inpe/Cptec, 1996

Cavalcanti, I.F.A., Kousky, V. E. *"Cold fronts over Brazil. Weather and climate in Brazil."* São Paulo, Texts Workshop. 133-148, (2009).

Carlson, T.N. *"Structure of a steady-state cold low."* Monthly Weather Review 95 (11), 763-777, (1967).

Costa Micejane da Silva, *"High-level cyclonic vortex associated with the northeastern Brazilian jet stream."* M.Sc. thesis, Universidade Federal de Alagoas, Maceio, 2010.

Costa, M.S., Levit, V., Fedorova, N. "Atmospheric circulation patterns in northeastern Brazil from high-level cyclonic vortex events with jet streams." Revista Brasileira de Geografia Física *6, no. 4, 794-804, (2013).*

Coutinho M.D.L., Gan M.A., Rao, V. B. *"Objective method of identifying high-level cyclonic vortices in the Tropical South region: validation."* Revista Brasileira de Meteorologia 25, 311-323, (2010).

Coutinho M.D.L., Lima K.C., Silva C.M.S. *"Regional climate simulations of the changes in the components."* International Journal of Climatology 36, 1170-1183, (2016).

Cordeiro Edwans Silva *"Synoptic Analysis of Thunderstorm Events for the State of Alagoas during the 15-Year Period (1998-2012)."* MSc thesis, Universidade Federal de Alagoas, Maceio, 2014.

Dean, G. A. *"Three-dimensional wind structure over South America and associated rainfall over Brazil",* MSc thesis, National Institute for Space Research (INPE), 1971.

Fedorova Natalia. *Sinótica IV: Sistemas e processos sinóticos atuantes na américa do sul*. Maceió: Ed. Universitária/ UFAL, 192 p. ISBN: 978-85-7177-422-3; CDU: 551.5, 2008

Figueroa, S. N. "*Study of summer circulation systems over South America and their simulations with numerical models.*" Dr. thesis, National Institute for Space Research (INPE), (INPE 7121-TRD/672), 1997.

Ferreira Solange Aragao. *"Interaction between Frontal Zones and High-Level Cyclonic Vortices and Weather Influence in Northeast Brazil."* MSc thesis, Universidade Federal de Alagoas, Maceio, 2005.

Ferreira, N.J.; Lacava, C.I.V., Sobral, Z.R. *"A climatological study of convective cloudbands in northeastern Brazil Part I: preliminary analysis.*" Australian Meteorological Magazine 50, 105–113, (2001).

Ferreira, R.R., Thermodynamic Aspects of a High Level Cyclonic Vortex over Northeast Brazil. M. Sc. thesis, Federal University of Campina Grande, 85p. 2017.

Ferreira, R.R., Souza, E.P. *"Thermodynamic Aspects of a High-Level Cyclonic Vortex over the Northeast Region of Brazil: a Case Study."* Yearbook of the Institute of Geosciences - UFRJ, 42 (3), 504-513, (2019).

Frank, N. L. *"The weather distribution with upper tropospheric cold lows in the tropics.*" U.S. Weather Bureau, Southern Region, Oct. (Technical Memorandum #28), 1966.

Frank, N. L. *"On the energetics of cold lows."* In: Symposium Tropical Meteorology, American Meteorological Society, Boston, p.6, (1970).

Gan Manoel Alonso. "*An Observational Study of the Upper Troposphere Cold Lows in the Subtropical Latitudes of the South and East Atlantic of Brazil.*" National Institute for Space Research. INPE - 2685-TDL/126, 1982.

Gan, M. A., Kousky, V. E. *"Upper Troposphere Cyclone Vortexes in the South Atlantic Ocean."* Revista Brasileira de Meteorologia 1, 19-28, (1986).

Gomes Heliofábio Barros *"Study of the Jet Stream near the State of Alagoas."* MSc thesis, Universidade Federal de Alagoas, Maceio, 2013.

G1. GAZETA DE ALAGOAS, 2008. Alagoas. Available at: <http://www.gazetadadedalagoas/> Accessed in February 2008.

Glickman, T.S., *Glossary of meteorology.* Boston: American Meteorological Society. 855 p., 2000.

Holton, James R. *An Introduction to Dynamic Meteorology.* 4° ed. 2004. Washington. Ed. Elsevier Science. 2004.

Kousky, V., Gan, M.A. *"Upper troposhperic cyclonic vortices in the tropical South Altantic.*" Tellus 33, 538–551, (1981).

Kulicov, V.A., Rudnev, G.V. "*Agrometeorología tropical.*" Científico-Técnica: La Habana, Cuba, 130-134, (1980).

Knight, C. A., Knight, N. C. "*Hailstorms."* In: Severe Convective Storms

28, n. 50, 223–254, Boston, MA: American Meteorological Society (2021).

Levit Vladimir, Cordeiro Edwans Silva, Cavalcante Lucas Carvalho Vieira *"Capitulo 2. Método da previsão de trovoadas no Nordeste do Brasil"* In: Meteorology in Topics – v. 3. San Paulo, Authors Club., 79-132, (2016)

Melo, D. L., Fedorova, N., Levit, V. *"Analysis of hail cases in the hinterland of Alagoas."* In: Interdisciplinary Formation of Exact Sciences: Knowledge and Research 2. Atena Editora, 139-155, (2022), ISBN 978-65-258-0197-1, doi 10.22533/at.ed.971222006.

Morais, M.D.C., Gan, M.A., Yoshida, M.C., *"Features Of The Upper Tropospheric Cyclonic Vortices Of Northeast Brazil"*. In: Life Cycle Stages. International Journal of Climatology, (2020), DOI: 10.1002/joc.6839.

Morais, M.D.C. "High-level cyclonic vortices operating in Northeast Brazil: observational and numerical study." PhD. Thesis, National Institute for Space Research of Brazil, 2016.

Molion, L.C.B., Calheiros, A.J.P. *"Critical events in northeast Brazil: the January 2004 event."* In: XIII Brazilian Congress of Meteorology, Fortaleza, (2004).

Mishra, S.K., Rao, V.B., Gan, M.A. *"Structure and evolution of the large-scale flow and an embedded upper-tropospheric cyclonic vortex over Northeast Brazil."* Monthly Weather Review 129, 1673-1688, (2001).

Mishra S.K., Rao V. Brahmananda, Franchito Sergio H. *"Genesis of the Northeast Brazil Upper Tropospheric Cyclonic Vortex: A primitive equation barotropic instability study."* Journal of the Atmospheric Sciences 64-4 (April 2007).

Muanza Gomes Antônio. *"Padrões de circulação atmosférico associados á ligação da extremidade frontal com a ZCIT"*, MSc. Thesis, Federal University of Alagoas, Maceio, Brazil, (2018).

Palmen, E. *"Origin and structure of high-level cyclones south of the maximum westerlies."* Tellus 1, 22-31, (1949).

Palmer, C. E. *"On high-level cyclones originating in the tropics."* Transactions of American Geophysics Union 32 (5), 683-695, (1951).

Paixao, E. B., Gandu, A. W. *"Caracterização do Vórtice Ciclônico de Ar Superior no Nordeste brasileiro (Parte I)."* In: XI Congresso Brasileiro de Meteorologia. (in Rio de Janeiro, Brazil), (2000).

Palharini, R.S.A., Vila, A.D. *"Climatological Behavior Precipitating Clouds in the Northeast Regions of Brazil."* Advances in Meteorology 11, 1-12, (2017).

Pontes da Silva B.F., Fedorova N., Levit V., Brito B.M., Santos A.G.,

Raimundo C.C. *"Mechanisms associated with intense rainfall observed between January 13th and 14th, 2005 in the sertão of the state of Alagoas."* In: XV Brazilian Congress of Meteorology, Sao Paulo, Brazil, (2008).

Pontes da Silva B.F.P., Fedorova N., Levit V., Brito B.M., Peresetsky, A. *"Synoptic systems associated with heavy rainfall in the State of Alagoas."* Revista Brasileira de Meteorologia 26 (3), 295-310, (2011).

Ramirez, M.C.V., Kayano, M.T., Ferreira, N.J. *"Statistical analysis of upper tropospheric vortices in the vicinity of northeast Brazil during the 1980-1989."* Atmosphere 12, 75–88, (1999).

Ramirez, M. C. V., Ferreira, N. J., Gan, M. A. *"High-level cyclonic vortices originating in the eastern tropical South Pacific - part I: synoptic aspects related to their formation."* In: Brazilian Congress of Meteorology, Rio de Janeiro, Brazil. (2000).

Rao, V. B. & Bonatti, J. P. *"On the origin of upper tropospheric cyclonic vortices in the South Atlantic ocean and adjoining Brazil during the summer."* Meteorology and Atmospheric physics 37, 11-16 (1987).

Rao, V.B., Franchito, S.H., Barbosa, T.F. (*Impact of highpotential vorticity intrusions into the tropical upper troposphere in South Atlantic on precipitation over Northeast Brazil."* Geophysical Research Letters 34, USA, 1-4, (2007).

Reboita, M.S., Campos, B., Santos, T., Gan, M.A., Carvalho, V.S.B. "*Synoptic and numerical analysis of a VCAN in Northeast Brazil."* Brazilian Journal of Physical Geography 10, 041–059, (2017).

Reboita, M.S., Gan, M.A., Rocha, R.P., Ambrizzi, T. *"South American rainfall regimes: A literature review."* Revista Brasileira de Meteorologia 25, 185-204, (2010).

Reis Jean Souza "Dynamic and thermodynamic characteristics of High-Level Cyclonic Vortices in Northeast Brazil." M.Sc.Thesis, Federal University of Rio Grande do Norte, 77p., 2018

Repinaldo Henrique Fuch Bueno **"***Upper Tropospheric Cyclonic Vortex and Brazilian Northeast Jet Stream in El Niño and La Niña Years."* M. Sc. Thesis, Federal University of Alagoas, Maceio, 2010.

Repinaldo, H. F. B., Fedorova, N., Levit, V., Rocha Repinaldo, C. R. *"Upper Tropospheric Cyclonic Vortex and Brazilian Northeast Jet Stream over Alagoas state: circulation patterns and rainfall." Revista Brasileira de Meteorologia* 35, n. Especial, 745-754, 2020. doi: http://dx.doi.org/10.1590/0102-7786355000003.

Silva, L. A. *"The influence of the high-level cyclonic vortex (VCAN) on the precipitation of Northeast Brazil (NEB) and the associated characteristics."* Dissertation (master's in meteorology), National

Institute for Space Research (INPE), São José dos Campos. (INPE-14488-TDI/1169) 2005.

Silva, L.A., Valverde Ramírez, M.C., Satyamurty, P., "*Spatial Distribution of the Cyclonic Vorticity in the Northeast of Brazil and Adjoining Ocean and its Relations with the ENSO.*" In: The 8th International Conference On Southern Hemisphere Meteorology And Oceanography, Foz do Iguaçu - Brazil. Proceedings of the 8th ICSHMO, 2006.

Silva G. C. O., Fedorova N., Levit V. "*Case Study of Thunderstorms with Electric Discharges in Alagoas.*", Yearbook of the Institute of Geosciences, 1-12, (2022) DOI: 10.11137/1982-3908_45_38477 ISSN: 0101-9759, e-ISSN 1982-3908.

Valverde Ramírez, M. C. *"Climatic patterns of high-level cyclonic vortices in Northeastern Brazil".* São José dos Campos. M.Sc. Thesis, National Institute for Space Research, Brazil, (INPE-6408-TDI/618), 1996.

Valverde Ramírez, M.C., Kayano, M.T., Ferreira, N.J. *"Statistical analysis of upper tropospheric vortices in the vicinity of northeast Brazil during the 1980-1989."* Atmosphere 12, no. 2, 75-88, (1999).

CHAPTER I.5

MIDDLE TROPOSPHERIC CYCLONIC VORTEXES

NATALIA FEDOROVA AND VLADIMIR LEVIT

1. Identification

Information on Middle Tropospheric Cyclonic Vortexes (MTCV) in the tropical South Atlantic was first published by Fedorova et al. (2006), and the association of MTCV with precipitation was described in Pontes Da Silva et al. (2011).

Middle Tropospheric Cyclonic Vortex (MTCV) is a cyclonic vortex located only at the middle levels in the tropical region of the South Atlantic and South America.

The criteria for MTCV identification (Fedorova et al., 2017) are the next:

1) streamlines convergence,
2) relative vorticity equal to or less than $-1x10^{-5}$ s^{-1}, and the average vorticity is $-3x10^{-5}s^{-1}$,
3) center of cyclonic circulation is located between 700 and 400 hPa,
4) diameter of the cyclonic circulation area equal to or more than 500 km.

If the daily displacement of the MTCV center is less than or equal to 5° of latitude or longitude, and if the air currents were the same, then the vortexes on two consecutive maps are considered to be the same vortex. Special attention was paid to events with UTCVs (*Upper Tropospheric Cyclonic Vortex*, Chapter I-4), which have more intensive centers at 200 hPa level and even can reach the middle levels.

An example of MTCV identification is presented in figures I-5-1, I-5-2 and I-5-3. Figure I-5-2 shows MTCV location at the middle levels of 500, 600 and 700 hPa. The layers above and below confirmed that vortex absences at the high (200 and 400 hPa, figure I-5-1) and low (800 and 1000 hPa, figure I-5-3) levels.

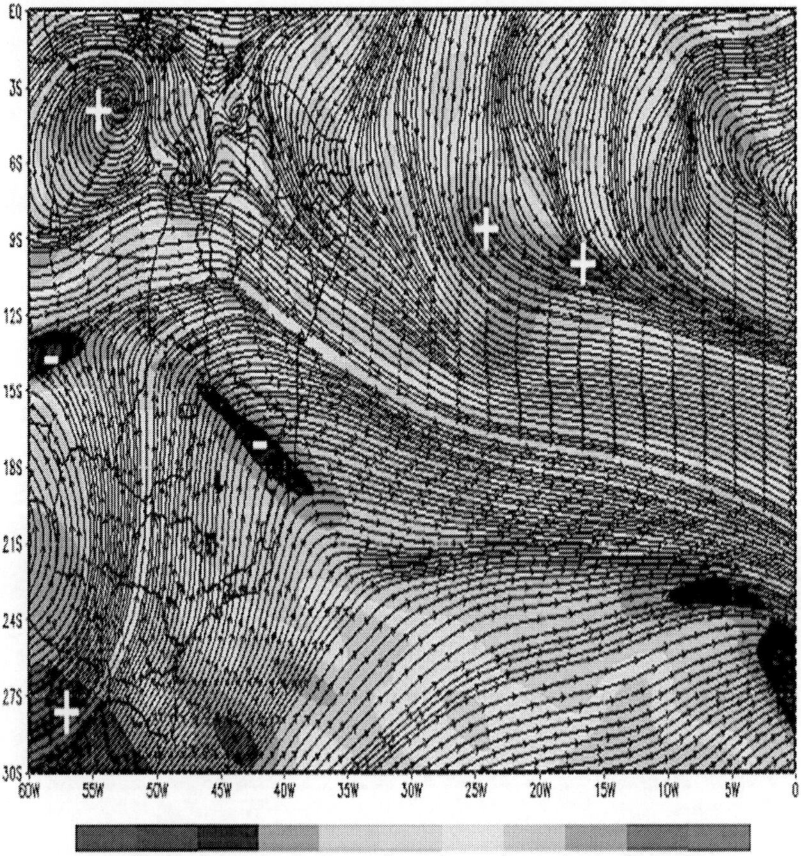
EQ
3S
6S
9S
12S
15S
18S
21S
24S
27S
30S
60W
55W
50W
45W
40W
35W
30W
25W
20W
15W
10W
5W
0
-10
-8
-6
-4
-2
0
2
4
6
8

a) 200 hPa

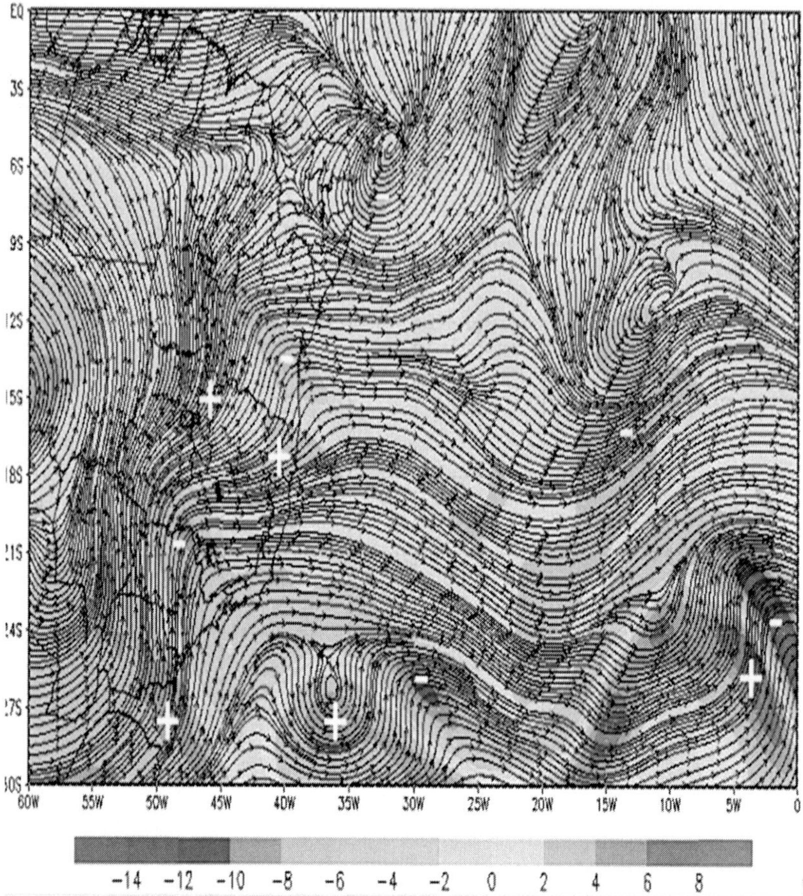

b) 400 hPa

Figure I-5-1 – Example of MTCV identification by streamlines and vorticity (color, x10^{-5} s^{-1}): Part I: vortex absences at high levels 200 hPa (a) and 400 hPa (b) on 03 July 2010, 18 UTC

Source: Silva, 2015; Fedorova & Silva, 2016.

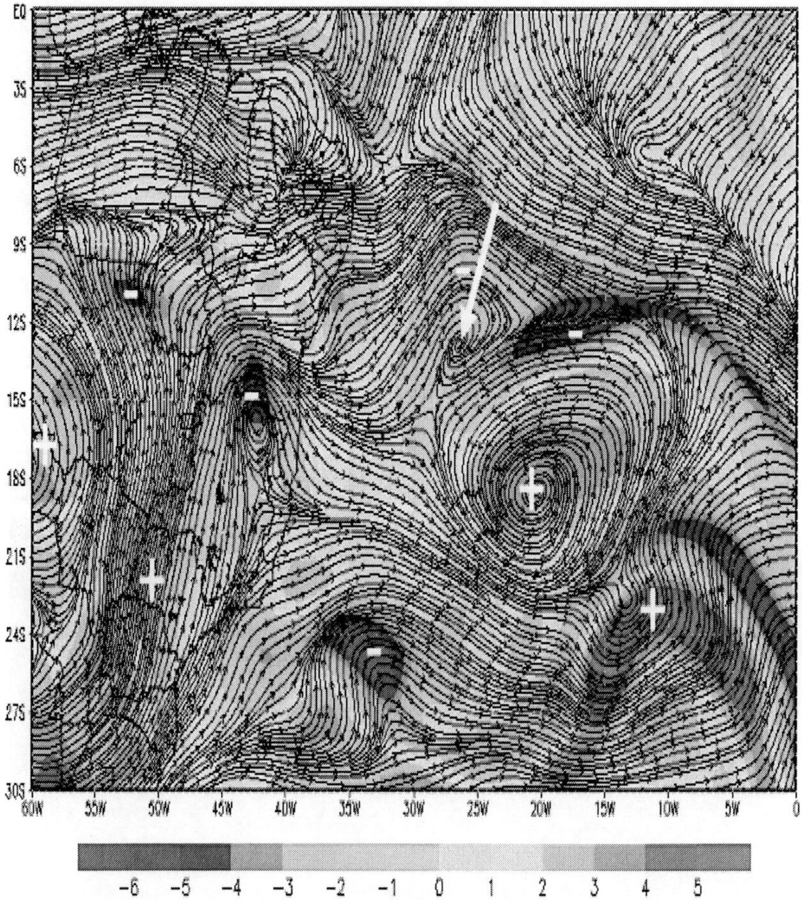
EQ
3S
6S
9S
12S
15S
18S
21S
24S
27S
30S
60W
55W
50W
45W
40W
35W
30W
25W
20W
15W
10W
5W
0
-6 -5 -4 -3 -2 -1 0 1 2 3 4 5

a) 500 hPa

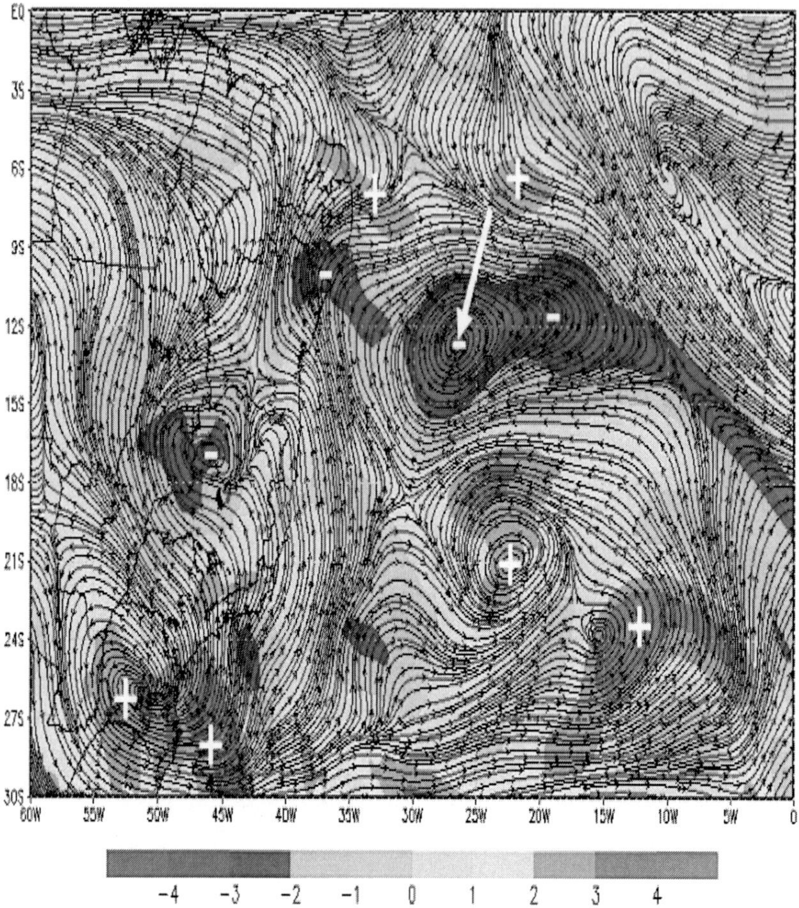
EQ
3S
6S
9S
12S
15S
18S
21S
24S
27S
30S
60W
55W
50W
45W
40W
35W
30W
25W
20W
15W
10W
5W
0
-4
-3
-2
-1
0
1
2
3
4

b) 600 hPa

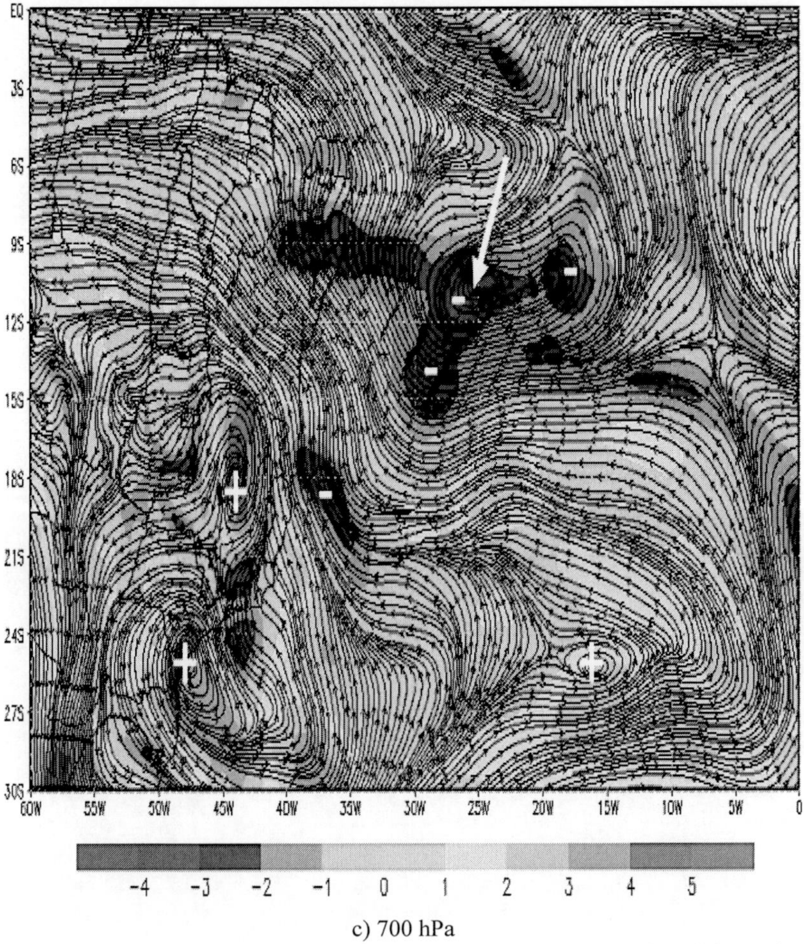

c) 700 hPa

Figure I-5-2 – Example of MTCV identification by streamlines and vorticity (color, x10^{-5} s^{-1}): Part II: vortex location (*arrow pointed in the vortex center)* at the middle levels of 500 hPa (a), 600 hPa (b) and 700 hPa (c) on 03 July 2010, 18 UTC
Source: Silva, 2015; Fedorova & Silva, 2016.

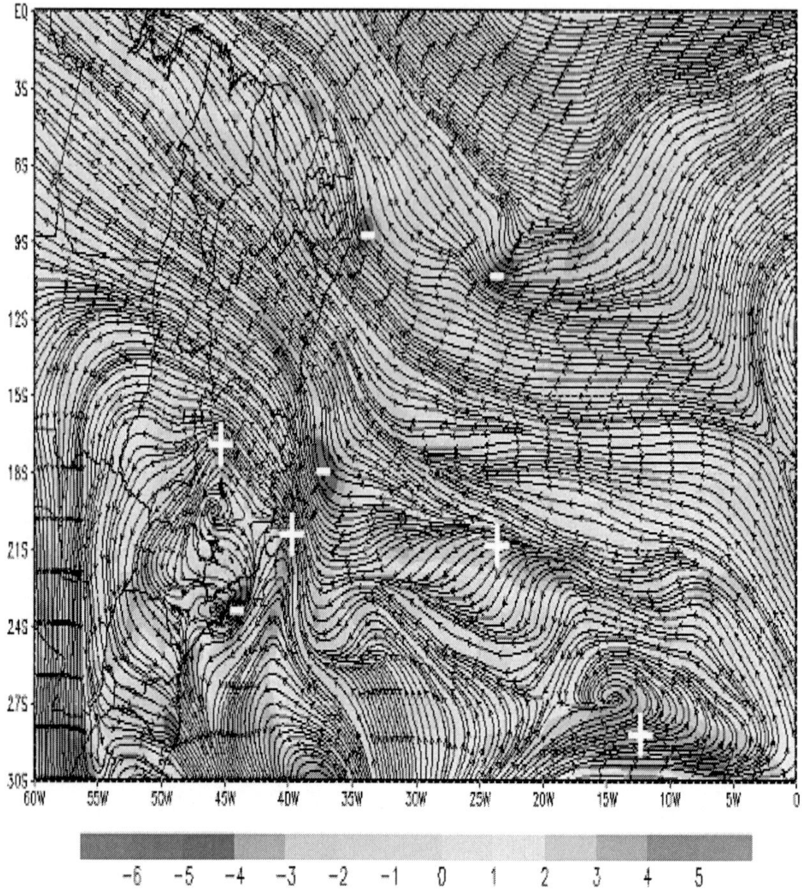
EQ
3S
6S
9S
12S
15S
18S
21S
24S
27S
30S
60W
55W
50W
45W
40W
35W
30W
25W
20W
15W
10W
5W
0
-6
-5
-4
-3
-2
-1
0
1
2
3
4
5

a) 800 hPa

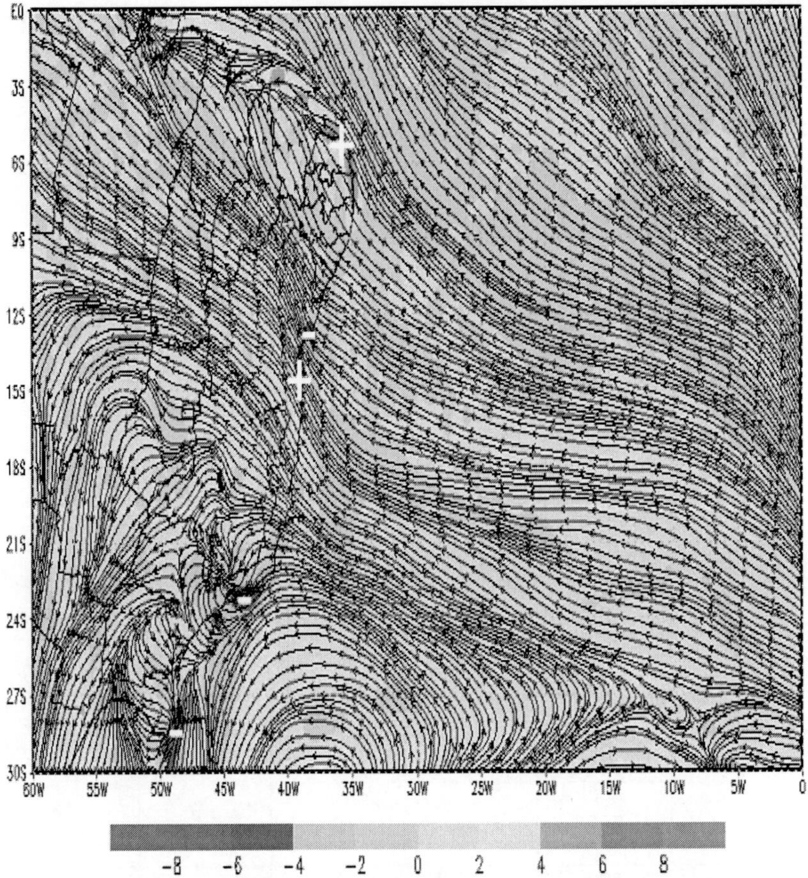

b) 1000 hPa

Figure I-5-3 – Example of MTCV identification by streamlines and vorticity (color, x10^{-5} s^{-1}): Part III: vortex absences at the low levels of 800 hPa (a) and 1000 hPa (b) on 03 July 2010, 18 UTC.
Source: Silva, 2015; Fedorova & Silva, 2016.

2. Location, duration, vertical structure

Location, duration

MTCV are observed frequently in the tropical regions of the South Atlantic and South America (0-60°W, 0-30°S), 200 - 250 vortices per year were detected (Santos, 2012; Fedorova et al., 2017). More frequently, MTCV centers were observed at the level of 700 hPa level (on 40% of days with vortices) and between 700 hPa and 500 hPa (on 90% of days with vortices) (Table I-5-1).

Table I-5-1 – Frequency (number of events - N) of MTCV location in 1, 2, 3 or 4 levels for three years (2008 - 2010 - N3) and average for one year (N1)
Source: Santos, 2012; Fedorova et al., 2017.

<table>
<tr><th></th><th></th><th>N3</th><th>N1</th><th>Σ</th><th>%</th></tr>
<tr><td rowspan="4">1
level</td><td>400</td><td>82</td><td>27</td><td rowspan="4">1122</td><td rowspan="4">80</td></tr>
<tr><td>500</td><td>256</td><td>85</td></tr>
<tr><td>600</td><td>269</td><td>90</td></tr>
<tr><td>700</td><td>515</td><td>171</td></tr>
<tr><td rowspan="4">2
levels</td><td>400
500</td><td>33</td><td>11</td><td rowspan="4">238</td><td rowspan="4">17</td></tr>
<tr><td>500
600</td><td>53</td><td>18</td></tr>
<tr><td>600
700</td><td>122</td><td>40</td></tr>
<tr><td>700
800</td><td>30</td><td>10</td></tr>
<tr><td rowspan="3">3
levels</td><td>400
500
600</td><td>3</td><td>1</td><td rowspan="3">41</td><td rowspan="3">3</td></tr>
<tr><td>500
600
700</td><td>28</td><td>9</td></tr>
<tr><td>600
700
800</td><td>10</td><td>3</td></tr>
<tr><td rowspan="2">4
levels</td><td>500
600
700
800</td><td>2</td><td>0.7</td><td rowspan="2">5</td><td rowspan="2">0</td></tr>
<tr><td>400
500
600
700</td><td>3</td><td>1</td></tr>
<tr><td>Σ</td><td></td><td>1406</td><td>469</td><td></td><td></td></tr>
</table>

MTCVs were located predominantly (86% of events) over the ocean between 6 and 18 °S. Nevertheless approximately 10 vortexes per year were registered over the continent in the BNE.

MTCV duration was short between 6 and 12 hours (70% of events). Approximately 15 events per year last 30-36 h and can last up to 42 h.

Vertical structure

In general, the divergence/convergence values in the MTCV region were weak between $-2x10^{-5}$ s^{-1} (convergence) and $0,5x10^{-5}$ s^{-1} (divergence), higher values were during formation (Fedorova et al., 2015). A weak baboclinic zone was identified in the MTCV layer in most events (Fedorova & Silva, 2016; Fedorova et al., 2017). The MTCVs over the continent were formed below the warm wave and above the weak baroclinic zone.

One example shows an intensive MTCV event with the duration of 66 h (from 03 July 2010, 12 UTC to 06 July 2010, 06 UTC), which was observed around the 600 hPa level during its entire lifetime. Figures I-5-4 and I-5-5 show maps of the vorticity, divergence, omega at the 600 hPa and thickness at the layer 500-700 hPa on 05 July 2010, 12 UTC. Value of the vorticity was between -3 and $-1x10^{-5}$ s^{-1} and highest values were in the center and rear side of the vortex (Figure I-5-4 a). The convergence of air current ($-0.6x10^{-5}s^{-1}$) was identified in the forefront part and the center of the vortex, while divergence ($0.5x10^{-5}$ s^{-1}) in the rear part (Figure I-5-4 b).

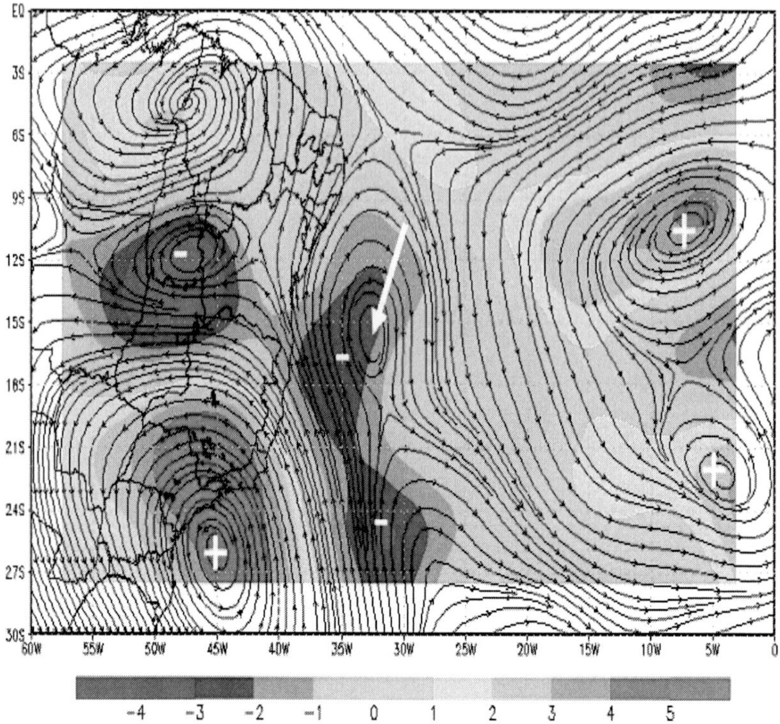
EQ
3S
6S
9S
12S
15S
18S
21S
24S
27S
30S
60W
55W
50W
45W
40W
35W
30W
25W
20W
15W
10W
5W
0
-4
-3
-2
-1
0
1
2
3
4
5

a)

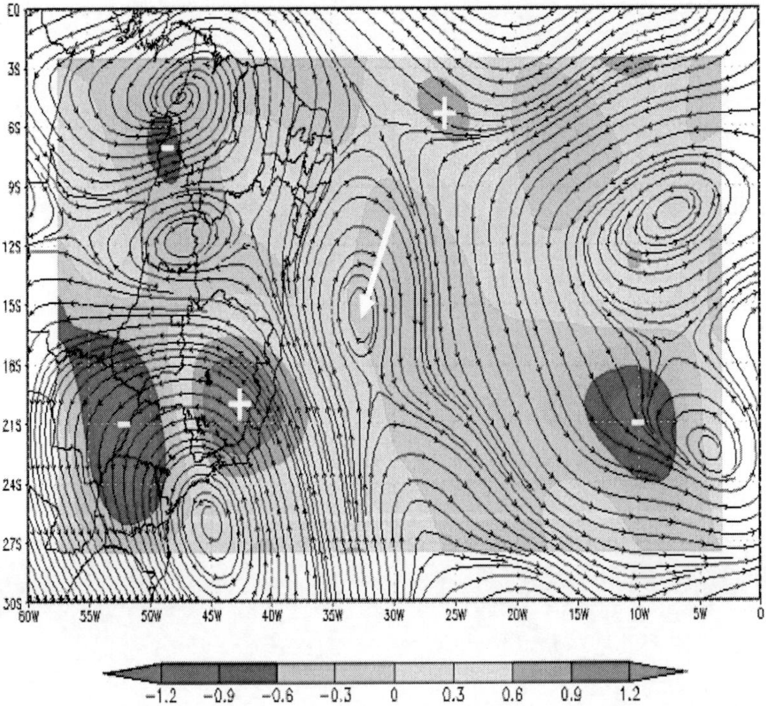

b)

Figure I-5-4 – MTCV structure: streamlines with vorticity (a) and divergence (b) at 600 hPa, on 05 July 2010, 12 UTC. *The arrow pointed to the vortex center*
Source: Fedorova et al., 2015.

A more intense lifting (up to 25x10^{-2} Pa/s) was determined at the vortex vanguard in the eastern vortex region and a weak sinking in the western region (Figure I-5-5 a). Thickness gradients showed the existence of baroclinic zone at the vortex level (Figure I-5-5 b).

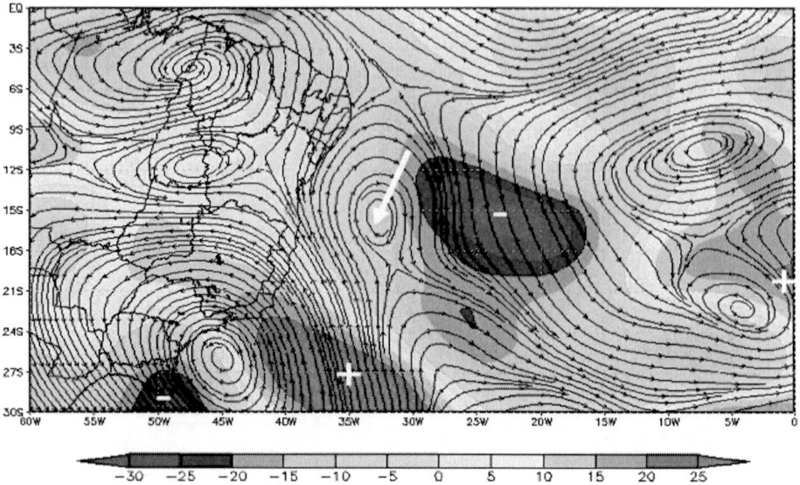

a)

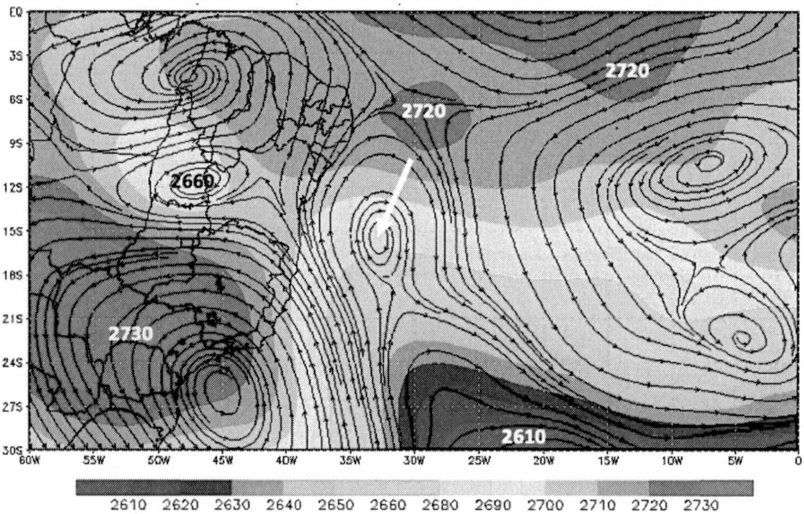

b)

Figure I-5-5 – MTCV structure: streamlines with omega at 600 hPa (a, Pa s^{-1}) and thickness in the layer of 500-700 hPa (b) on 05 July 2010, 12 UTC. *Arrow pointed to the vortex center.*

Source: Fedorova et al., 2015; Silva, 2015; Fedorova & Silva, 2016.

Vertical motions

The vertical motion in the MTCV, in general, are weak, predominantly sinking and are changed during vortex's life. Omega values are usually -10 and $5x10^{-2}$ Pa/s at the MTCV center and -25 and $5x10^{-2}$ Pa/s at the periphery (Fedorova et al., 2015, 2017; Fedorova & Silva, 2016).

No pattern in the vertical structure of the vertical motion have yet been identified. At the same time, in all events there was no uniform motion in the vortex region (different vertical motions were observed in different parts of the MTCV). The usual vertical motions during the vortex's life are: 1) lifting at all levels during the formation, with the maximum values being above the MTCV; 2) the lifting was observed in the northern part of the vortex and sinking in the southern part during the vortex development; 3) a weak lifting was found in a certain vortex region and a more intensive sinking in most of the vortex before its dissipation.

An example of vertical motion is presented in figure I-5-6. At the initial stage of the vortex's life, lifting above the MTCV center and sinking on the equatorial side of the vortex predominated (Figure I-5-6 a). During the vortex development, lifting on the equatorial side of the vortex and sinking on the polar side were observed (Figure I-5-6 b). An area with the lifting in the final period of the vortex life increased and entered the vortex center (Figure I-5-6 c).

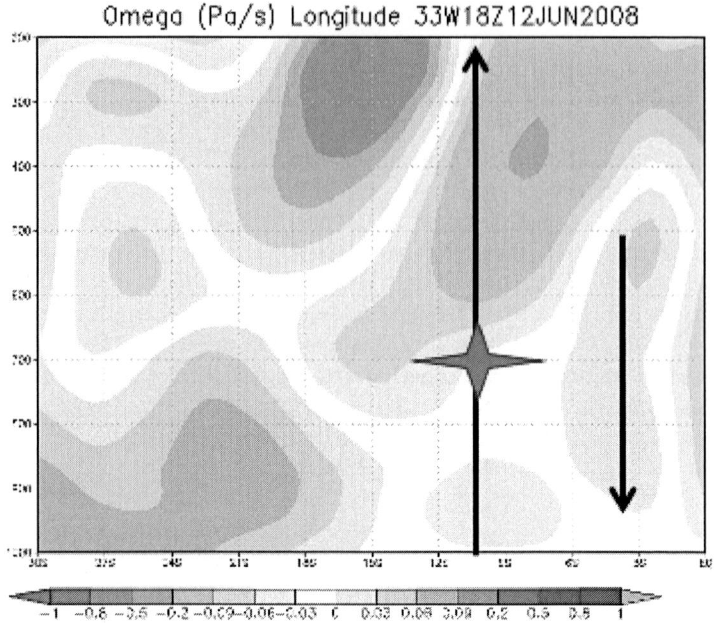
Omega (Pa/s) Longitude 33W18Z12JUN2008
−1 −0.8 −0.5 −0.2 −0.09 −0.06 −0.03 0 0.03 0.06 0.09 0.2 0.5 0.8 1

a)

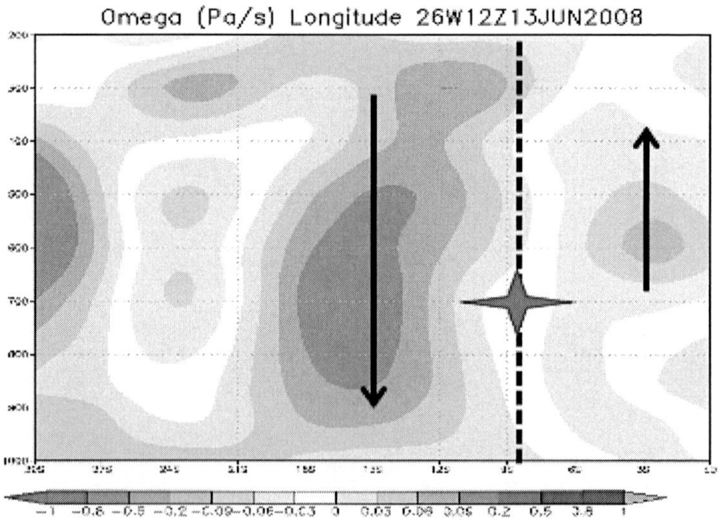
Omega (Pa/s) Longitude 26W12Z13JUN2008

b)

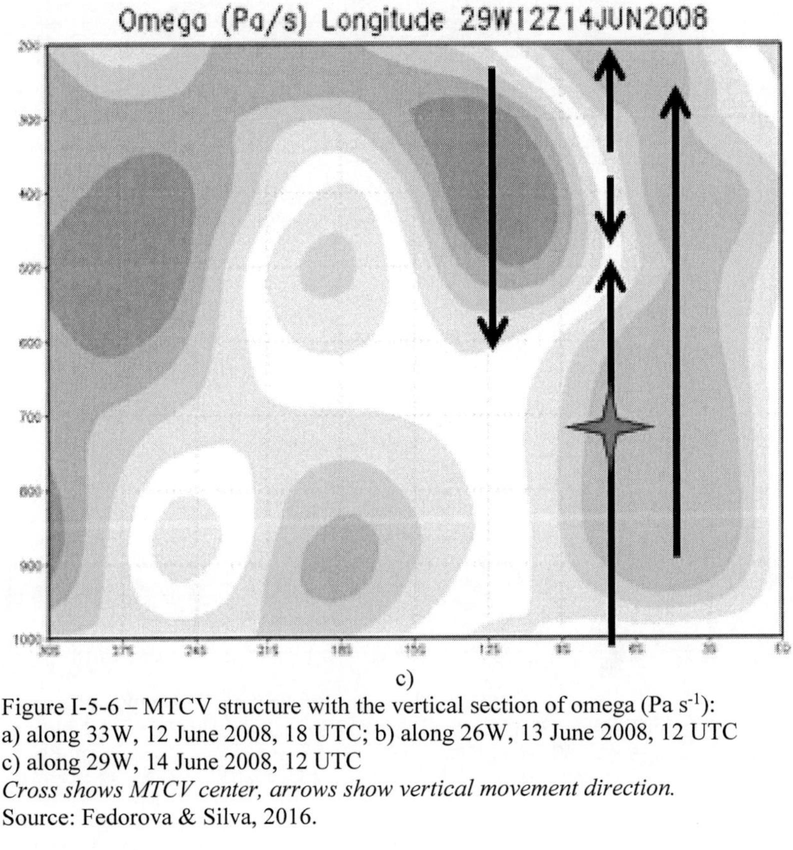

c)

Figure I-5-6 – MTCV structure with the vertical section of omega (Pa s^{-1}):
a) along 33W, 12 June 2008, 18 UTC; b) along 26W, 13 June 2008, 12 UTC
c) along 29W, 14 June 2008, 12 UTC
Cross shows MTCV center, arrows show vertical movement direction.
Source: Fedorova & Silva, 2016.

Potential vorticity

A vertical section of potential vorticity that intersects crossing the MTCV center shows the vortex location (Fedorova et al., 2017; Fedorova & Silva, 2016). As an example, a vertical section of the MTCV center along 33°W longitudes, on 12 January 2008, 18UTC is presented (Figure I-5-7). The closed isoline of the potential vorticity equal to 1 PVU around the 700 hPa level shows the location of the MTCV center. The nearly closed isoline of the potential vorticity equal to -0,5 PVU between 500 and 850 hPa represents the spatial distribution of the cyclonic circulation.

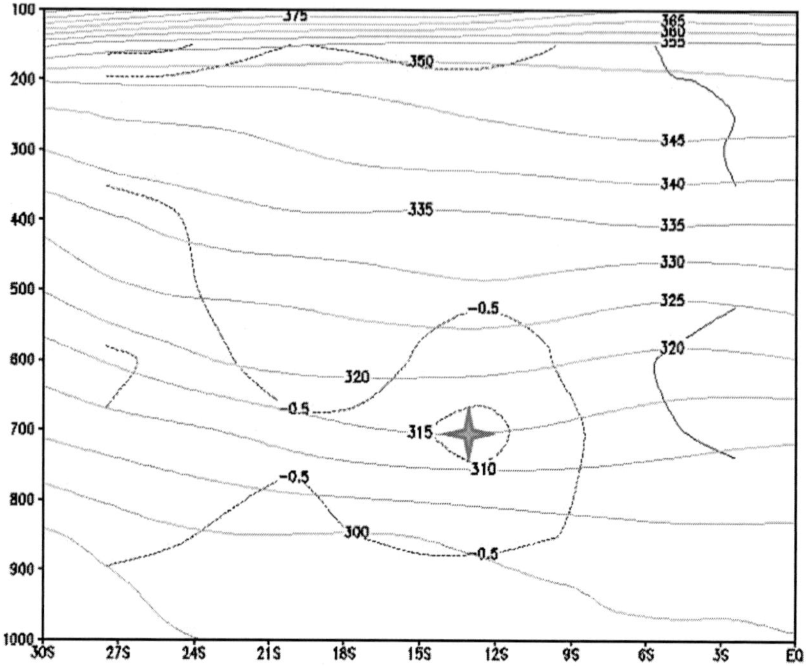

Figure I-5-7 – MTCV structure: vertical sections of the potential temperature and potential vorticity along 33°W longitudes, on 12 June 2008, 18UTC.
Potential temperature (K) – black full line;
potential vorticity (PVU, 1PVU=$10^{-6}m^2s^{-1}Kkg^{-1}$) dotted line, cross – MTCV center by PVU.
Source: Fedorova et al., 2017; Fedorova & Silva, 2016.

3. MTCV trajectories

The MTCV trajectories are very different; so far, no pattern has been described (Fedorova & Silva, 2016). Examples of the trajectories for three events are shown in Figure I-5-8.

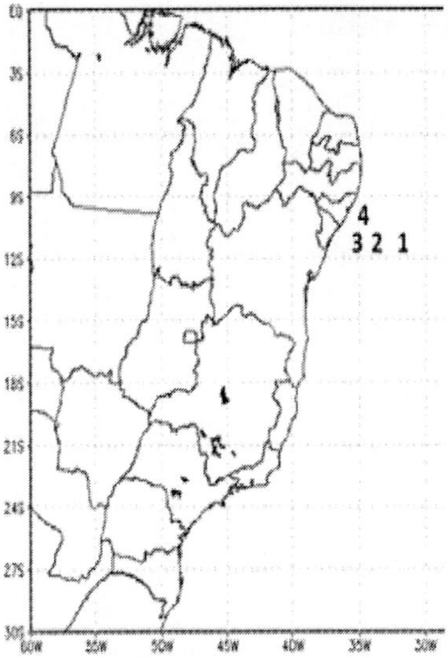

a) 2008 June from 12, 12UTC to 13, 06 UTC.

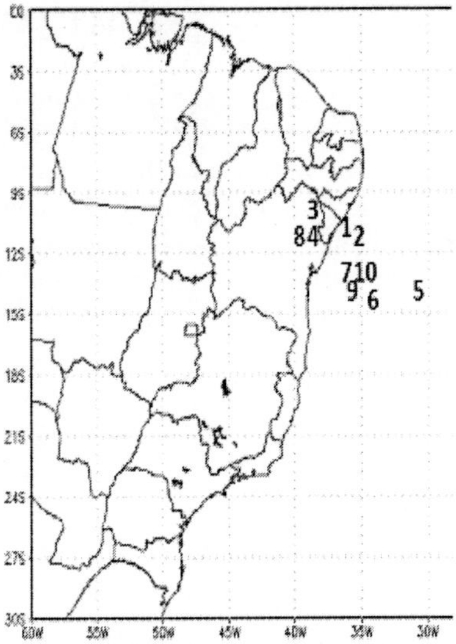

b) 2009 March from 13, 06UTC to 15, 18UTC.

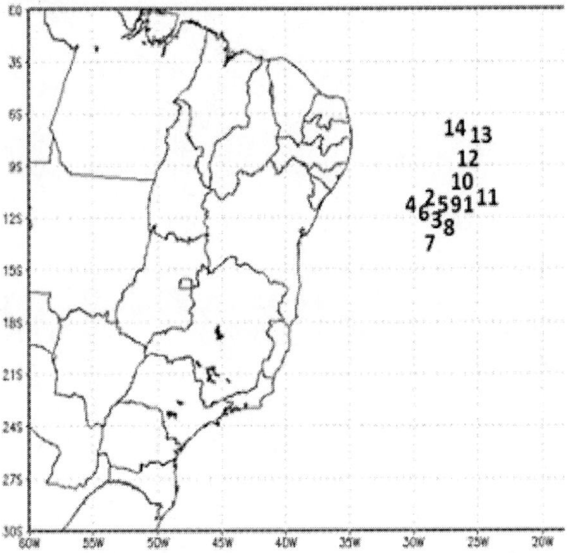

c) 2009 November from 09, 06UTC to 12, 12UTC.

Figure I-5-8 – Trajectories of MTCVs

The numbers are the serial number of the MTCV center location.

Source: Fedorova & Silva, 2016.

4. MTCV formation

MTCVs are formed in the air current from the west (Type I, Figure I-5-9 a, b, c), from the east (Type II, Figure I-5-9 d, e, f) and in the meridional current from both hemispheres (Type IV, Figure I-5-9 g, h, i, j) (Silva T.G., 2017; Fedorova & Silva, 2016). The axis of the trough before the MTCV formation can have a different position: meridional, NW - SE and SW - N (Figure I-5-9 a, b, c, d, e, f). The meridional flow shows the confluence of the air currents of both hemispheres (Figure I-5-9 g), which is similar to the ITCZ formation. Furthermore, MTCVs are formed between currents from the northern and southern hemispheres (Figure I-5-9 h). In addition, MTCVs are formed in the air current from the northern or southern hemispheres, which passed through the equator (Figure I-5-9 i, j).

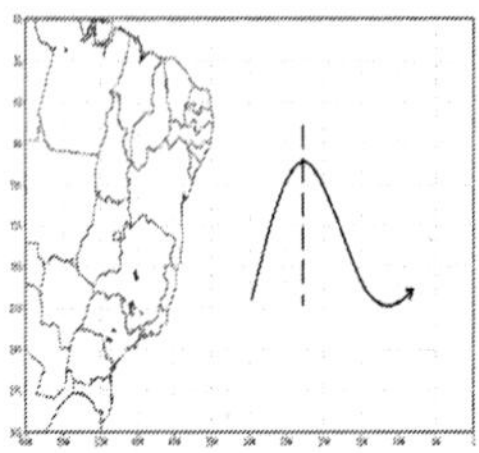

a) Type Ia

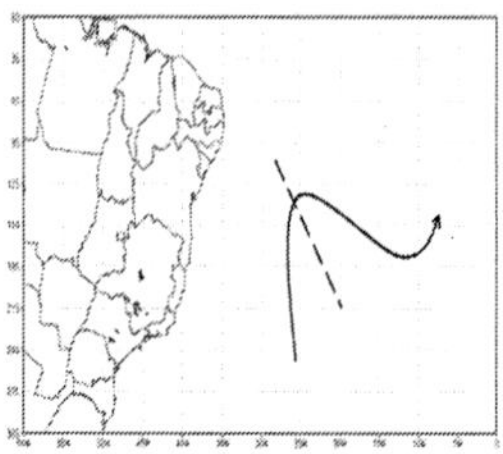

b) Type Ib

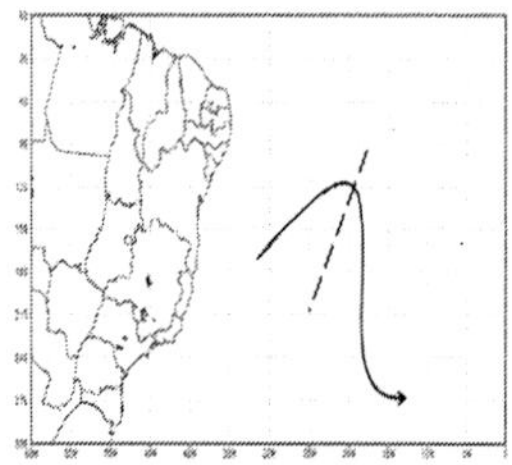

c) Type Ic

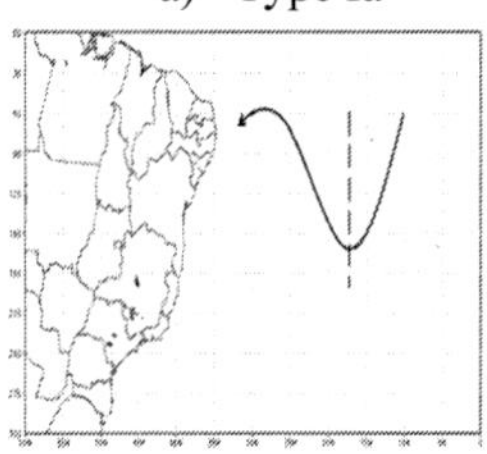

d) Type IIa

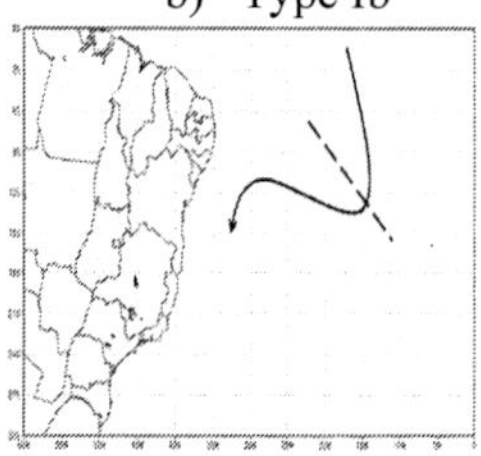

e) Type IIb

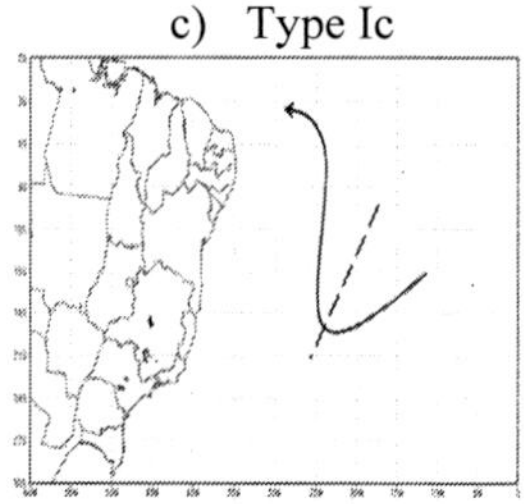

f) Type IIc

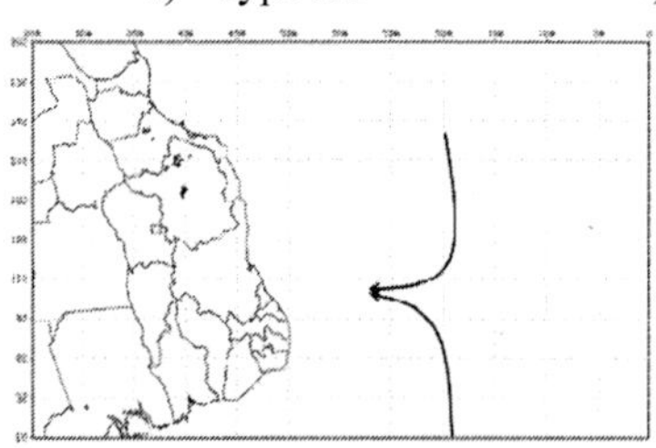

g) Type IIIa

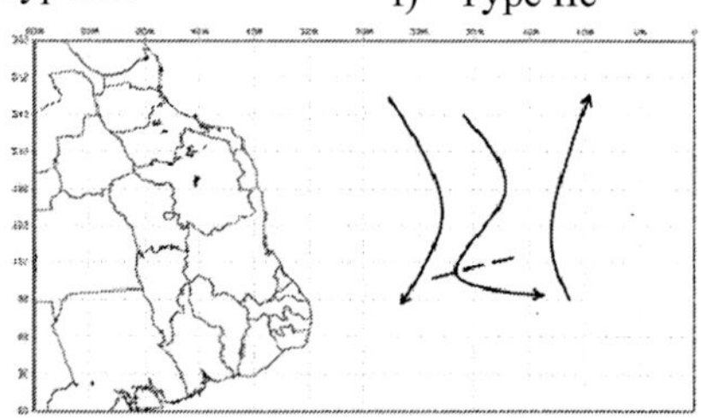

h) Type IIIb

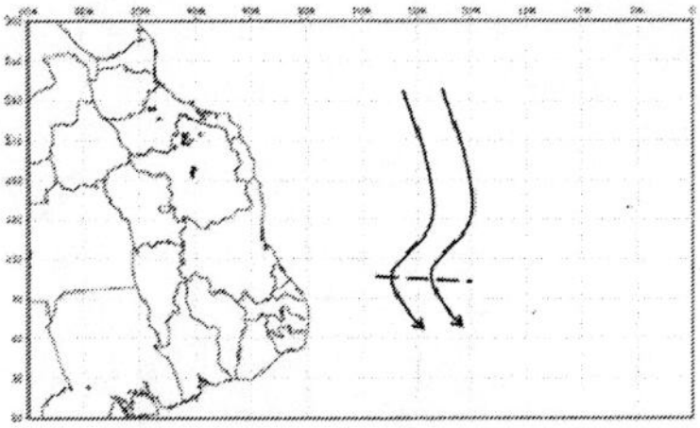
i) Type IIIc

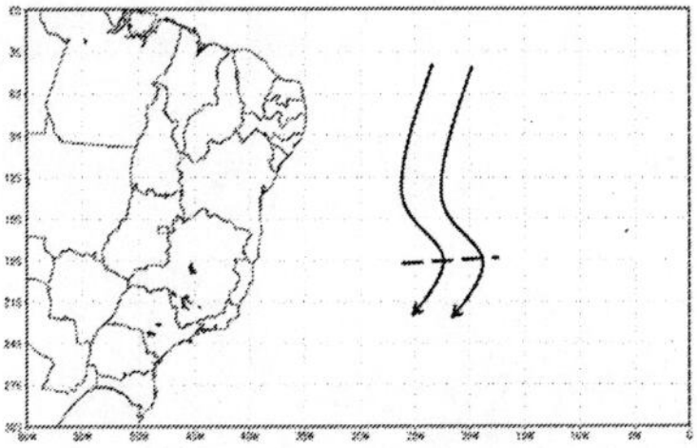
j) Type IIId

Figure I-5-9 – Types of MTCV formation in the air current from: the west - Type I a), b) c); the east - Type II d), e) f) and in the meridional currents Type III (g, h, I, j). Source: Silva T.G., 2017; Fedorova & Silva, 2016.

Types I and II were more frequently observed (Table I-5-2) (Silva T.G., 2017; Fedorova & Silva, 2016). As a rule, MTCV were formed in the air flow from the west (Type I, 29 events) and from the east (Type II, 21events). Usually, in the air current from the west, the axis of the trough was meridional before the MTCV formation (Type Ia, 17 events).

Table I-5-2 – Frequency (number of events) of the different types of MTCV formation
Source: Silva T.G., 2017; Fedorova & Silva, 2016.

Type I: from the west			**Type II: from the east**			**Type III: meridional currents**			
Ia	Ib	Ic	IIa	IIb	IIc	IIIa	IIIb	IIIc	IIId
17	8	4	4	13	2	5	2	5	2
29			**21**			**14**			

In the air current from the east (Table I-5-2), the axis of the trough was predominantly NW - SE. It is important to note the influence of the air currents from the Northern Hemisphere in Type III.

Preliminary results show that the MTCV formation was associated with the intrusion of stratospheric air along the surface of a cold front located south of this vortex (Silva N.M., 2017). This invasion triggered/caused the MTCV formation. This process can be seen in the vertical section of the potential vorticity (Figure I-5-10). Once formed during its lifetime, the vortex was associated with tropospheric air without any connection to the stratospheric air. The example shows that the cold front located at the left edge of the figure I-5-10, which was identified by the vertical distribution of potential vorticity gradients. The MTCV was formed at the level of 600 hPa and this process is visible from the increase in the area of -0.6 PVU ($1PVU=10^{-6}m^2s^{-1}Kkg^{-1}$) to the north (to the right side of the figure) of the frontal zone at the same level (Figure I-5-10 a and b). A clearly defined MTCV region in the potential vorticity section is traced throughout the lifetime of the vortex in figures I-5-10 c and d.

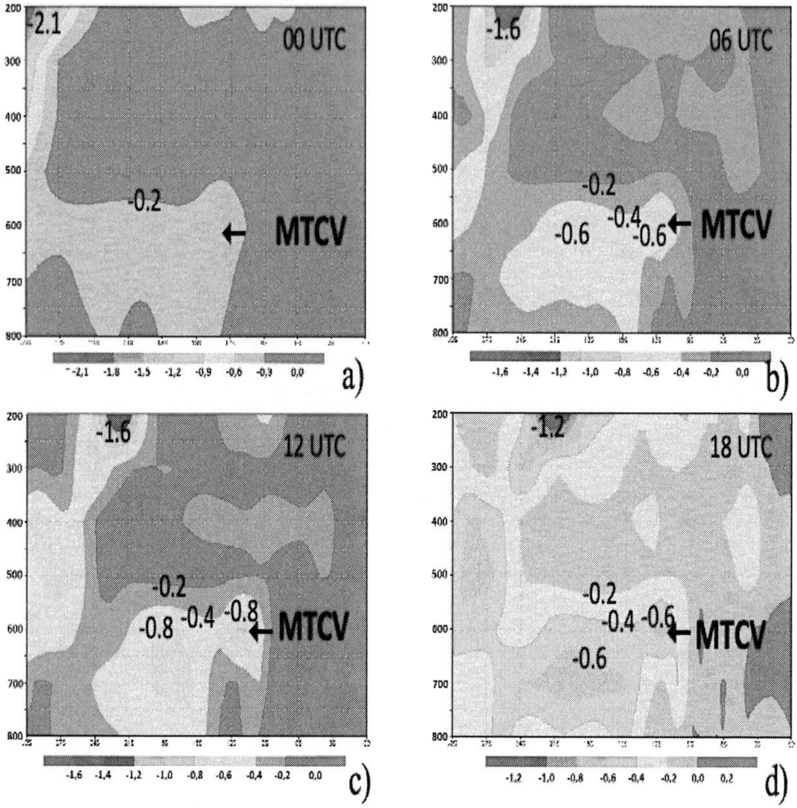

Figure I-5-10 – Vertical sections of the potential vorticity along 32, 35, 37 and 37°W longitudes, on 12 January 2008
Source: Silva N.M., 2017.

Association with the synoptic systems. All MTCVs were located above the trade winds at the low levels (figure I-5-1, 1000, 800 hPa; Chapter I-3). Also, MTCV formation due to convergence of the trade winds of the northern and southern hemispheres (Type IIIa of MTCV formation, Figure I-5-9; Chapters I-1 and I-3) shows similarities with ITCZ.

5. Seasonal variation

MTCVs were observed during each season (on average 19 MTCVs per month in the tropical regions of the South Atlantic and the South America 0-60°W, 0-30°S) with low seasonal variation (Santos, 2012; Fedorova et al., 2017; Figure I-5-11). In spring, their frequency was lower (average 49 vortices per spring season or 16 vortices per month) than in other seasons (average 20 events per month) (Table I-5-3). The number of MTCVs with a duration of more than 42h was 13-15 per year. Vortices with a lifetime of more than 30h were detected more frequently in summer (9 events per season) and rarely (4-5 events per season) in autumn (Table I-5-3).

Table I-5-3 – Frequency (numbers of events) and duration of MTCV during seasons in 2008, 2009 and 2010.
Source: Santos, 2012; Fedorova et al., 2017.

Duration (h)	Summer	Autumn	Winter	Spring	Year
6-12	126	134	126	106	492
18-24	31	35	32	22	120
30-36	16	7	11	10	44
≥42	12	7	11	10	40
Σ	185	183	180	148	696

Over the continent at the region between 9 - 21°S in summer and winter about 9 events were observed, in autumn and spring from 4 to 11 (Santos, 2012) (Figure I-5-11).

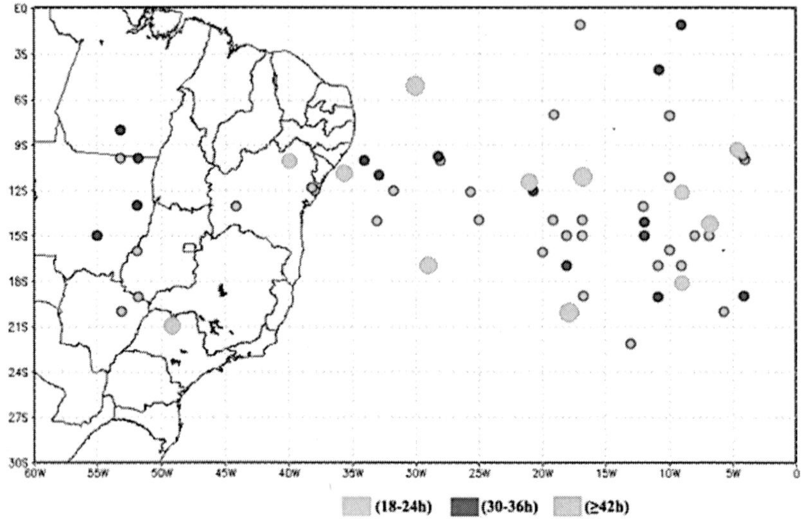

a) summer

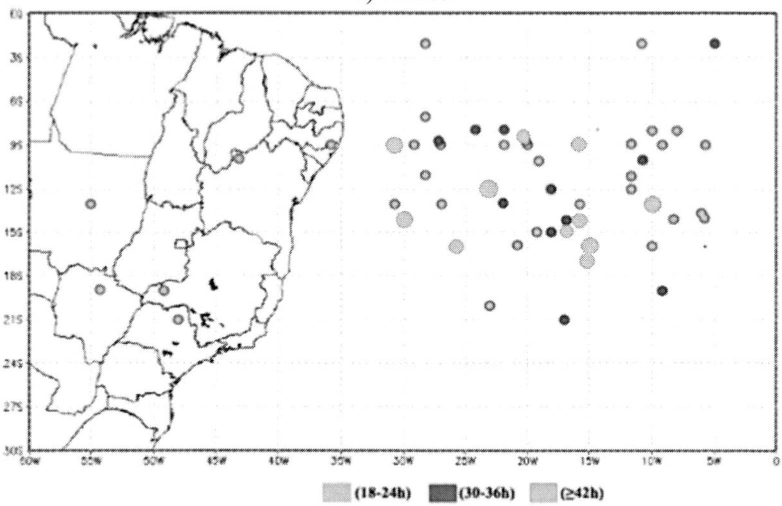

b) winter

Figure I-5-11 – MTCV location in summer (a) and winter (b) in 2008, 2009 and 2010

Big grey circle shows the location of MTCV of a long duration (≥ 42 hours)

Source: Fedorova et al., 2017; Fedorova & Silva, 2016.

6. Relationship between MTCV and precipitation

Knowledge of large-scale systems and their relationship to adverse phenomena is important for weather forecasting. Dry conditions in the MTCV and cloudiness at low levels can be confirmed by vertical sections, satellite images and humidity maps from vortex formation to dissipation (Silva et al., 2014; Fedorova et al., 2017). Precipitation was detected in all MTCV events and thunderstorms were observed rarely at the vortex periphery.

Usually, low-level clouds are observed throughout the vortex lifetime in the NEB. Precipitation occurred in typical events was about 12mm / 24h. Convective clouds, organized as instability lines at the low and middle levels, are identified from satellite images at the vortex periphery (Fedorova et al., 2017).

The example from 5 July 2010 shows the position of the vortex center at 15°S / 34°W. The vertical profiles (Figure I-5-12 a and b) show high humidity at low levels up to about 900 hPa and dry air at middle and high levels with a wet layer at 400 hPa at the vortex center. Low instability was confirmed by low CAPE+ at low and middle levels. Satellite images showed cloudiness at low levels and rarely clouds reaching middle levels (Figure I-5-12 c and d).

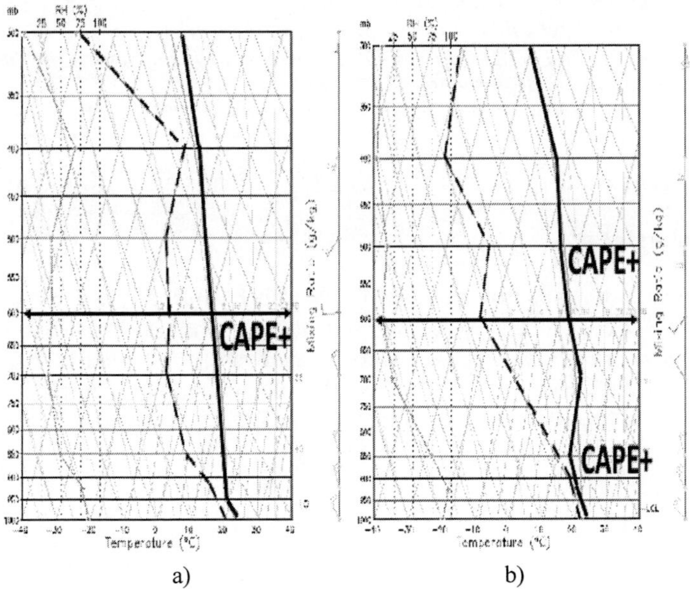

a) b)

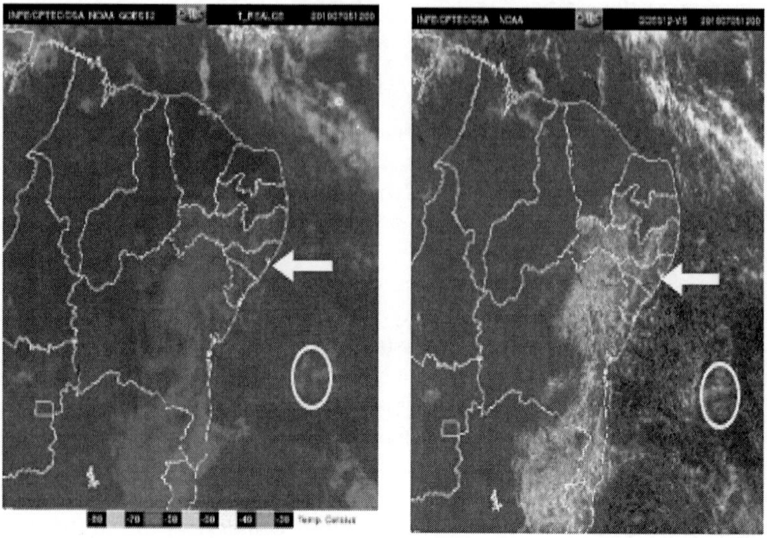

c) d)

Figure I-5-12 – Vertical profiles of temperature (*black line*) and dew point (*dotted line*) at the MTCV center (a), over the coastal region (b), satellite images in IR (c) and VIS (d) channels on 5 July 2010, 12 UTC.
Circles in the figures (c) and (d) show the MTCV center;
the double arrow in the figures (a and b) show MTCV level;
the white arrows in the figures (c and d) show the location of vertical profile over the coastal region in the figure (b).
Source: CPTEC, Fedorova et al., 2015, Fedorova & Silva, 2016.

7. Relationship between MTCV and adverse phenomena

Thunderstorm

Thunderstorm formation in the MTCV has been confirmed by several events. Thunderstorms were revealed at the periphery of the cyclonic circulation in all events. These phenomena are rare, so all kinds of data were used to obtain information about the thunderstorm occurrence. The examples below show the use of the traditional METeorological Aerodrome Report -METAR and information received from Tropical Rainfall Measuring Mission -TRMM satellite. This satellite was operated from 1997 to 2015 in coordination with other NASA Earth Observing System satellites.

Example 1. The first MTCV was observed at 600 hPa level over the South Atlantic near the BNE (Figure I-5-13) (Fedorova & Silva, 2016). The western periphery of this vortex was located over the continent and the thunderstorm was captured by METAR data.

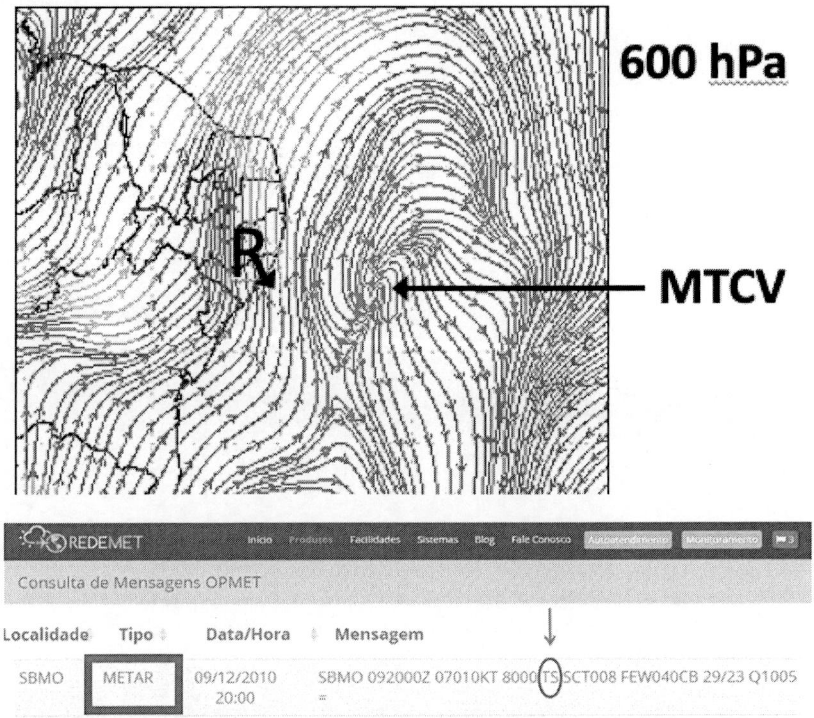

Figure I-5-13 – Thunderstorm formation in the MTCV on 9 December 2010:
Above: Streamlines at 600 hPa with MTCV center location.
Below: METAR data with an information about thunderstorm (TS)
Meteorological symbol of thunderstorm shows the location this adverse phenomenon, black arrow pointed to MTCV center. Source: Fedorova & Silva, 2016.

Example 2. The center of the second MTCV was more intense at the level of 700 hPa over the Atlantic Ocean at the latitude of the central BNE region (Figure I-5-14) (Fedorova & Silva, 2016). The western part of this vortex affected much of the continental region. The TRMM satellite data showed many places with thunderstorms.

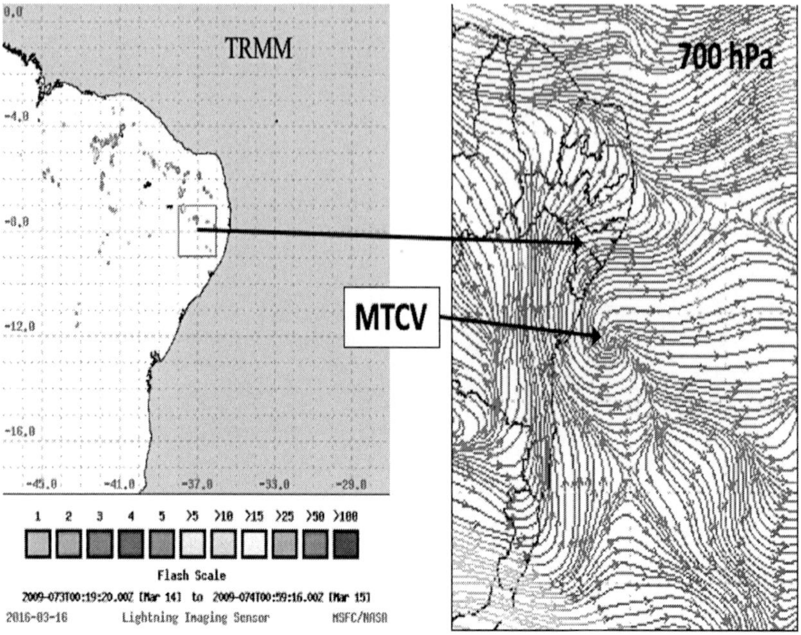

a) b)

Figure I-5-14 – Thunderstorm in the MTCV on 15 March 2009:
a) TRMM satellite data, *the dots in the square show a thunderstorm*
b) Streamlines at 700 hPa with MTCV location; *black arrow pointed to MTCV center*
Source: MSFC/NASA, Silva, 2015; Fedorova & Silva, 2016

Fog event

A rare fog event in the southern BNE was identified at the northern periphery of the MTCV on 12 May 2009 (Figure I-5-15). The minimum visibility was 800 m and was recorded at dawn at 6 and 7 am local time.

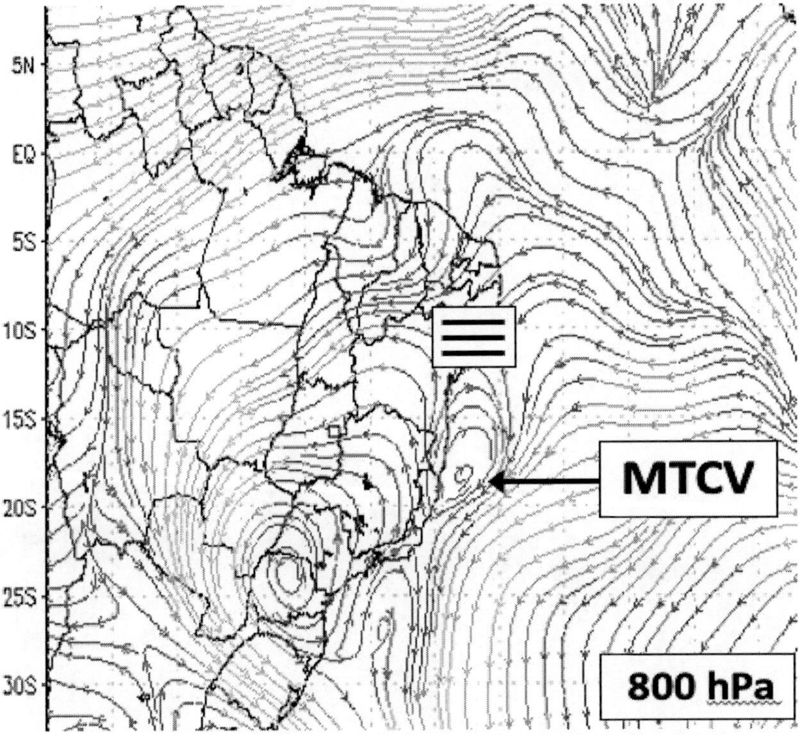

a)

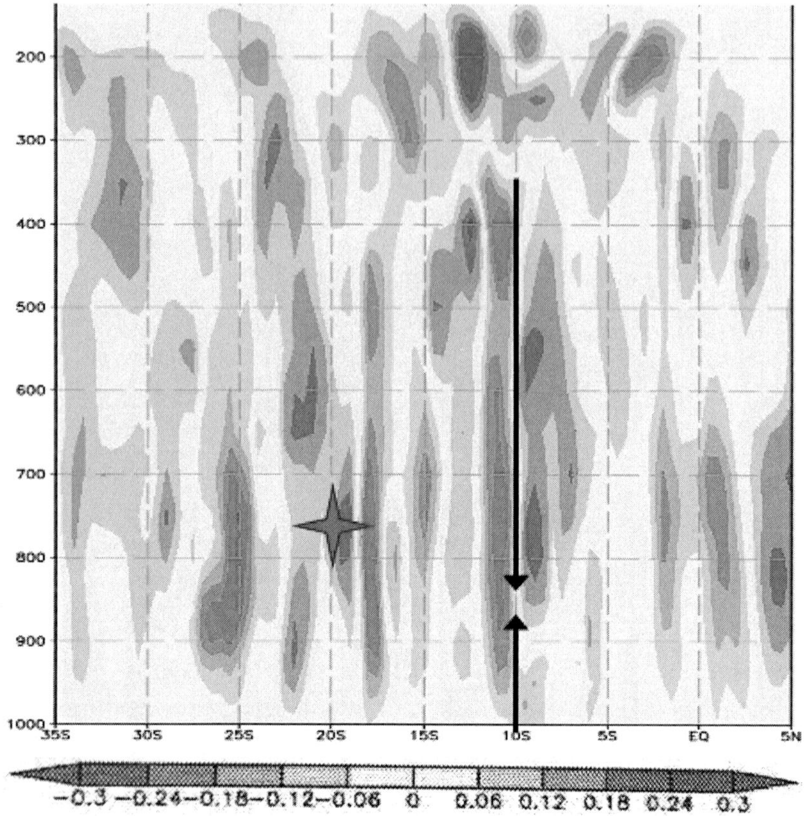

b)

Figure I-5-15 – An example of fog formation in MTCV on 12 May 2009:
a) Streamlines at 800 hPa (by Reanalysis 2) with MTCV location; *black arrow pointed to MTCV's center;*
meteorological symbol of fog with tree horizontal lines shows local of the fog;
b) Vertical section of vertical motion (omega, Pa s^{-1}; by CFSR) along 32°W longitudes; c*ross shows MTCV center at 789 hPa level, arrows show vertical movement direction over the fog.*
Source: Fedorova & Levit, 2016.

Mesoscale Convective Complex (MCC) in the MTCV

The MTCV center can be seen by streamlines at 700 hPa level (Figure I-5-16b). On infrared satellite images, the MCC is visible almost in the coastal zone, on the northeastern periphery of this vortex (Figure I-5-16a). Lifting from the surface up to 700 hPa was observed in the MCC area (Figure I-5-16c); these upward movements contributed to the formation of the MCC with precipitation. Precipitation was detected only on the MCC periphery, on the western side, and amounted to 17 mm/24h.

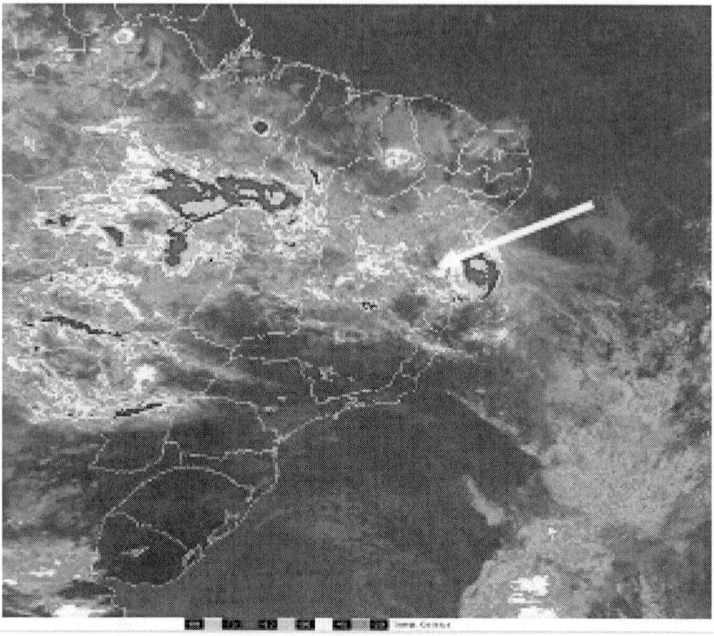

a)

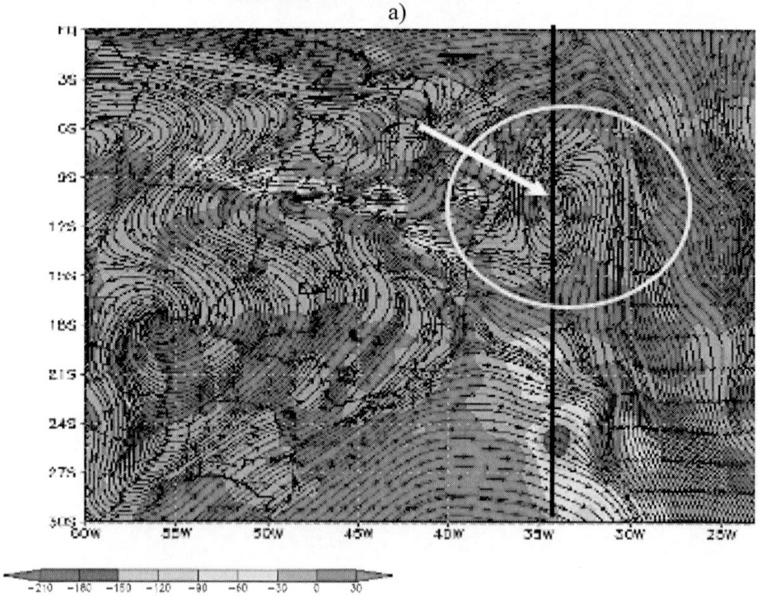
EQ
3S
6S
9S
12S
15S
18S
21S
24S
27S
30S
60W
55W
50W
45W
40W
35W
30W
25W
-210 -180 -150 -120 -90 -60 -30 0 30

b)

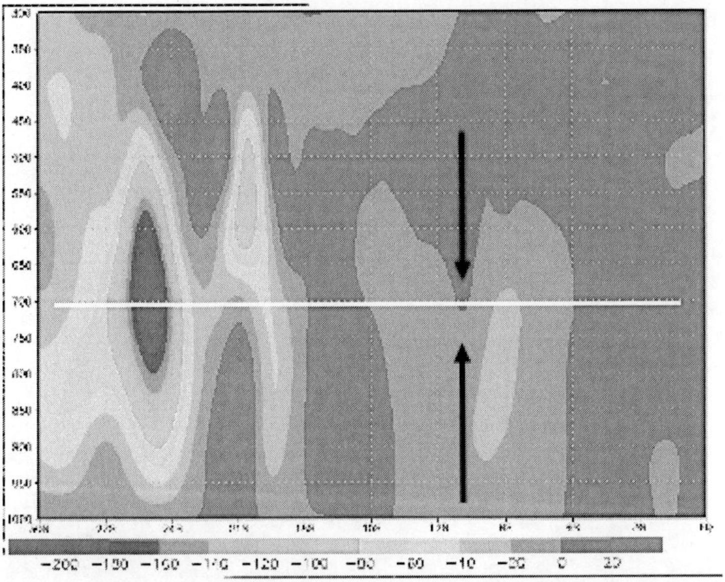

c)

Figure I-5-16 – MTCV with MCC on 9 December 2010, 12 UTC
a) satellite images in IR channel; *black arrow pointed to the western side of the MCC with precipitation information*
b) maps of vertical motion (omega, Pa s^{-1}, colors) and streamlines; *the white arrow points to MTCV's center; the white circle shows the MTCV area*
c) vertical section of the vertical movement (omega, Pa s^{-1}) along 34°W longitudes; *horizontal white line shows the MTCV center at the 700 hPa level, black arrows show vertical movement direction over the MCC region.*
Source: Fedorova & Silva, 2016; Silva, 2015.

8. MTCVs over the BNE continent

Location, duration and structure

MTCVs over the BNE continental region were analyzed during 2017, and 8 vortexes were identified (Silva & Fedorova, 2019). MTCVs were located at the 700hPa level and were deeper (up to 600hPa) in two events. Two models (NCEP and ECMWF) show the same spatial location of the MTCV. The vortices persisted for an average of 12 hours with a maximum of 30 hours.

The vorticity was weak between -1 and $-3x10^{-5}s^{-1}$ in all cases. The vertical movement was also week, predominantly downward, up to 0,2 $Pa.s^{-1}$.

Examples of the vorticity and vertical motion in the trough before the vortex formation and in the vortex are presented in figure I-5-17.

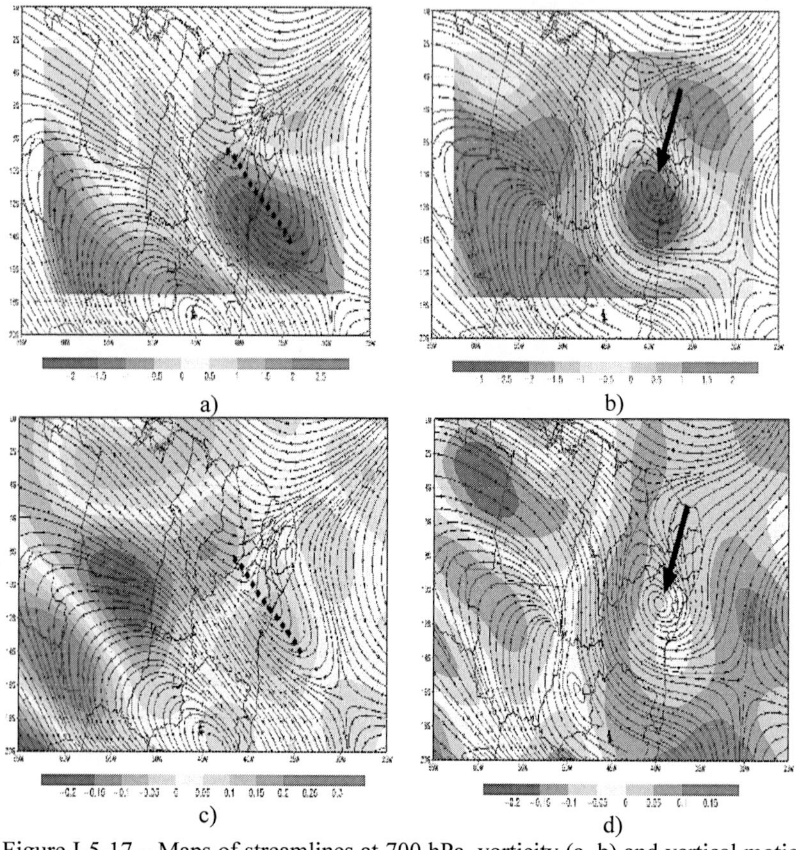

Figure I-5-17 – Maps of streamlines at 700 hPa, vorticity (a, b) and vertical motion (c, d) in the trough before the vortex formation (a, c; dotted line) and in the vortex (b, d; arrows) on 16 August 2017.
Source: Silva & Fedorova, 2019.

One example of the MTCV over the continent is presented below. The trough formed at 800 hPa, after 6 hours it became more intense and was located between 700 and 500 hPa (Table I-5-4). The vortex was observed at the levels of 700 and 600 hPa at 12 and 18 UTC.

Table I-5-4 – Location of MTCV at different levels over the BNE on 26 June 2017.
V- vortex center, T- trough.

Level (hPa)	00UTC	06UTC	12UTC	18UTC
200	-	-	-	-
300	-	-	-	-
400	-	-	-	T
500	-	T	-	-
600	-	T	V	V
700	-	T	V	V
800	T	-	-	-
900	-	-	-	-
1000	-	-	-	-

Formation

MTCV over the continent were formed predominantly (7 event from 8) from a trough in the current from the east (Type II, figure I-5-9) (Silva & Fedorova, 2019). This formation process can be seen on the example of 16 August 2017 (Figure I-5-18).

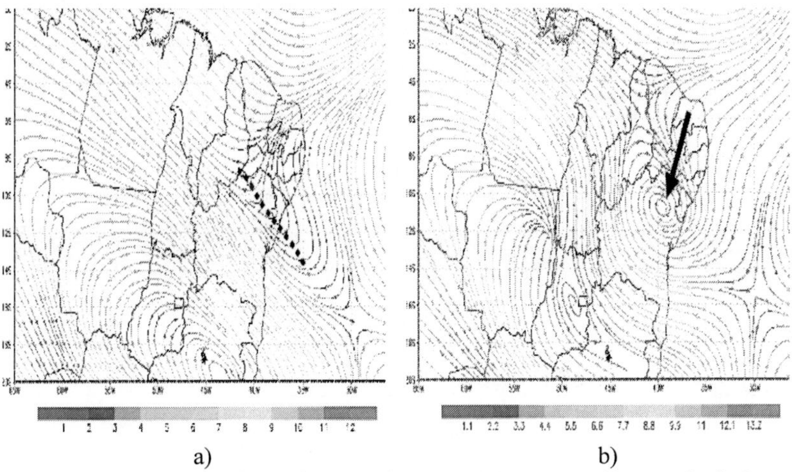

Figure I-5-18 – MTCV formation on 16 August 2017, streamlines and wind speed (m/s) on 00 UTC (a) and 06 UTC (b) at the 700 hPa level.
The dotted line shows the trough axis before vortex formation (a) and arrow pointed to the vortex center.
Source: Silva & Fedorova, 2019.

One atypical MTCV formation process is presented in Figure I-5-19; MTCV was formed in the trough between two Highs (Silva & Fedorova, 2019).

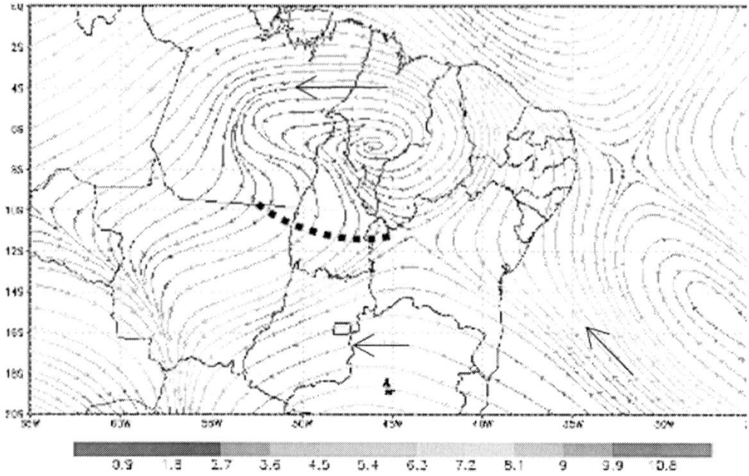

a)

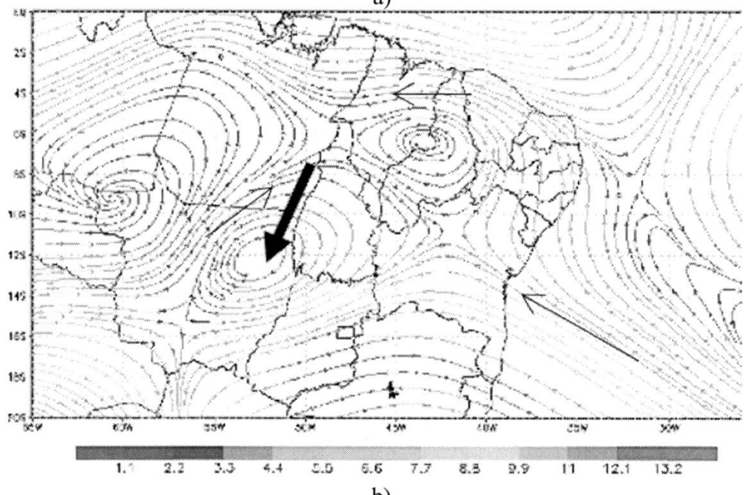

b)

Figure I-5-19 – MTCV formation on 26 June 2017, streamlines and wind speed (m/s) on 06 UTC (a) and 12 UTC (b) *The dotted line shows the trough axis before vortex formation (a) and arrow pointed to the vortex center.*

⟶ *Shows the circulation direction.*

Source: Silva & Fedorova, 2019.

Thermal structure

The MTCVs were formed more frequently (in five events) below (Figure I-5-20 a, b) and inside the warm wave (Figure I-5-20 c, d) (Silva & Fedorova, 2019). It is important to note, that all MTCVs were created above the weak baroclinic zone (Figure I-5-20 e, f). This weak baroclinic zone was deeper and reach the vortex layer in three events.

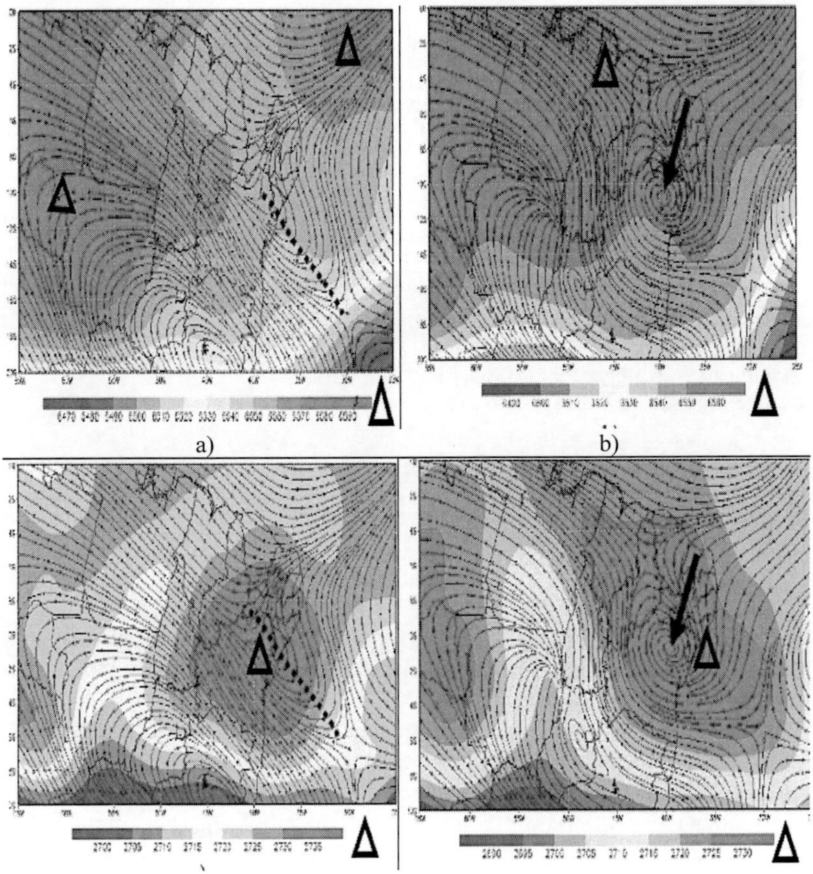

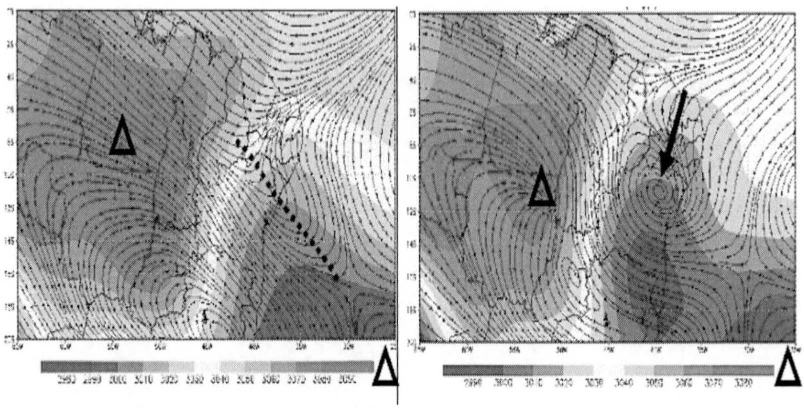

e) f)

Figure I-5-20 – Maps of streamlines at 700 hPa, on 16 August 2017 with
(a, b) thickness in the layer 200-500 hPa above the MTCV
(c, d) thickness in the layer 500-700 hPa at the level of the MTCV
(e, f) thickness in the layer 700-1000 hPa below the MTCV
in the trough before the vortex formation (a, c, e; dotted line) and in the vortex (b, d, f; arrows)
▲ indicates the higher scale number in the figures.
Source: Silva & Fedorova, 2019.

Humidity, cloudiness and precipitation.

Dry conditions in the MTCV layer were confirmed by vertical sections and humidity maps from vortex formation to dissipation (Silva & Fedorova, 2019). For example, dry air (relative humidity < 30%) at 700 hPa can be seen in the MTCV zone (Figure I-5-21a). An example of vertical profile shows high humidity ($1^{o}C<T-Td<3^{o}C$) in the lower layer up to 850 hPa and dry air above (Figure I-5-21d). The stable layer between 850 and 700 hPa creates conditions for the humidity accumulation below and, as a result, stratus and stratocumulus clouds formation. These types of cloudiness are confirmed by the infrared (IR) satellite images (Figure I-5-21 b). Low-level clouds were observed throughout the lifetime of the vortex. These clouds are associated with weak precipitation. An example of precipitation that occurred in a typical event was about 12mm / 24h (Figure I-5-21 c).

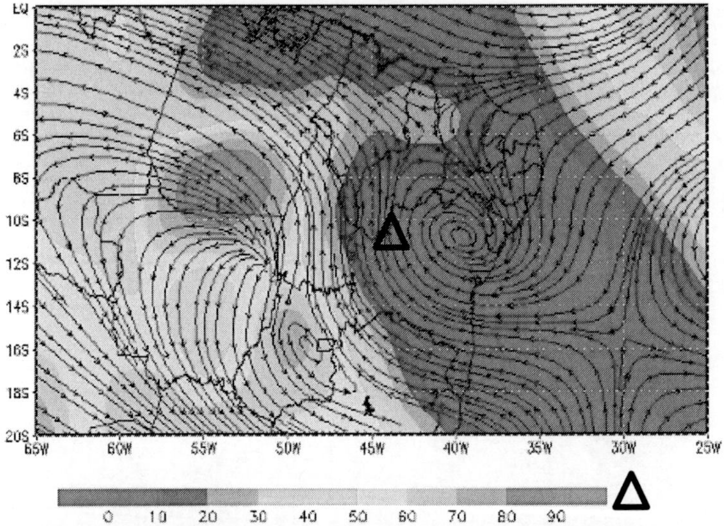

a)

Δ shows higher number of the scale at the figure

b)

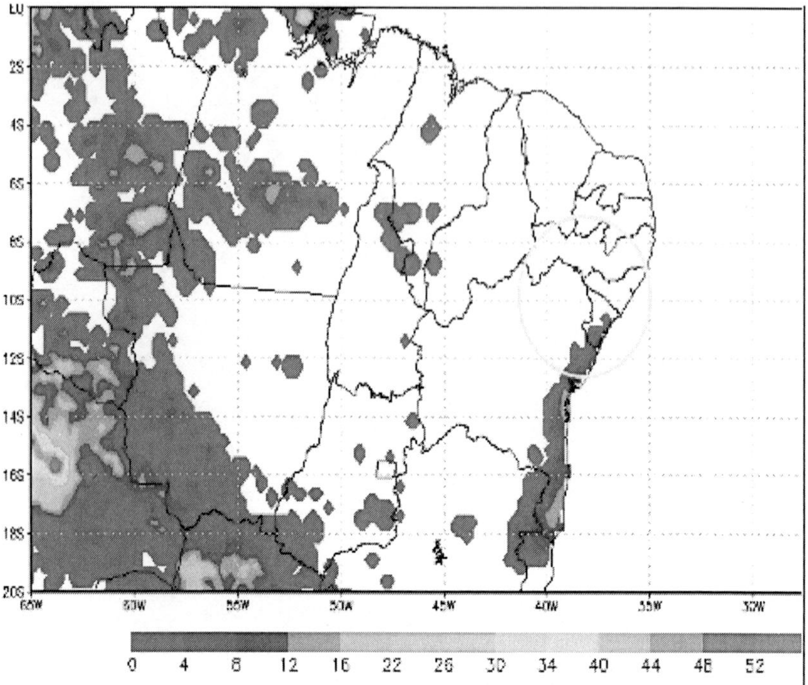
EQ
2S
4S
6S
8S
10S
12S
14S
16S
18S
20S
0 4 8 12 16 22 26 30 34 40 44 48 52

c)

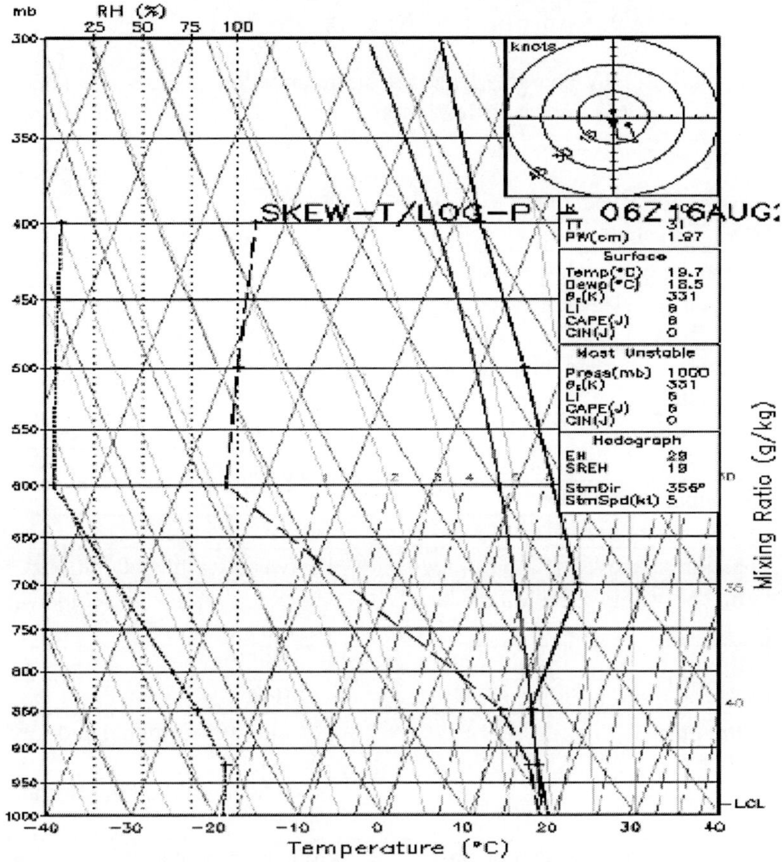

d)

Figure I-5-21 – Humidity, cloudiness and precipitation in the MTCV on 16 August 2017, 06 UTC:

a) streamlines and relative humidity (%) at level of 700 hPa;

b) satellite infrared (IR) image, *white circle shows the MTCV location;*

c) precipitation (mm/24h), *black circle shows the MTCV location;*

d) vertical profile of the temperature (*right full line*) and dew point (*dashed line*) at the MTCV center (10°S, 40°W).

Source: Silva & Fedorova, 2019; NCEP; CPTEC; CHIRPS.

9. Summary of MTCV characteristics

The main MTCV characteristics are summarized in Table I-5-5. This table can assist in MTCV identification and forecasting of the precipitation and adverse phenomena.

Table I-5-5 - Principal characteristics of MTCV

Characteristic	**Value or description**
Levels of a more intensive circulation	700-500 hPa
Lowest level reached by a cyclonic circulation	800 hPa
Highest level reached by a cyclonic circulation	400 hPa
Trajectories	varied
Frequencies	~200-250 per year
Seasonal variation	little variation
Thickness at the MTCV levels	weak baroclinic
Thickness above the MTCV levels	between warm and cold waves
Thickness below the MTCV levels	between warm and cold waves or only warm wave
Minimum vorticity at the MTCV levels	$-6x10^{-5}$ s^{-1}
Vertical motions at the MTCV levels	Varied
Cloudiness	Clouds at the middle and low levels, Rarely cumulonimbus at the periphery
Precipitation and Phenomena	Weak precipitation predominant Rare thunderstorms at the periphery

10. References

Fedorova, N.; Gemiacki, L.; Carvalho, L.C.; Levit, V.; Rodrigues, L.R.L.; Costa, S.B. *"Frontal Zone on the North-East of Brazil."* In: International Conference on Southern Hemisphere Meteorology and Oceanography (ICSHMO), 8. Foz do Iguaçu. São José dos Campos: INPE, 1765-1768. 2006.

Fedorova, N.; Santos, D.M.B.; Segundo, M.M.L.; Levit, *V. "Middle Tropospheric Cyclonic Vortex in Northeastern Brazil and the Tropical*

Atlantic." Pure Appl. Geophys, 174, 397-411, (2017). DOI: 10.1007/s00024-016-1381-1

Fedorova, N., Silva T.G. *Capítulo 3. Vórtice ciclônico de médios níveis e sua influência no tempo do nordeste brasileiro.* In: Meteorologia em tópicos – volume 3, UFPel – Pelotas, 133-178, 2016. ISBN 978-85-68891-02-5

Fedorova, N.; Levit, V; Silva T.G. *Capitulo 9. VÓRTICE CICLÔNICO DE MÉDIOS NÍVEIS ESTÁ INFLUENCIANDO NO TEMPO DE ALAGOAS?* In: Atmosfera e Sociedade: Meio Ambiente e Elementos Meteorológicos, Volume IV, Edufal, Maceio –Al, 139-149, 2015. ISBN 978-85-7177-919-8; CDU 551.5.504

Fedorova, N.; Levit, *Fog in the Tropical Region. Fog formation in the tropical region of the northeast of Brazil.* LAP LAMBERT Academic Publishing, Saarbrücken, Germany, 82p, 2016, ISBN 978-3-659-87098-9

Pontes Da Silva, B.F.; Fedorova, N.; Levit, V; Peresetsky, A; Brito, B.M. *"Sistemas sinóticos associados às precipitações intensas no Estado de Alagoas."* Revista Brasileira de Meteorologia 26, 323-338, (2011).

Santos Deydila Michele Bonfim "*Vórtices Ciclônicos de Médios Níveis (VCMN): Uma análise de frequência e estrutura.*", 76p., M. Sc. Teses, Universidade Federal de Alagoas, Maceió, 2012.

Silva, T.G.; Fedorova, N.; Levit, V. "*Influências dos Vórtices Ciclônicos de Médios Níveis no tempo de Alagoas.*" In: Congresso Brasileiro de Meteorologia, 18, SBMET: Recife, 2014.

Silva Thaíse Gomes. "*Vórtices Ciclônicos de Médios Níveis: estrutura, processos de formação e tempo associado*", 44p., Teses, Universidade Federal de Alagoas, Maceió, 2015.

Silva Thaíse Gomes. "*Formação dos vórtices ciclônicos de médios níveis.*" M. Sc. Teses, Universidade Federal de Alagoas, Maceió, 2017.

Silva Nídia Maria. "*Análise da estrutura de vórtices ciclônicos de médios níveis usando vorticidade potencial*", 87p, Teses, Universidade Federal de Alagoas, Maceió. 2017.

Silva M. C., Fedorova N. "*Processo de formação do vórtice ciclônico de médios níveis na região continental e fenômenos associados.*" Revista Brasileira de Geografia Física, 12, n.02, 365-379, (2019).

CHAPTER I.6

BRAZILIAN NORTHEAST JET STREAM

NATALIA FEDOROVA, VLADIMIR LEVIT,
BRUCE FRANCISCO PONTES DA SILVA,
BRUNO MIRANDA DE BRITO,
EDWANS SILVA CORDEIRO AND
MARCIO HENRIQUE DOS SANTOS SILVEIRA

1. Identification

The climatological location of the Subtropical Jet Stream (SJS) in South America is approximately at 30°S throughout the year (Taljaard, 1972; Satyamurty et al., 1998). Data on the maximum wind speed at 250 hPa between 25 and 30°S (Conaty *et al*., 2001) confirmed this climatological position of SJS.

Climatology SJS (1979-2008) was performed using ERA-Interim reanalysis (Zimmermann, 2017). The SJS core was identified by daily vertical profiles of horizontal wind speed and potential temperature. The SJS core has a horizontal wind speed of 30 to 50 m s^{-1} in the summer and 40 to 60 m s^{-1} in the winter, usually greater over the ocean than over the continent. SJS has always been between 200 and 250 hPa and between latitudes 18-29°S in winter and 16-35°S in summer.

The information cited above shows the lack of jet stream in the Brazilian Northeast (BNE). Virji (1981), indicated strong winds (with a wind speed of more than 20 m s^{-1}) between the Bolivian High and the UTT.

The Brazilian Northeast Jet Stream (BNEJS) has been recently described (Fedorova et al., 2018a and 2018b). According to this study, BNEJS are defined as a jet current between 20°W– 50°W and 0° – 20°S with a maximum wind speed in their core above 20 m s^{-1}. The average BNEJS wind speed was 37 m s^{-1}, and the maximum reached 64 m s^{-1}.

2. Vertical Structure

A typical example of the vertical structure of a BNEJS event (05 January 2019, 00 UTC) is shown in Figure I-6-1a. Streamlines show the BNEJS location over the Brazilian Northeast and over the North Atlantic Ocean from Maranhão to Sergipe States. The maximum wind speed in the core reached about 52 $m \cdot s^{-1}$.

A radiosonde in São Luís, the capital of Maranhão (Figure I-6-1b), was located at the entrance of the jet core (Figure 1a). Using these data, it can be seen that the maximum wind speed was observed at levels of 170-200 hPa. The vertical section of the wind speed (Figure I-6-1c) at the entrance of jet core along 44°W from 05°N to 05°S shows a maximum wind speed of 52 $m \cdot s^{-1}$ at a level of 200 hPa.

The variation in potential temperature near the jet stream axis is shown in Figure I-6-1d. A different isentrope slope in the 100 - 300 hPa layer near the jet axis is typical for jet streams (Namias & Clapp, 1949, Djuric, 1994; Fedorova 1999, 2008; Fedorova et al., 2018a).

A vertical section of vertical motions (Figure I-6-1e) shows a transvers circulation at the entrance of jet core with an upward movement on the equatorial side (the warmer side) and a downward movement on the polar (colder) side.

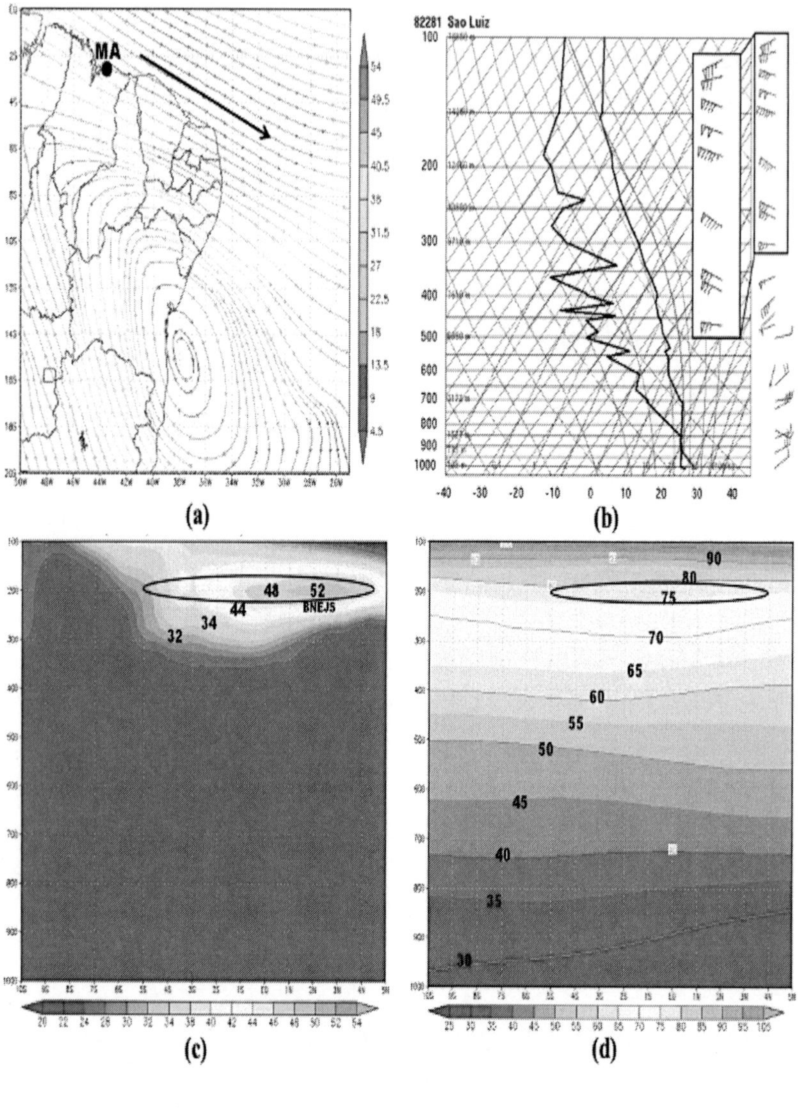
MA
82281 Sao Luiz
BNEJS
(a)
(b)
(c)
(d)

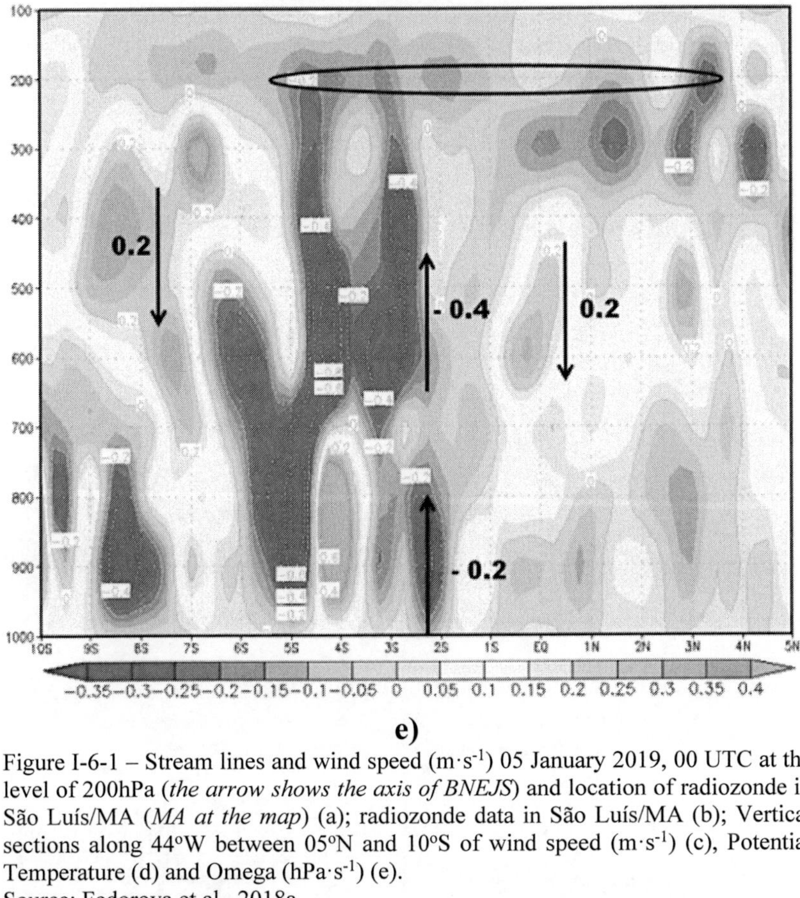

e)

Figure I-6-1 – Stream lines and wind speed (m·s^{-1}) 05 January 2019, 00 UTC at the level of 200hPa (*the arrow shows the axis of BNEJS*) and location of radiozonde in São Luís/MA (*MA at the map*) (a); radiozonde data in São Luís/MA (b); Vertical sections along 44°W between 05°N and 10°S of wind speed (m·s^{-1}) (c), Potential Temperature (d) and Omega (hPa·s^{-1}) (e).
Source: Fedorova et al., 2018a

3. Association with synoptic systems and BNEJS direction

The main synoptic-scale systems that are observed in this region at the high levels are: *Bolivian High (BH), Upper Tropospheric Cyclonic Vortexes (UTCVs), Upper-Tropospheric Troughs (UTTs), South Atlantic Subtropical High (SASH) and UTCV in the Northern Hemisphere (UTCV-NH)* (Fedorova et al., 2018b).

BNEJS is located between these systems and the presence of systems determines the direction of the jet stream:

a) a southerly (southwest and southeast) BNEJS is located between the BH and the upper tropospheric cyclonic vortex (UTCV) (and/or upper-tropospheric trough (UTT)) (Figure I-6-2a and I-6-2b);
b) a northerly and northwest BNEJS is located between the South Atlantic Subtropical High (SASH) and the UTCV (and/or UTT) (Figure I-6-2d and I-6-2c);
c) a westerly BNEJS is located between the UTCV in the Northern Hemisphere and the UTT of the Southern Hemisphere (Figure I-6-2e).

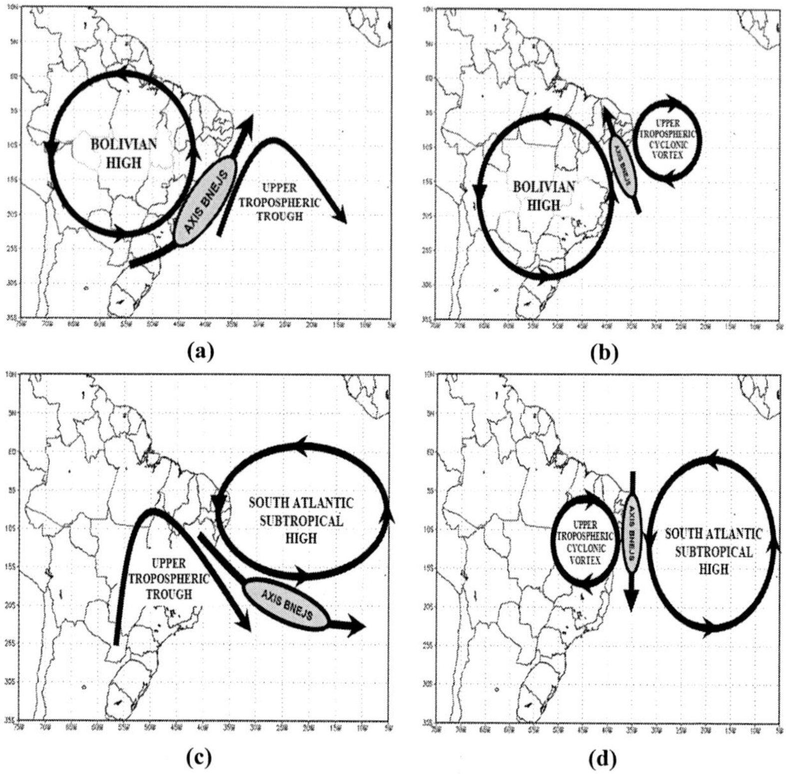

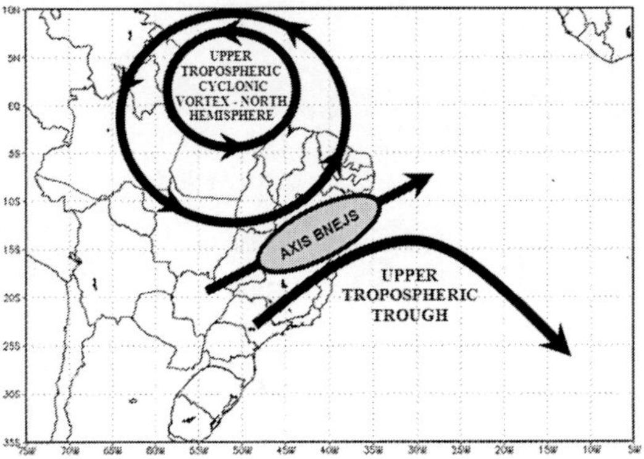

(e)

Figure I-6-2 – Types of BNEJS: southerly BNEJS between BH and UTT (a) and UTCV (b), (a – from SE and S; b – from SW); northerly BNEJS between SASH and UTCV (d) and UTT (c), (d – from N; c – from NW); westerly BNEJS between UTCV-NH and UTT –SH (e).
Source: Fedorova et. al., 2018b.

Examples of southerly, northerly and westerly BNEJS are presented in figures I-6-3, I-6-4 and I-6-5, respectively.

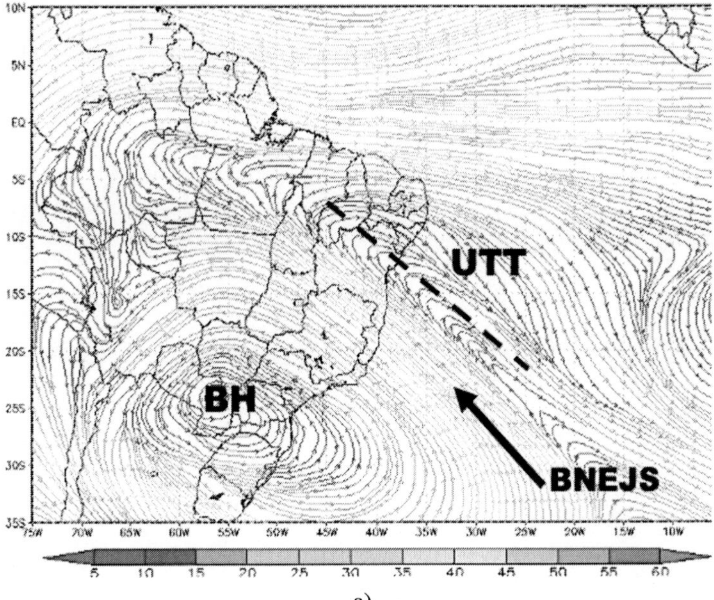

a)

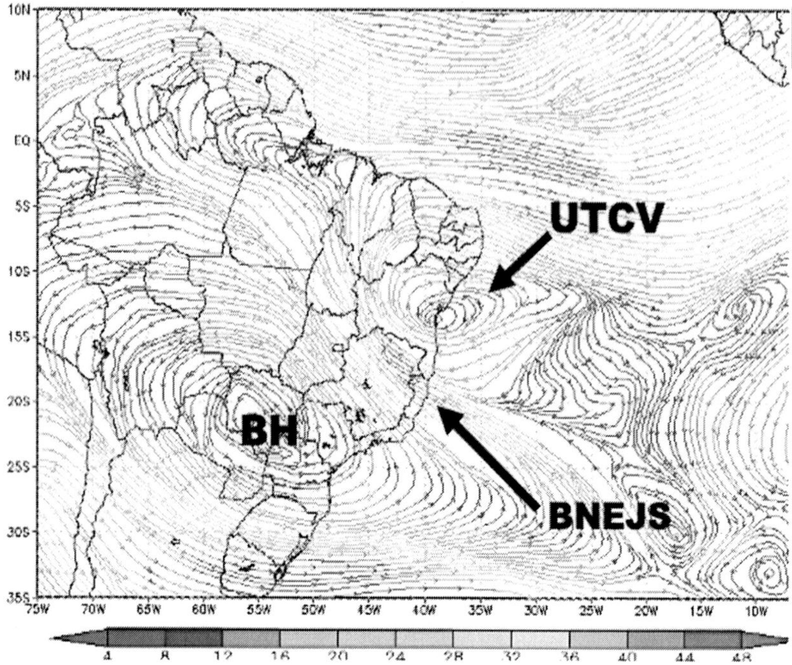

b)

Figure I-6-3 – Examples of BNEJS: southerly BNEJS between BH and (a) UTT 03 January 2019 and (b) UTCV 06 January 2019.

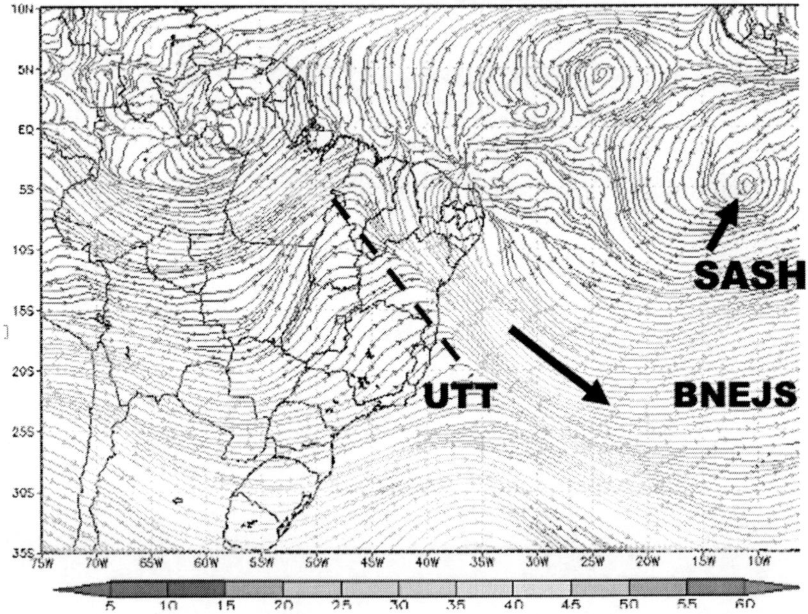

a)

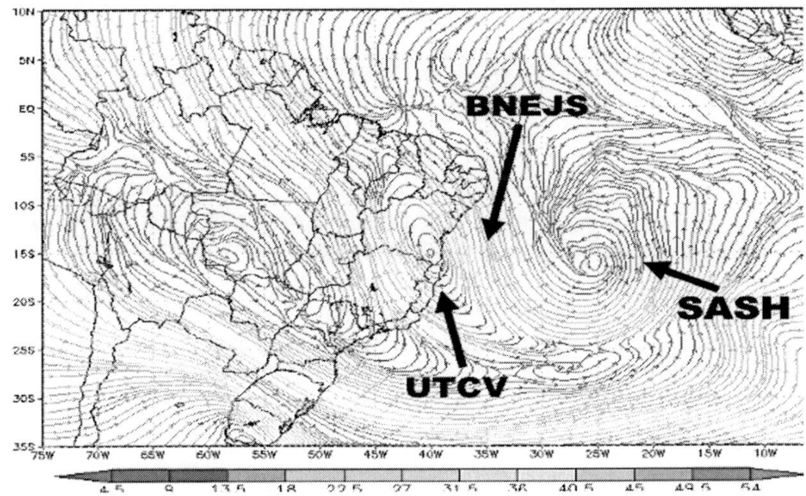

b)

Figure I-6-4 – Examples of BNEJS: northerly BNEJS between SASH and (a) UTT 07 May 2019 and (b) UTCV 07 April 2019.

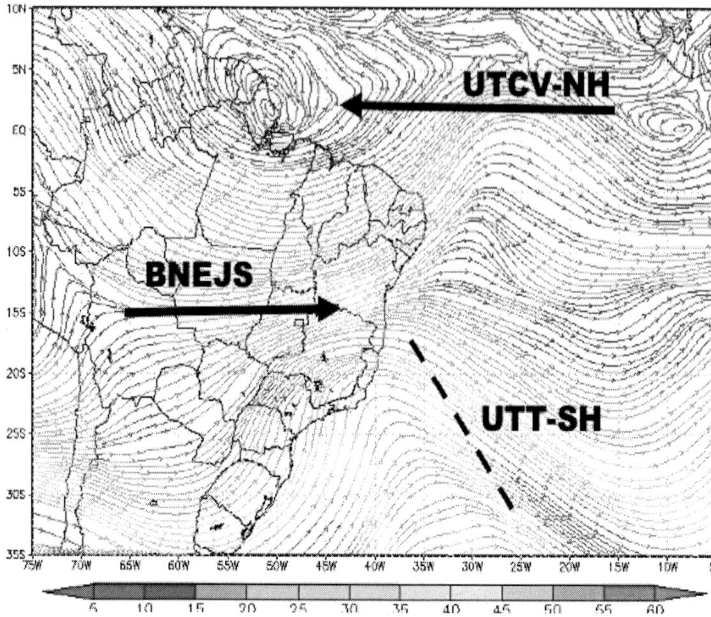

Figure I-6-5 – Example of BNEJS: westerly BNEJS between UTCV-NH and UTT–SH 21 June 2019.

According to Fedorova et. al., 2018a and 2018b, the year can be divided into the following periods depending on the characteristics of the BNEJS:

a) *December to February* (a dry period in the coastal region of BNE) is characterized by a relatively small number of fast BNEJSs with a low wind speed in the core, with a dominant south-easterly direction and BNEJS location between the UTCV and the Bolivian High.
b) *April to October* (a rainy period and transition to a dry season in the coastal region) is characterized by a rather high number of fast BNEJSs with a high wind speed in the core, a predominant northwesterly direction and BNEJS location between the UTT and the SASH.
c) *March and November* (transitional periods) has a moderate number of BNEJSs without a high wind speed in the core, with a predominant southwesterly direction and BNEJS location between the UTT and the Bolivian High.

4. Association of BNEJS with tropical and polar jet streams

The following jet streams of the southern and northern hemisphere together with BNEJS affect BNE whether (Figure I-6-6) (Vaz, 2014):

- polar jet stream of Southern Hemispheres – PJSSH;
- polar jet stream of Northern Hemispheres – PJSNH;
- subtropical jet streams of Southern Hemispheres – STJSSH;
- subtropical jet streams of Northern Hemispheres – STJSNH.

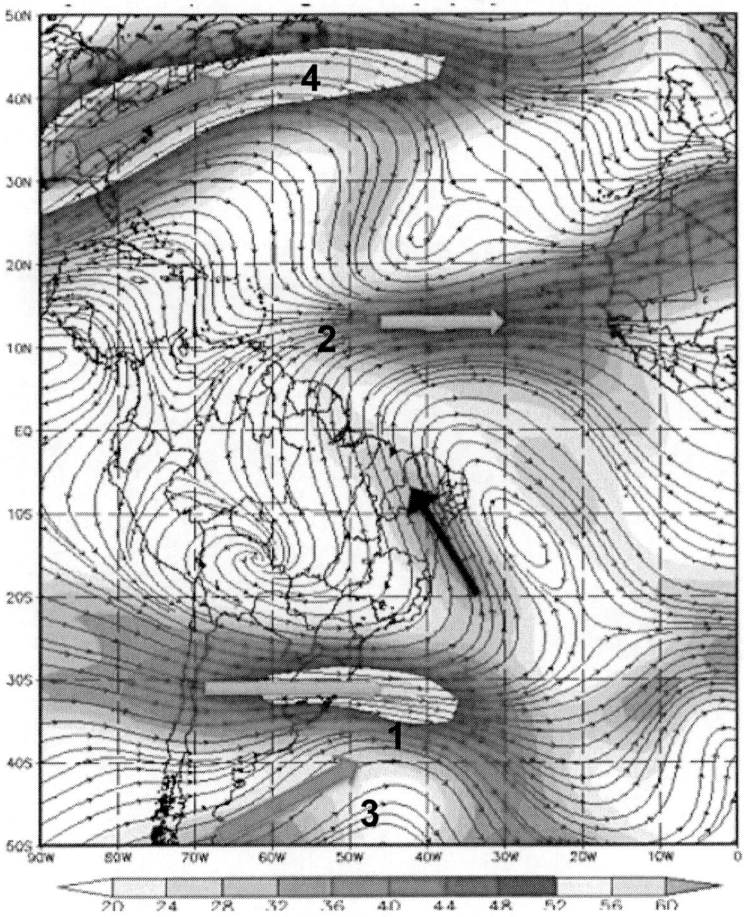

Figure I-6-6 – Jet streams at the level of 200 hPa 04 December 2009:
BNEJS - black arrow; 1-STJSSH, 2-STJSNH; 3-PJSSH, 4- PJSNH.
Source: adapted from Vaz, 2014.

Figure I-6-6 shows jet streams in just one day. Similar jet streams were observed throughout January 2020 (Figure I-6-7). The pattern of synoptic scale systems in the summer season shows the location of the Bolivian High (BH) over northern Paraguay 22°S/60°W and Upper Tropospheric Cyclonic Vortex (UTCV) over the Atlantic Ocean, almost 17°S/23°W. BNEJS with a wind speed above 20m/s was located between the BH and UTCV systems, above the BNE (Bahia State).

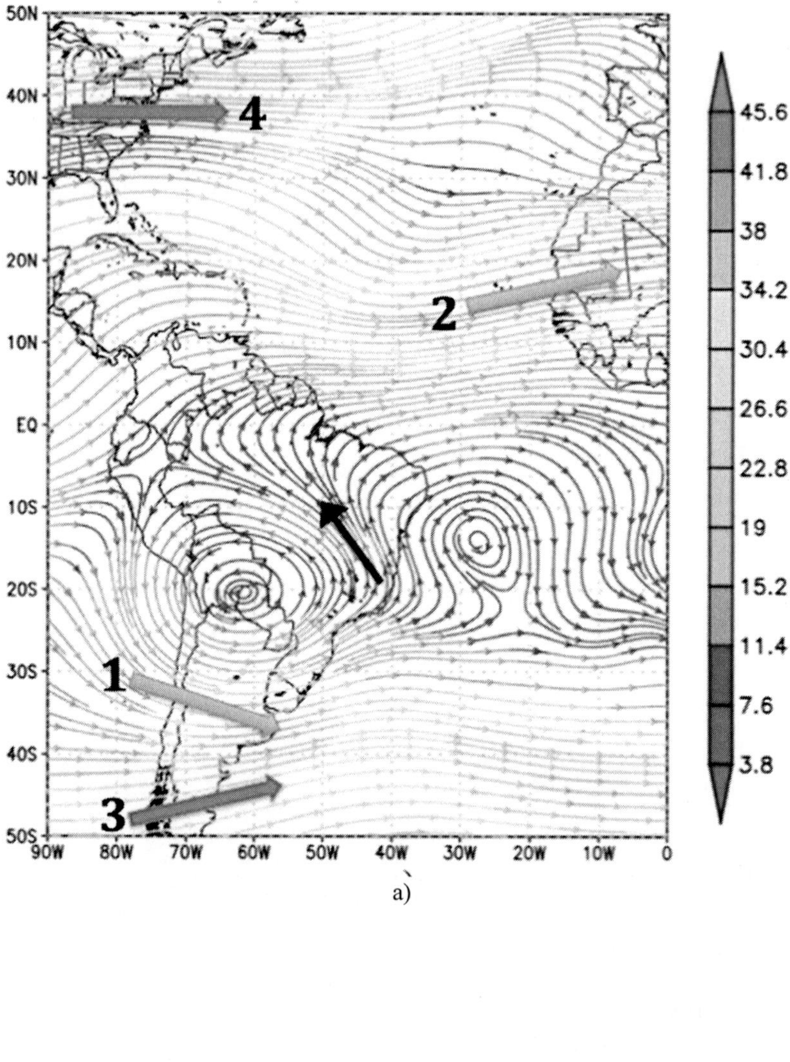

a)

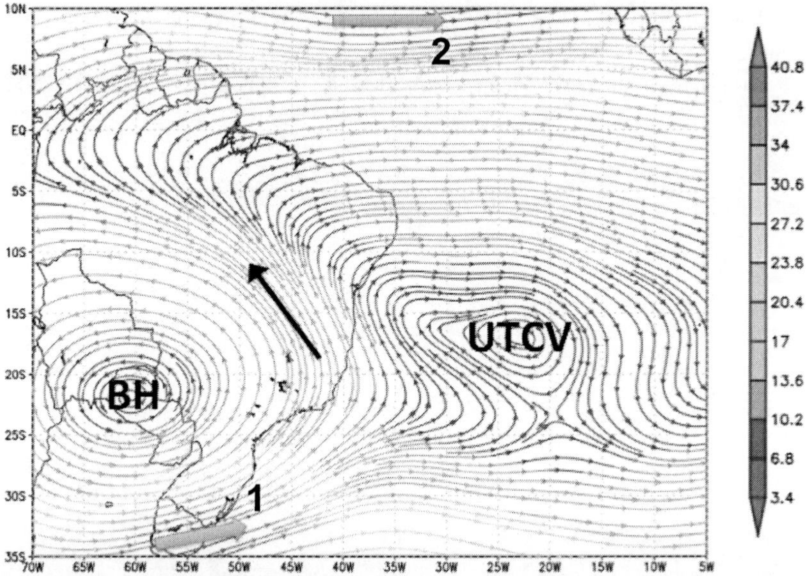

b)

Figure I-6-7 – Jet streams at the level of 200 hPa (*PJSSH, PJSNH, STJSSH, STJSNH and* BNEJS*)* (a) and (b) Jet streams nearly the BNE (*STJSSH, STJSNH and* BNEJS) together with synoptic systems (BH and UTCV) in January 2020.
BNEJS - black arrow; 1-STJSSH, 2-STJSNH; 3-PJSSH, 4- PJSNH.

The frequencies (annual and seasonal) of all jet streams, given above, were studied for 10 years (2000-2009) (Vaz, 2014). These frequencies were analyzed using jet stream days as well as events. As an event it was called a period with the same CJ.

4.1 Annual frequencies

Annual frequencies are shown in Figure I-6-8 (Vaz, 2014). The largest number of events from all 290 cases was detected when there was a relationship between BNEJS, STJSSH and PJSSH (Figure I-6-8a, *line 6*). These events were well distributed by year. The number of days was also large (Figure 8b, *dashed line*).

A connection with the northern hemisphere has been detected in many cases. Eighty-four days of a relationship between BNEJS, STJSNH and PJSNH were found (Figure I-6-8b*, line 1*). As regards the relationship between CJNEB and STJSNH, 16 events were found (Figure 8a, *line 1*). The greatest number of cases, reaching 6, was observed in 2002.

The relationship between BNEJS and STJSSH was found in 21 events, mainly in 2002 and 2008 (Figure I-6-8a, *line 5*).

Practically no relationship was found between BNEJS, STJSNH, STJSSH, and PJSNH (Figure I-6-8a, *line 3*), since only 4 events were diagnosed in 2000 and 2001 out of 10 years studied.

Relations between BNEJS, STJSNH, STJSSH and PJSSH were noted only in 7 cases (Figure I-6-8a, *line 7*).

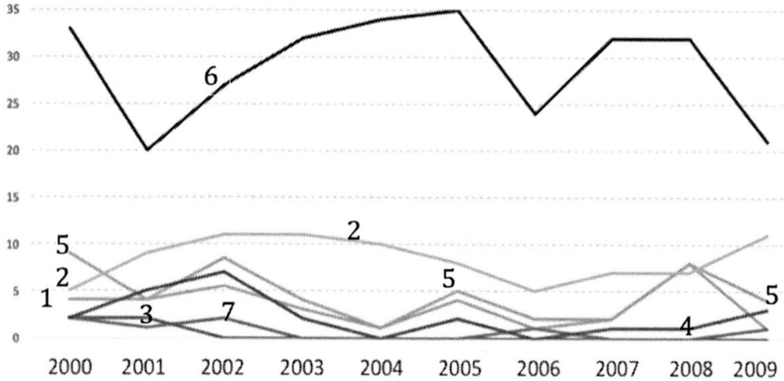

a)

1-BNEJS, STJSNH
2-BNEJS, STJSNH, STJSSH
3-BNEJS, STJSNH, STJSSH, PJSNH
4-All jet streams
5-BNEJS, STJSSH
6-BNEJS, STJSSH, PJSSH
7-BNEJS, STJSNH, STJSSH, PJSSH

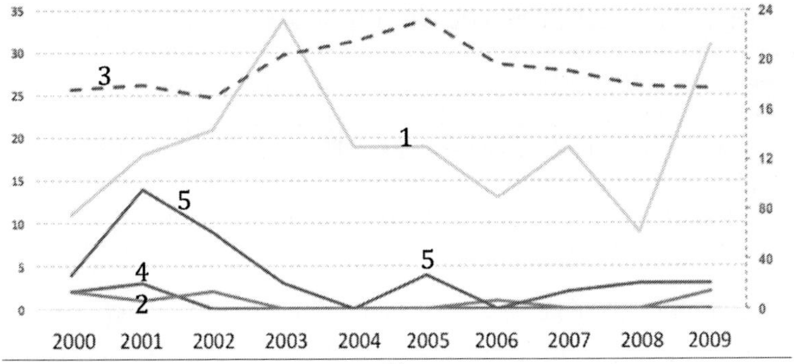

b)

1- BNEJS, STJSNH, PJSNH,
2-BNEJS, STJSNH, STJSSH, PJSSH
3-BNEJS, STJSSH, PJSSH
4-BNEJS, STJSNH, STJSSH, PJSNH
5-All jet streams

Figure I-6-8 – Annual frequencies (2000-2009) of BNEJS relationship with other jet streams.
Axis Y (left side) – days or events number.
Axis Y (right side) and dashed line– day number for the dashed line
(a) Events of BNEJS relationships and (b) *Days* of BNEJS relationships.
Source: adapted from Vaz, 2014.

4.2 Seasonal frequencies

The relationship between jet streams by months between 2000 and 2009 is shown in Figure I-6-9 (Vaz, 2014). Relations between BNEJS and STJSNH were observed in January, February, and December (Figure 9a, *line 1*). These events occurred in the winter in the northern hemisphere because STJSNH is observed closer to HS. The largest number of days (7 days) with this type of relations was observed in December. December is the winter month of the Northern Hemisphere, the climatic characteristics of which are more favorable for this type of relationship.

The relationship between BNEJS and STJSSH was mainly observed in January and February (Figure I-6-9a, *line 5*). Unlike the previous case, this type of connection is better distributed throughout the year, except for September.

Connections between BNEJS, STJSNH and STJSSH are absent in the Northern Hemisphere in summer and are more common in winter (Figure I-6-9a, *line 2*). In January and December there were more days with this type of connection. And its climatological characteristics were favorable for the connection between BNEJS and CJs of the Northern Hemisphere. The relationship between BNEJS, STJSSH and PJSSH showed both a large number and an excellent monthly distribution (Figure I-6-9b, dashed line). More days were observed in the cold season (May – August), and fewer in the hot period. This is due to the "steering current" location, which is further north in the cold period, and further south in the warm period.

Relations between BNEJS, STJSNH, STJSSH and PJSSH occur preferably during the warm period (mainly in December) of the Southern Hemisphere (Figure I-6-9a, *line 7*). This can be explained by the fact that in the cold period of the Northern Hemisphere, STJSNH is located further south.

When all CJ studied were evaluated, events were recorded rarely in all seasons (Figure I-6-9a, *line 4*; Figure I-6-9b *line 5*). Priority was in the warm period of the Southern Hemisphere or in December. Fourteen events were observed in 2001. Also in the fall of 2005, a rare case of 4 days was observed. This rare episode was recorded when it was not expected, due to the location of the CJs of the Northern Hemisphere more to the north.

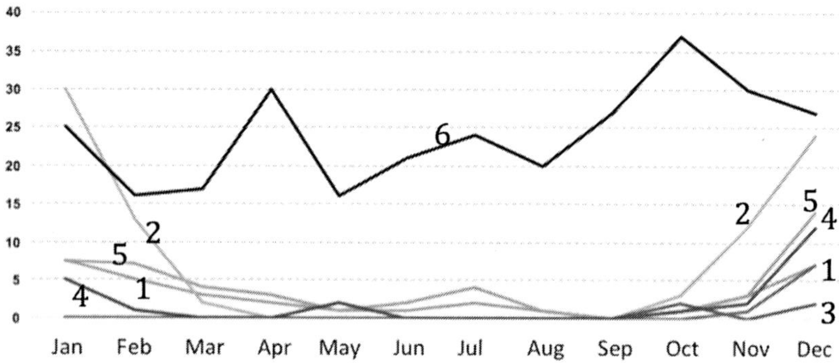

a)

1-BNEJS, STJSNH
2-BNEJS, STJSNH, STJSSH
3-BNEJS, STJSNH, STJSSH, PJSNH
4-All jet streams
5-BNEJS, STJSSH
6-BNEJS, STJSSH, PJSSH
7-BNEJS, STJSNH, STJSSH, PJSSH

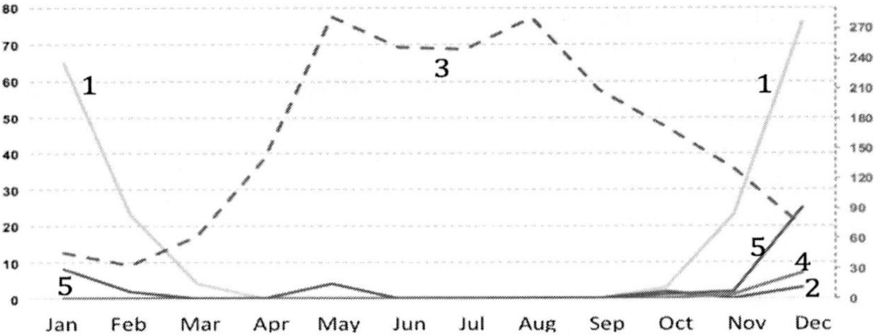

b)

1-BNEJS, STJSNH, PJSNH,
2-BNEJS, STJSNH, STJSSH, PJSSH
3-BNEJS, STJSSH, PJSSH
4-BNEJS, STJSNH, STJSSH, PJSNH
5-All jet streams

Figure I-6-9 – Seasonal frequencies (months) of BNEJS relationships with other jet streams.
Axis Y (left side) – days or events number.
Axis Y (right side) and dashed line– day number for the dashed line
(a) Events of BNEJS links and (b) *Days of* BNEJS links.
Source: adapted from Vaz, 2014.

5. BNEJS and precipitations relationship

Until now no description of precipitation patterns associated with the presence/absence of BNEJS has been published. Still, according to Djuric D. (1994), and Fedorova and Lyra (2017). BNEJS modifies vertical movements around UTCV. This factor, theoretically, contributes to the deep convection development and occurrence of intense precipitation over some areas under UTCV. However, without more detailed analysis, it is not possible to determine, whether deep convection appears or not in BNEJS' absence.

In most cases, BNEJS does not appear without other synoptic systems. It makes analysis more complex, as synoptic systems make different contributions to the total rainfall. In addition, Pontes da Silva et al. (2011) found that Wave Disturbances in the Trade Winds (WDTW, Chapter I-3), BNEJS and Cold Front Extremities (CFE - the western edge of a frontal cloud band, Chapter I-7) were associated with heavy precipitations in the coastal region of the Brazilian Northeast between 2003 and 2006. Also, they show that WDTW and BNEJS were the prevailing systems in events of moderate and low intensity precipitation.

BNEJS contribute to the lifting process, facilitating clouds formation and heavy rainfall (in 38% of heavy precipitation events). Besides, BNEJS was found as one of the main systems associated with intense precipitation in the coastal region of the Brazilian Northeast (Pontes da Silva et al., 2011). It is important to note that almost always BNEJS acts together with another synoptic system, such as WDTW (Chapter I-3), CFE (Chapter I-7), ITCZ (Chapter I-1), UTCV (Chapter I-4) or their combination.

It has been documented that most of the BNEJS events were observed from May to August (eastern BNE rainy season) near the coastal zone of the BNE (Fedorova et al., 2018a). However, this zone is very narrow near the ocean, compared to the entire BNE region.

Summarizing the above information, it is impossible to state that BNEJS is the main synoptic system associated with precipitation over BNE. On the other hand, climatological location of the BNEJS in the rainy season is observed over the BNE. However, the presence of BNEJS does not allow us to conclude whether the day will be rainy or not.

Two cases are presented to illustrate the difficulty associated with checking the contribution of BNEJS to monthly precipitation over BNE. These cases occurred during two different seasons (summer and winter), period of which was detected in a study by Pontes da Silva et al. (2011). In both cases, BNEJS persisted long enough to analyze the average wind and anomalies without significant interference from other upper-level systems for the selected months (January 2004 and July 2006) (Figures. I-6-10 and I-6-11).

5.1 BNEJS and precipitations in summer

The first example shows BNEJS in January 2004, the summer of the Southern Hemisphere. BNEJS appeared along the western side of UTCV (BNEJS Type I, according to Fedorova et al., 2018b).

Climatological data (Figure I-6-10a) show the BNEJS axis over the BNE, and wind speeds reaching 17.5 $m.s^{-1}$. The average wind speed is about 12 $m.s^{-1}$, according to Kalnay et al. (1996).

The UTCV center and the BNEJS axis in January 2004 (Figure I-6-10b) were located more east, above the Atlantic Ocean, in comparison with the climatological position (Figure I-6-10a). Strong winds (about 20 $m.s^{-1}$) in the BNEJS core this month reached double climatological values. Anomalies in wind speed confirmed an atypical situation (a square above the ocean next to the BNE, Figure I-6-10c).

It can be seen that negative precipitation anomalies were observed in the BNEJS region (Figure I-6-10d), but high positive anomalies were detected in almost entire BNE. This means that heavy precipitation was observed at the western edge of UTCV in the region with BNEJS (Figure I-6-10b).

Some areas of the BNE have been affected by severe flooding.

According to a previous study (Kousky and Gan, 1981), a stronger Bolivian High (BH) creates intense UTCV (a classic vortex close to BNE). In January 2020, the BH location was northeast of its typical location, and at its core the winds were weaker. This means that there were no conditions for the intensification of UTCV. However, UTCV was stronger than usual, possibly under the influence of an anomalous subtropical jet stream (in the southern part of BH) together along with BNEJS.

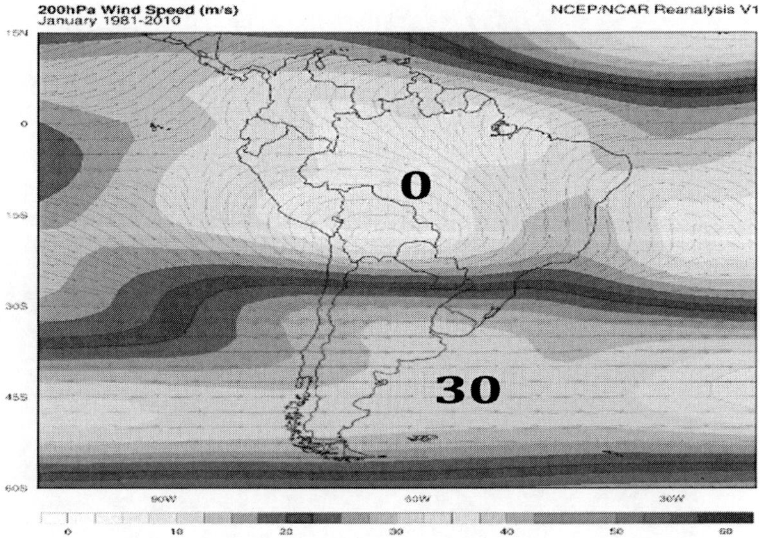

a)

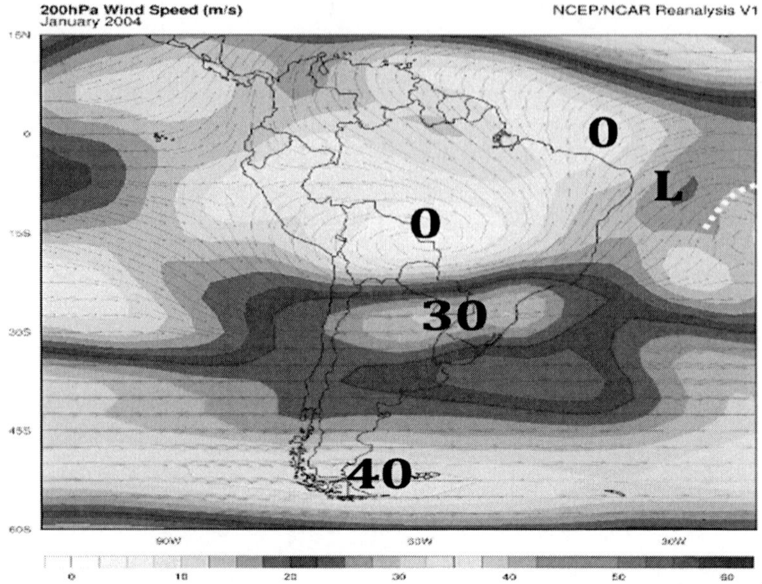
200hPa Wind Speed (m/s)
January 2004
NCEP/NCAR Reanalysis V1
0
L
0
30
40

b)

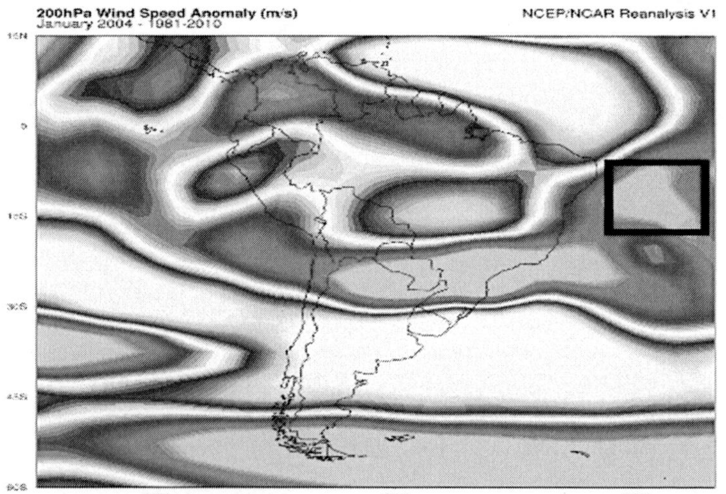
200hPa Wind Speed Anomaly (m/s)
January 2004 - 1981-2010
NCEP/NCAR Reanalysis V1
ClimateReanalyzer.org | Climate Change Institute | University of Maine

c)

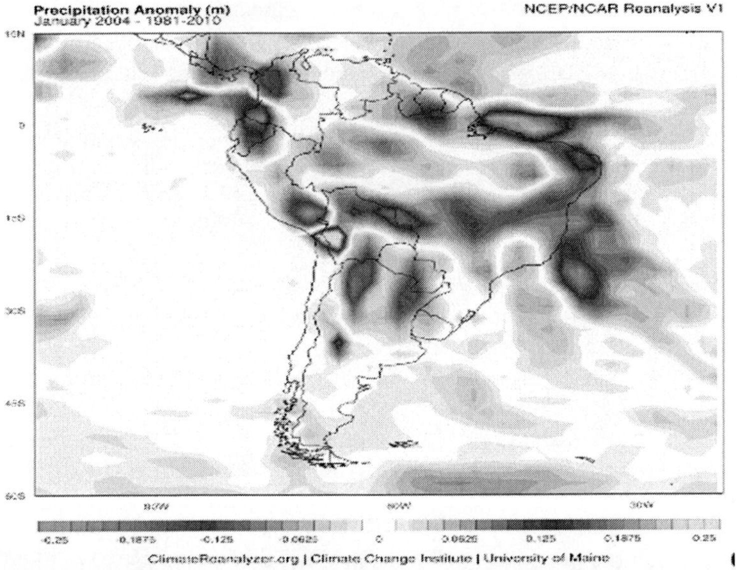

d)

Figure I-6-10 – An example of the relationship between BNEJS (Type I) and summer precipitations: a) and b) wind direction and speed ($m.s^{-1}$) in January (climatology from 1981 to 2010) (a) and in January 2004 (b)
c) wind speed anomalies in January 2004 compared with 1981 to 2010
d) precipitation anomalies in January 2004 compared with 1981 to 2010.
The white dotted line (b) illustrates the position of the *BNEJS axis, L (b)* indicates *the* core of *UTCV and square (c) shows the wind speed anomaly associated* with *BNEJS.*
Source: NCEP, Climate Reanalyzer (https://ClimateReanalyzer.org), Climate Change Institute, University of Maine, USA.

5.2 BNEJS and precipitations in winter

The second example shows BNEJS and winter precipitation anomalies (July 2006). BNEJS was classified as Type III (according to Fedorova, et al., 2018b), with BNEJS between UTCV in the Northern Hemisphere (UTCV-NH) and the Upper-Tropospheric Trough (UTT-SH). BNEJS was stronger compared to the summer event (first example).

According to climatological data for July from 1981 to 2010 (Figure I-6-11a), wind direction and speed are represented by zonal westerly winds above the NEB with maximum values of about 20 ms^{-1} in the southern NEB. However, in July 2006, the NCEP wind reanalysis showed BNEJS location over the ocean near NEB with stronger winds of about 25 $m.s^{-1}$ in the BNEJS core (much higher than the climatological wind speed in the same region) (Figure I-6-11b). The wind speed anomalies confirm this difference (square above NEB, Figure 11c).

The precipitation anomalies observed in the second example were negative in the NEB and positive over the ocean next to this region (Figure I-6-11d).

In the area under the entrance/center of BNEJS there were negative precipitation anomalies, as in the eastern NEB, but there are some positive anomalies in its exit region (above the ocean near the NEB) and over the warmer side (north / northwest side) next to the BNEJS entrance (Figure I-6-11d).

Easterly Waves (EW) and WDTW are typical rainfall-related systems in this region. According to Asnani (1993), if the eastward stream is shallow (only low levels), shallow EWs are expected, but with a deeper eastward current (up to middle or high levels), deeper EWs are observed. Strong current at high levels stopped or blocked a typical displacement of EW (at the low and middle levels) or WDTW (at the low levels) from the east. Therefore, more precipitation fell over the ocean than in the east of the BNE.

The observational data were consistent with theory in this example. In July 2006 flow at the high levels (200 hPa) was much stronger than average (Figures I-6-11b-c). The abnormal position and intensity of BNEJS can explain the negative rainfall anomalies over the eastern NEB (July is the rainy season).

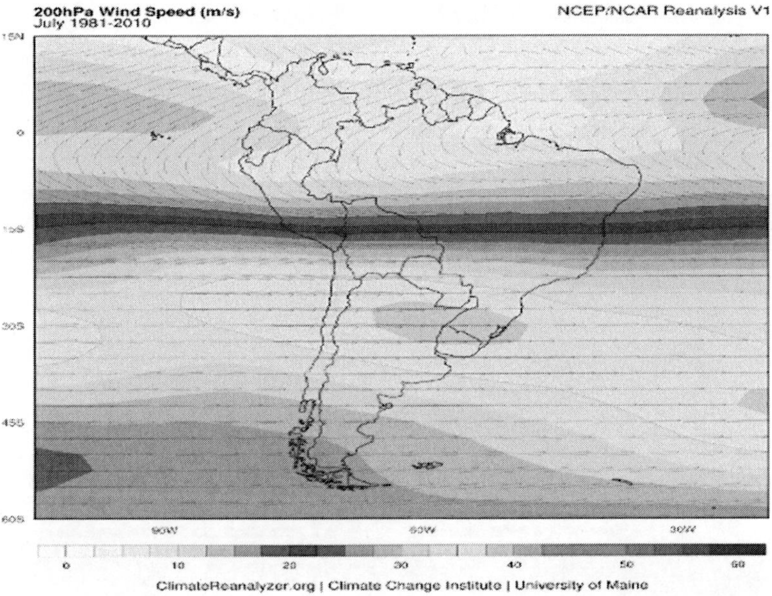
200hPa Wind Speed (m/s)
July 1981-2010
NCEP/NCAR Reanalysis V1
15N
0
15S
30S
45S
60S
90W
60W
30W
0
10
20
30
40
50
60
ClimateReanalyzer.org | Climate Change Institute | University of Maine

a)

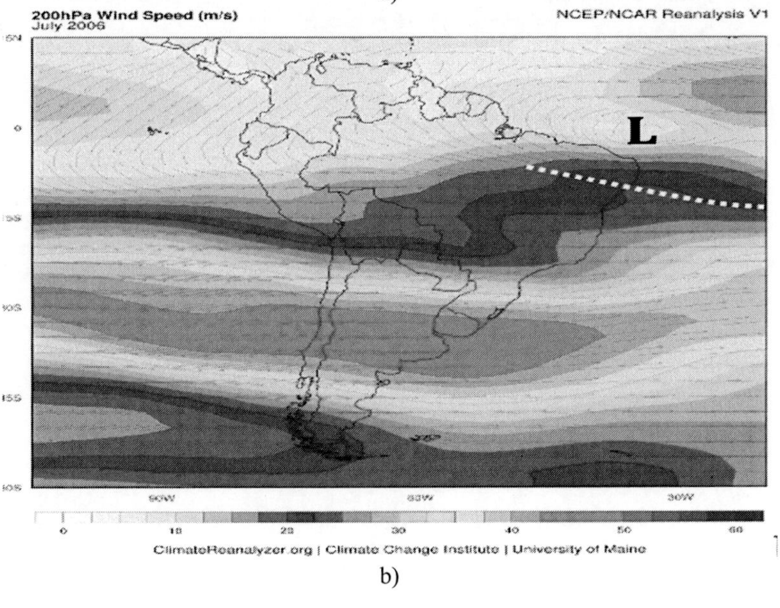
200hPa Wind Speed (m/s)
July 2006
NCEP/NCAR Reanalysis V1
L
0
90W
60W
30W
0
10
20
30
40
50
60
ClimateReanalyzer.org | Climate Change Institute | University of Maine

b)

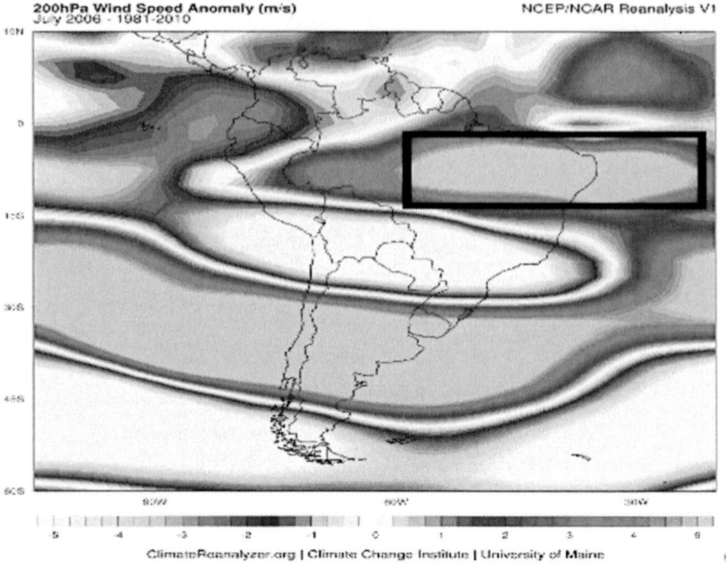

c)

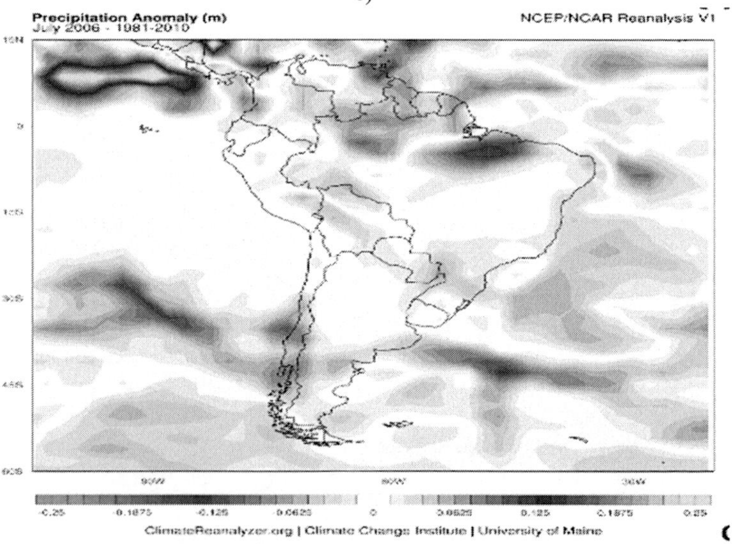

d)

Figure I-6-11 – An example of the relationship between BNEJS (Type III) and winter precipitations: a) and b) wind direction and speed (m.s^{-1}) in January (climatology from 1981 to 2010) (a) and in January 2004 (b)
c) wind speed anomalies in January 2004 compared with 1981 to 2010

d) precipitation anomalies in January 2004 compared with 1981 to 2010.
The white dotted line (b) illustrates the position of the *BNEJS axis, L (b)* indicates *the* core of *UTCV and square (c) shows the wind speed anomaly associated* with *BNEJS.*
Source: NCEP, Climate Reanalyzer (https://ClimateReanalyzer.org), Climate Change Institute, University of Maine, USA.

6. BNEJS and Mesoscale Convective Complexes relationship

The relationship between jet streams and adverse meteorological phenomena has been described in many studies, which usually assume that severe weather conditions are more likely in regions with upward movement under the quadrants of the right-entrance and left-exit of the jet (Djuric, 1994; Uccellini, L., Johnson D. 1979.).

This is due to mass continuity constraints that require vertical circulation to accompany the jet-level convergence-divergence pattern at the jet level when air parcels are accelerated in the jet inlet zone and decelerated in the jet outlet zone (Bluestein, 1993).

Lee and Galway (1956) studied flow characteristics at 200 hPa level and found a correspondence between the presence of a wind maximum at this level and tornado occurrence. Later by Whitney (1977), the effect of a subtropical jet stream on strong local storms, including tornadoes, was also revealed. A correlation between tornadoes and quadrants of upper-level jet streams was studied in detail by Rose et al. (2004) for use in short-term weather forecasting. The author found that tornadoes most often occurred below the jet exit area, with the left-exit being more favorable than the right one. In the entrance area, the occurrence of a tornado was strongly favored below the right-entrance quadrant, although this maximum was not as well defined as the maximum below the left- and right-exit quadrants.

Northeast Brazil (BNE) is not usually described as a region where deep convection occurs climatologically (Velasco and Fritsch, 1987; Romatschke and Houze, 2010). However, a recent analysis of extreme rainfall events (Pontes da Silva, 2011) indicated the emergence of mesoscale convective systems such as Mesoscale Convective Complexes (MCC).

The effect of the Brazilian Northeast Jet Stream (BNEJS) and Low-Level Jet Stream (LLJS) on MCC events was discussed in Milhahn (2013), and Fedorova & Lyra (2017). In this study, the majority of the reported MCCs were associated with jet streams, i.e., 53% were associated with LLJS and BNEJS accompanied MCC in 33% of cases. In addition, in 23% of cases, both BNEJS and LLJS are present. The absence of jet stream was observed only in 37% of cases (Figure I-6-12).

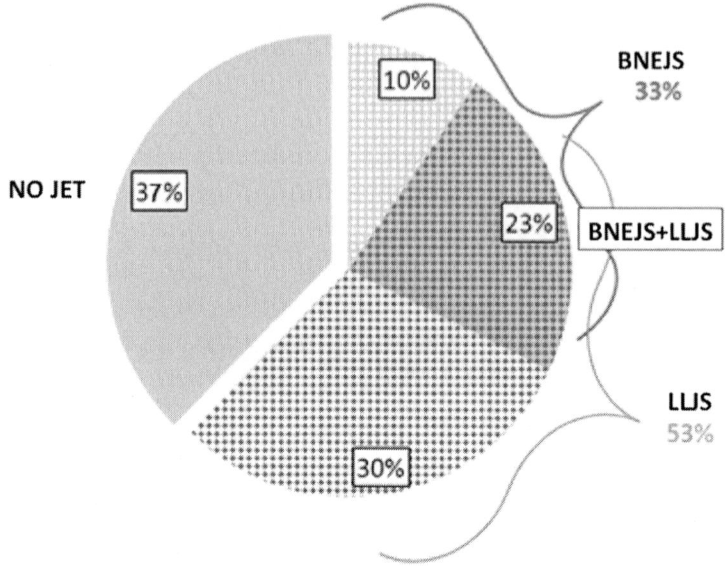

Figure I-6-12 – Frequencies of BNEJS and LLJS in MCC events in the BNE
Source: Milhahn, 2013, Fedorova & Lyra, 2017.

Given the seasonal distribution of BNEJS with MCC, they were more frequent in the fall (53%) and in the summer (37%), rare in the spring season (10%) and absent in winter. For LLJS cases, summer and fall were summed up to 43% and spring and winter - up to 12% and 2%, respectively.

Table I-6-1 – Percentage of MCC events and quadrants of BNEJS
Source: Milhahn, 2013, Fedorova & Lyra, 2017.

	Flow-related position		
	Entrance	Center	Exit
Left (warm side)	41%	12%	15%
Middle	6%	0%	6%
Right	12%	6%	3%

Both BNEJS and LLJS showed a zonal current in most cases (53%), while the meridional and transversal currents accounted for 28% and 19% of cases, respectively. The MCC locations along the BNEJS flow are not uniformly distributed between the input, output, left, middle and right sides (Table I-6-1). However, MCC was more frequent on the left entrance (41% of events), which is expected due to the upward movements in this region. The left exit was the second most frequent area, with 15%.

As an example, a well-developed convective system with MCC characteristics, observed May 8, 2019 over the coast of the BNE is shown in Figure I-6-13. The cut-out of satellite image in Figure I-6-13a shows a round-shaped cluster of cumulonimbus clouds about 200km in diameter along the BNE coast. At the upper levels, Figure I-6-13b shows the MCC located in the left-entrance region of the BNEJS in the area with anticyclonic circulation in the field of streamlines at 200hPa.

Figure I-6-13c shows a meridional section of vertical velocity, fixed in -34W and ranging from -4S (I) to -16S (II) latitudes, where negative omega values (up to $-0.3Pa.s^{-1}$) are visible along the entire atmosphere near the jet stream ($20m.s^{-1}$ isotach is shown), indicating upward movements that contributed to the MCC development.

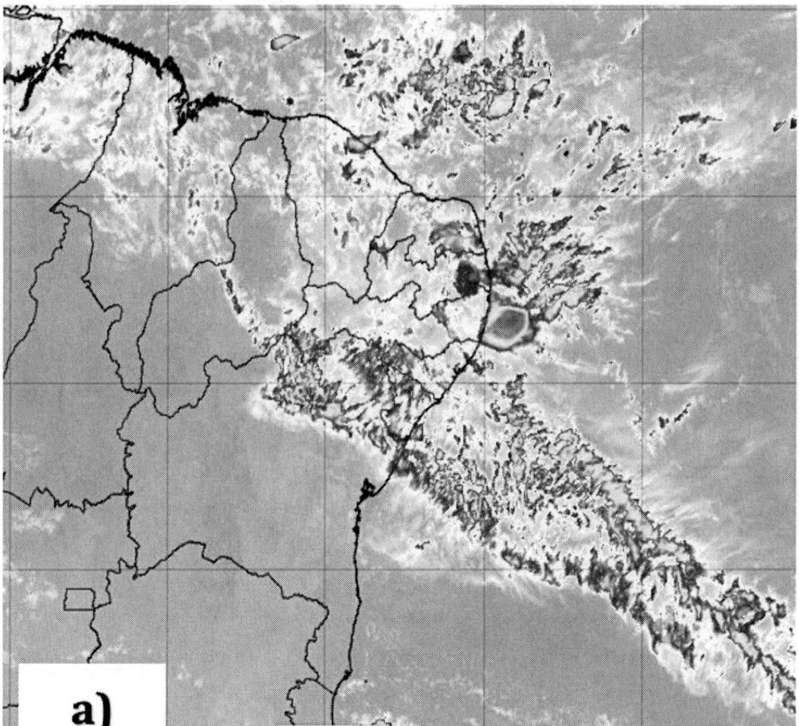

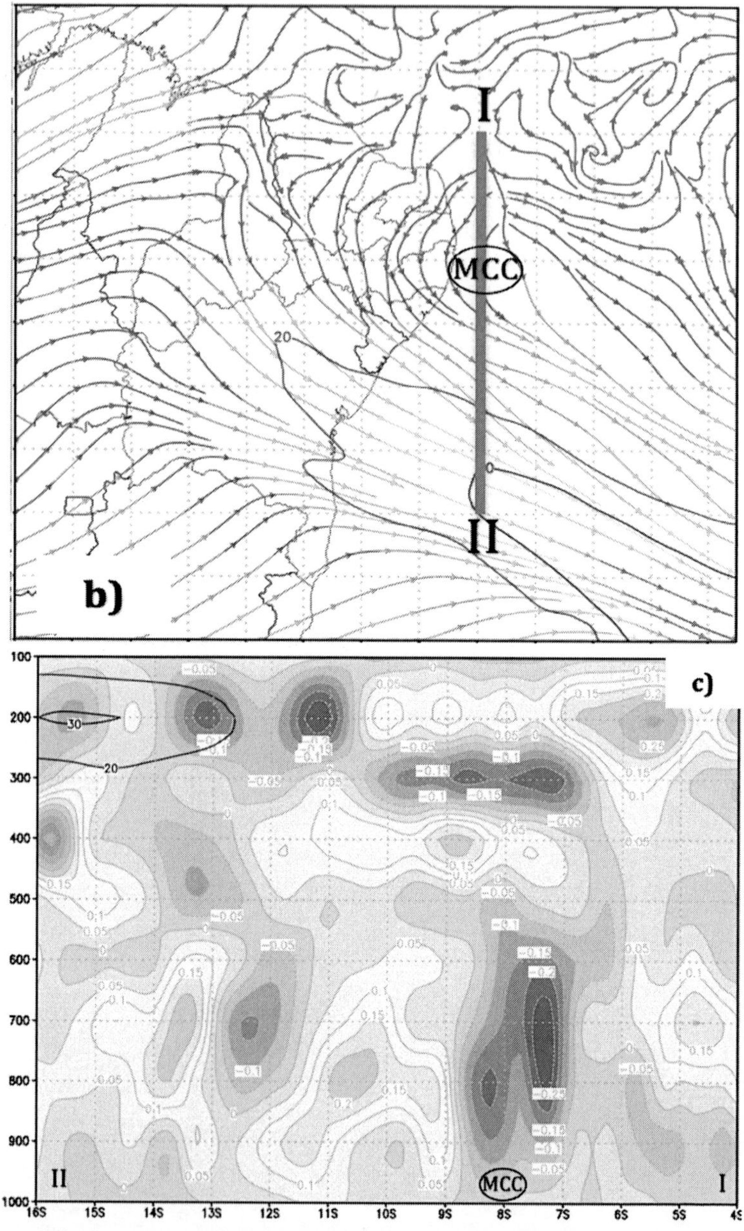

Figure I-6-13 – An example of a relationship between BNEJS and MCC 08 May 2019 at 10 UTC.

a) Cut-out of infrared satellite image, b) Streamlines at 200hPa and c) Omega meridional section (Pa s^{-1})
Sources: a) *INPE (Nacional Institute of Space Research)*, b) and c) Generated using Copernicus Climate Change Service Information, 2020.

7. References

Asnani G.C. *Tropical Meteorology*. In: Easterly Waves. Pashan, India. 1202p. 1993.

Bluestein H. B. *Observations and Theory of Weather Systems. Vol. II. Synoptic–Dynamic Meteorology in Midlatitudes*, Oxford University Press, 594p. 1993.

Conaty A.L., Jusem J.C., Takacs L., Keyser D., and R. Atlas. "*The Structure and Evolution of Extratropical Cyclones, Fronts, Jet Streams, and the Tropopause in the GEOS General Circulation Model*", Bulletin of the American Meteorological Society 82, No. 9, 1853-1867, (2001).

Copernicus Climate Change Service (C3S). *ERA5: Fifth generation of ECMWF atmospheric reanalysis of the global climate.* Copernicus Climate Change Service Climate Data Store (CDS). Accessed in 07 Feb 2020. Available at: https://cds.climate.copernicus.eu/cdsapp#!/home 2017.

Djuric Dusan, *Weather Analysis*. New Jersey: Prentice Hall, 304p, ISBN: 0-13-501149-3, 1994.

Fedorova N., Levit V., Campos A.M.V. *"Brazilian Northeast Jet Stream: frequency, wind speed and direction."* Meteorological Applications, 25: 254-260, DOI: 10.1002/met.1688 (2018a).

Fedorova N., Levit V., Campos A.M.V. "*Brazilian Northeast Jet Stream: Association with synoptic scale systems."* Meteorological Applications, 25: 261-268, DOI: 10.1002/met.1693, (2018b).

Fedorova Natalia. *Meteorologia Sinótica I.* Pelotas: Ed. Universitária/UFPel, 260p, ISBN: 85-7192-108-3, 1999.

Fedorova Natalia. *Sinótica III: Frentes, correntes de jato, ciclones e anticiclones.* Material didático: sinopses, figuras, equações Maceió: Ed. Universitária/ UFAL, 192 p, ISBN:978-85-7177-421-6, 2008.

Fedorova N., Lyra M.J.A. *"Capítulo 1. Corrente de jato e fenômenos associados."* In: Meteorologia em tópicos – volume 5, UFPel – Pelotas, 11-68, 2017. ISBN 978-85-68891-04-9

Kalnay et al. "*The NCEP/NCAR 40-year reanalysis project*", Bull. Amer. Meteor. Soc. 77, 437-470, (1996).

Kousky, Vernon E. & Gan Manoel Alonso. *"Upper tropospheric cyclonic vortices in the tropical South Atlantic"*. Tellus, 36(6), 538-551, (1981).

Lee, J. T., Galway J. G. *"Preliminary report on the relationship of the jet at the 200 mb level and tornado occurrence."* Bull. Amer. Meteor. Soc. 37, 327–332, (1956).

Milhahn Aribert Walter *"Correntes de jato e sistemas sinóticos associados aos complexos convectivos de mesoescala no nordeste Brasileiro."* MSc thesis., Universidade Federal de Alagoas, Maceió, 48p. 2013.

Namias, J., Clapp P. F. *"Confluence theory of the high tropospheric jet stream." J. Meteor.*, 6, 330–336, (1949).

National Institute of Space Research. Satellite Imagery Division. Brazil Northeast Image. 2019. Cachoeira Paulista, 2019. 1 Satellite Image. GOES 16 / Channel 13 (10.30 Microns). Available at: <http://satelite.cptec.inpe.br/repositoriogoes/goes16/goes16_web/ams_ret_ch13_baixa/2019/05/S11635388_201905081000.jpg>. Accessed in: 07 Feb 2020.

Pontes da Silva BF, Fedorova N, Levit V, Peresetsky A. "*Sistemas sinóticos associados às precipitações intensas no Estado de Alagoas (Synoptic systems associated with the heavy precipitations in the Alagoas State)."* Rev. Bras. Meteorol. 26: 295–310, (2011).

Romatschke, U., Houze, Jr. R. A. *"Extreme summer convection in South America."* J. Climate, 23, 3761-3791, (2010).

Rose SF, Hobbs PV, Locatelli JD, Stoelinga MTA *"10-yr climatology relating the locations of reported tornadoes to the quadrants of upper-level jet streaks."* Weather Forecasting, 19: 301–309, (2004).

Satyamurty P, Nobre CA, Silva Dias PL. "*South America.*" In Meteorology of Southern Hemisphere, Karoly DJ, Vincent DG (eds). American Meteorological Society: Boston, MA; 119–139, (1998).

Taljaard JJ. "*Synoptic meteorology of the southern hemisphere."* In Meteorology of Southern Hemisphere, Newton W (ed). American Meteorological Society: Boston, MA; 139–213, (1972).

Uccellini, L., Johnson D. *"The coupling of upper and lower tropospheric jet streaks and implications for the development of severe convective storms."* Mon. Wea. Rev., 107, 682–703, (1979).

Vaz Isabela Nogueira. *"Estudo sobre origem da corrente de jato do nordeste brasileiro e suas ligações inter-hemisféricas (Study on the origin of the jet stream in northeastern Brazil and its inter-hemispheric connections)",* MSc thesis. Teses, Universidade Federal de Alagoas, Maceió, (2014)

Virji H. "*A preliminary study of summertime tropospheric circulation patterns over South America estimated from cloud winds."* Mon. Weather Rev. 109: 599–610, (1981).

Velasco, I.; Fritsch, J. M. "*Mesoscale Convective Complexes in the Americas*." Journal of Geophysical Research: Atmospheres 92, n. 20, p. 9591–9613, (1987).

Whitney LF. *"Relationship of the subtropical jet stream to severe local storms."*, Mon. Weather Rev. 105: 398–412, (1977).

Zimmermann D.F.R. *"Subtropical jet climatology over South America."* MSc thesis. San Paulo, University of San Paulo, 89pp. (2017)

CHAPTER I.7

FRONTAL ZONES IN THE TROPICAL REGION

NATALIA FEDOROVA AND VLADIMIR LEVIT

1. Identification

The frontal zones of baroclinic cyclones are the main synoptic systems that change the weather in extratropical regions. Several studies show a displacement of these frontal zones to tropical regions (Parmenter, 1976; Kousky, 1979; Oliveira's 1986 research, which has been quoted by Satymurti et al., 1998; Cavalcanti & Kousky, 2009). Information about these frontal zones in the tropical regions of South America and South Atlantic has been published for the first time in recent years (for example, Fedorova et. al., 2016, Veber et al., 2020, Muanza Gomes et al., 2021).

The position of the frontal zone in baroclinic cyclones of the extratropical region is determined (Petterssen,1956; Bluestein, 1993; Smith et al., 1995; Djuric, 1994; Fedorova, 1999, 2008b,c; Fedorova & Carvalho, 2000; Simmonds et al., 2012; Fedorova et al., 2016) as follows:

1) *baroclinic cyclone trough*: from maps of pressure at the surface level and cyclonic relative vorticity at the low levels (1000 and 925hPa);
2) *confluence of streamlines* at the low level up to 850 hPa;
3) *high temperature gradient* at the surface level and in the layer of the frontal zone; the following maps are commonly used for this: a) temperature maps at 1000 hPa, b) thickness maps at 1000/500 hPa and c) thermal advection at 925 and 850 hPa; frontal zone located between areas with cold and warm advection.

All these characteristics have not been observed together in the tropical region. Usually, one of these maps shows the frontal zone in the tropical regions (Fedorova et al., 2016). Confluence of streamlines at the low levels and cyclonic vorticity are more typical of this tropical region in the frontal zone.

The best criterion for the frontal zone identification in the tropical regions is a variation in the equivalent potential temperature (θ_e) (Bolton,

1980; Gemiacki, 2005; Cruz, 2008; Fedorova et al., 2016; Muanza Gomes, 2018; Muanza Gomes et al., 2021). Therefore, horizontal distribution of θ_e and θ_e advection (Advθ_e) are used for the frontal identification. These maps show a heat wave ahead of the frontal zone and a cold wave behind; the frontal zone is located between positive and negative values of Advθ_e.

The example shows the identification of the frontal periphery in the BNE and the neighboring ocean. A cloudiness band from a baroclinic cyclone is visible on infrared satellite images (Figure I-7-1a). Pressure and relative vorticity maps (Figure I-7-1b, c) show the first baroclinic cyclone (L1, centered 50°S) and a new cyclone with one closed isobar (L2, centered 25°S). The trough of the cyclone L1 reaches the cyclone L2. Another trough from the equatorial region approaches the same cyclone L2 from the northwest. The confluence of air current along these two troughs can be seen from the streamlines (Figure I-7-1d). The highest negative Advθ_e values are associated with cold air advection behind the cold front and its periphery (Figure I-7-1e).

a)

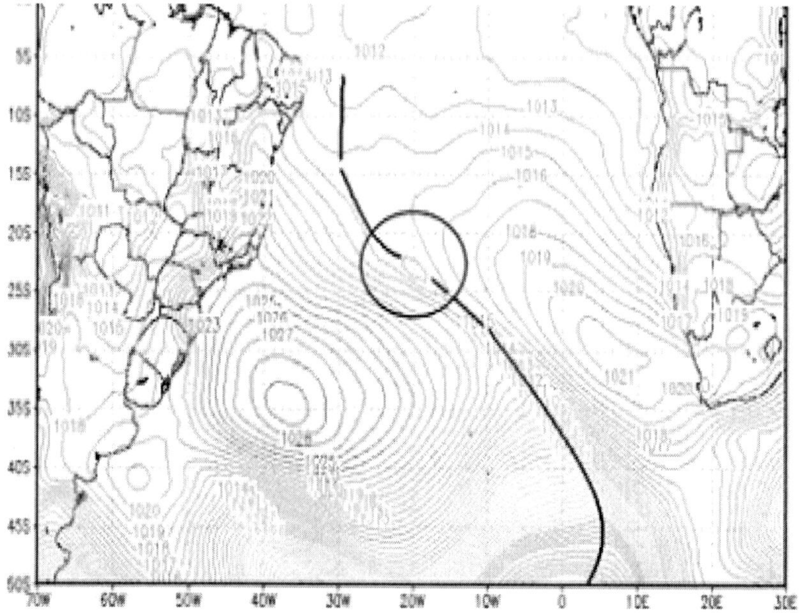
5S
10S
15S
20S
25S
30S
35S
40S
45S
50S
70W
60W
50W
40W
30W
20W
10W
0
10E
20E
30E

b)

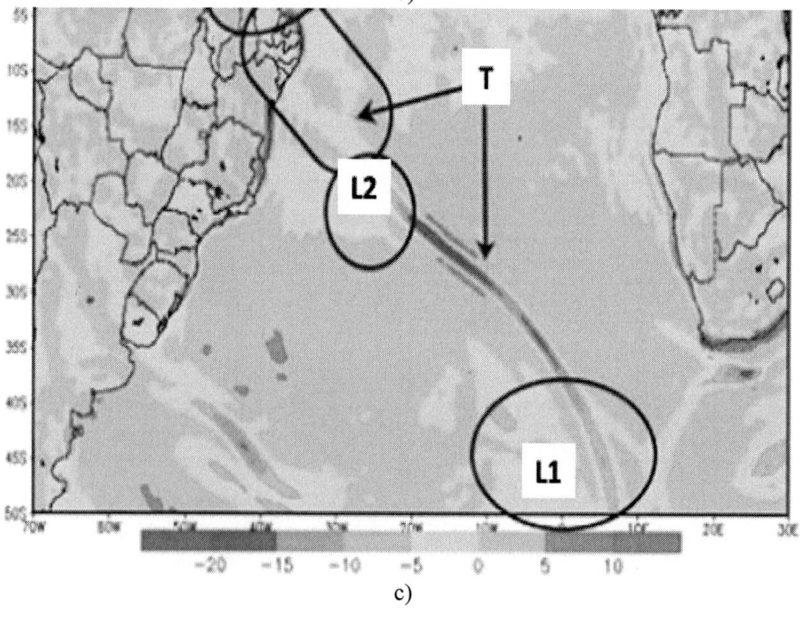
T
L2
L1
-20
-15
-10
-5
0
5
10

c)

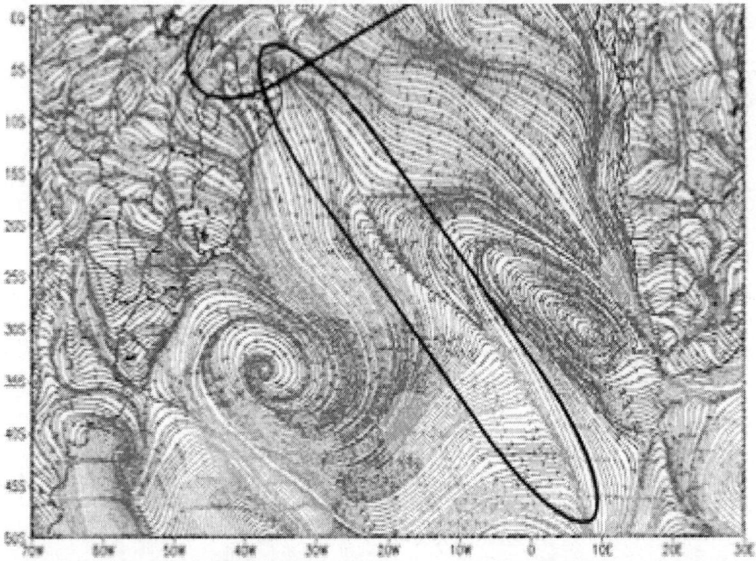

d)

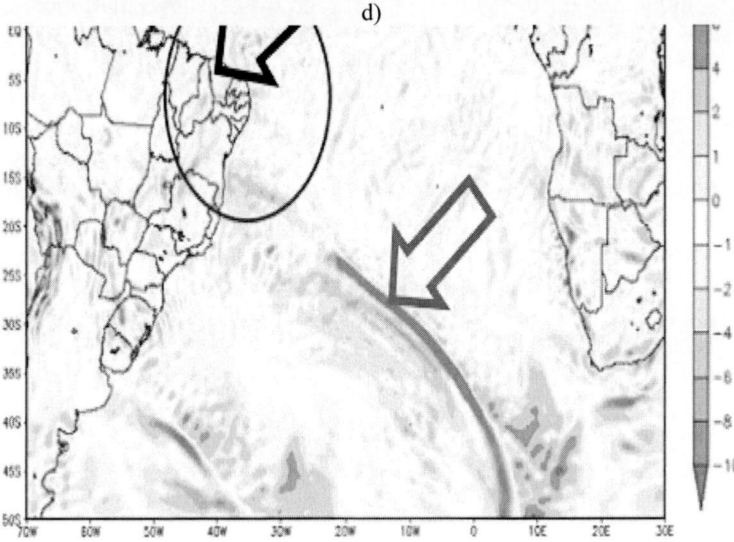

e)

Figure I-7-1 – Example of the frontal periphery identification in the BNE and neighboring ocean using satellite images and various variables:
a) infrared satellite images;
b) pressure maps at the surface level (black line - baroclinic cyclone trough);

c) relative vorticity at 1000 hPa (L1 and L2 show the centers of cyclones with negative numbers, T- trough);
d) streamlines at 1000 hPa (confluence of air current into black ellipse); and
e) Advθ_e map (Ks^{-1}) (arrows indicate the highest negative values).
Source: Muanza Gomes, 2018; Muanza Gomes et al., 2021; CPTEC/INPE, 2018

2. Vertical structure

The vertical structure of the frontal zone in the tropical region shows more intense gradients of several characteristics at the middle levels than at the low ones (Sinhori, 2015). These characteristics are as follows: temperature, equivalent potential temperature (θ_e) and relative vorticity. In general, temperature and wind speed variations in these frontal zones are very weak, so these characteristics are not recommended for frontal identification. In addition, the wind direction shows its slow change at low levels. The best characteristic for frontal identification on a vertical section is the equivalent potential temperatures, which present intense gradients at the middle level, usually in the 800-500 hPa layer.

Example of the IR satellite image (Figure I-7-2) shows the frontal zone location in the central and southern regions of the BNE. More intense gradients in the vertical temperature section are observed at the level of 1000-900 hPa around 10°S, which corresponds to the northern half of the frontal cloudiness band (Figure I-7-3a).

Vertical section of the θ_e shows an intensive gradient (5K per 5° de longitude) near the surface and at the low levels between 10 and 15°S (Figure I-7-3b). The most intense gradients (above 13K per 5° de longitude) were observed in the 680 - 880 hPa layer. It is important to note that the band of intensive θ_e gradient is inclined over cold air and at 300 hPa is observed at 23°S. Such a slope is characteristic of the vertical propagation of a cold front.

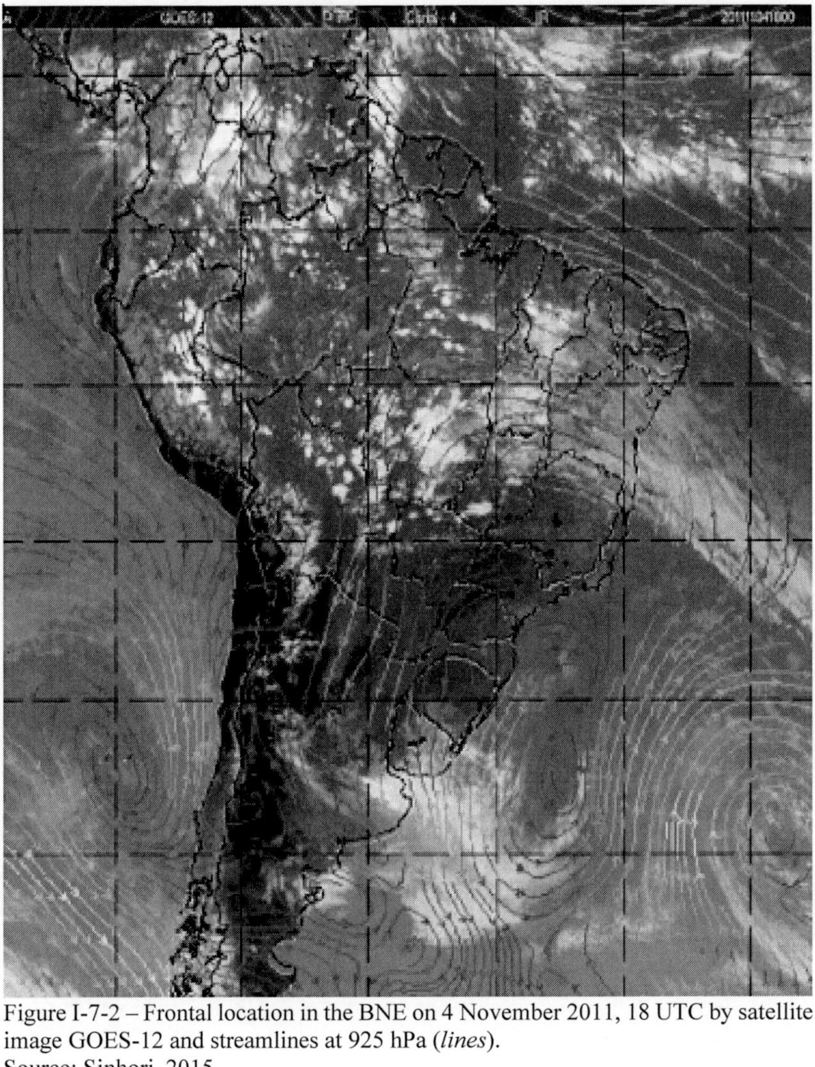

Figure I-7-2 – Frontal location in the BNE on 4 November 2011, 18 UTC by satellite image GOES-12 and streamlines at 925 hPa (*lines*).
Source: Sinhori, 2015

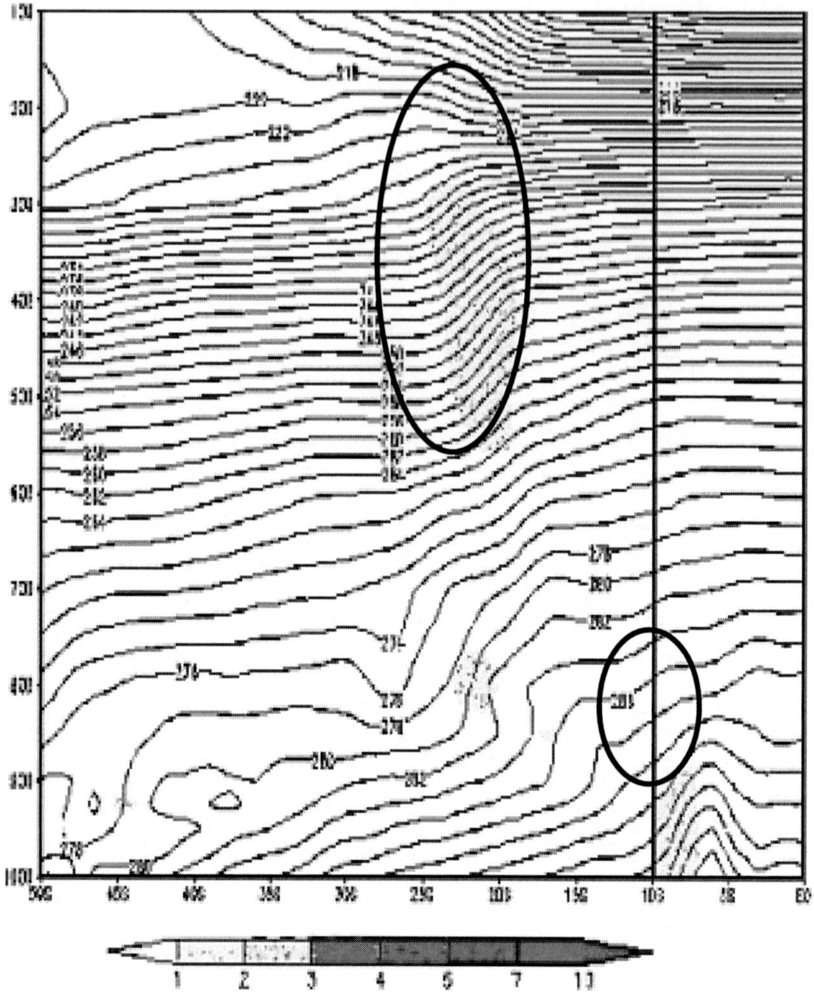

a) T (ºC)

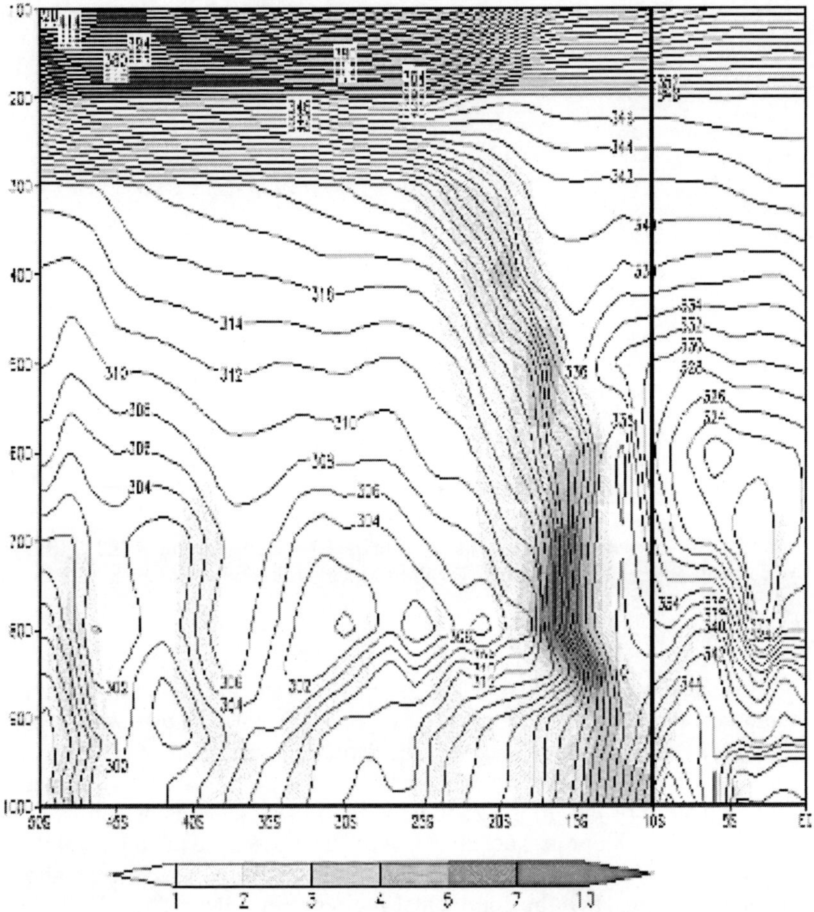

b) θ_e (K)

Figure I-7-3 – Structure of the frontal zone in the BNE on 4 November 2011, 18 UTC by vertical section along longitude 36°W of a) temperature and b) equivalent potential temperature (θ_e)
Source: Sinhori, 2015

The vertical section of the wind shows no change in wind speed in the frontal zone (Figure I-7-4a). However, there was a change in wind direction in front of the frontal zone from southeast to north (0-15°S) and behind the front from southwest to south (20-30°S). The vertical section of vorticity demonstrated the weak numbers of relative vorticity at the low levels (Figure I-7-4b). Very high values of the relative vorticity were observed at the high levels (300 hPa) over the frontal zone and were associated with jet stream (Sinhori, 2015).

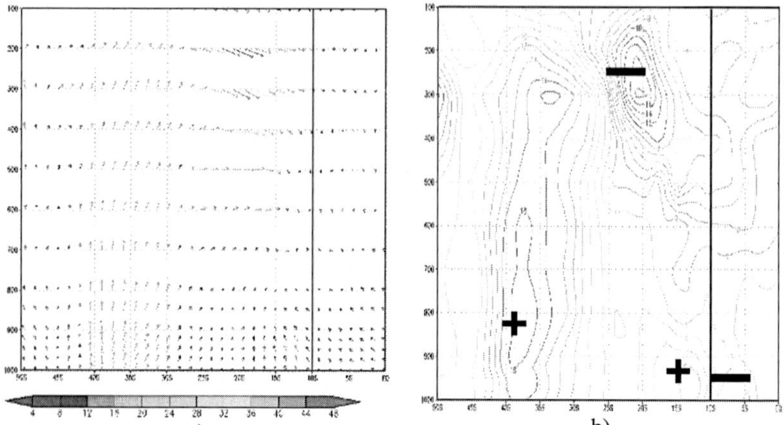

a) b)

Figure I-7-4 – Structure of the frontal zone in the BNE on 4 November 2011, 18 UTC by vertical section along longitude 36°W of a) wind (direction and velocity, m/s) and b) relative vorticity (s^{-1}).
Source: Sinhori, 2015.

Frontal zones in the BNE are associated with the jet streams of the southern and northern hemispheres: *Brazilian Northeast Jet Stream (BNEJS); Subtropical Jet Stream of the Southern Hemisphere (STJSSH) and Subtropical Jet Stream of the Northern Hemisphere (STJSNH)* (Chapter I-6). The pattern of the frontal zone relative to these jet streams can be seen in Figure I-7-5 (Albuquerque, 2022). *STJSNH* crosses the main frontal zone between 20 and 30°S. The periphery of this frontal zone was located above the ocean and reached the continental region of the BNE. The BNEJS and STJSNH entered exactly into this region of the frontal periphery.

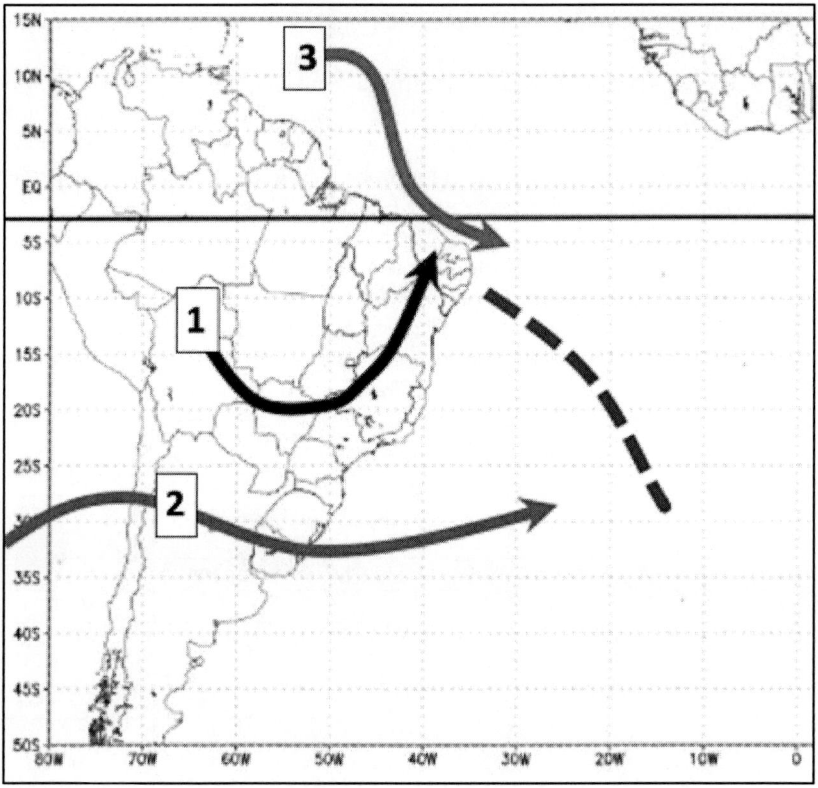

Figure I-7-5 – Pattern of the frontal zone (*dashed line*) in the BNE relative to the jet streams:
1-BNEJS; 2- STJSSH; 3- STJSNH
Source: Albuquerque, 2022

3. Seasonal variation

The number of frontal zones detected in southern BNE (Caravelas, BA) reached, approximately, 19 per year (56 over three years, 2004 - 2006) according to the methodology used in synoptic operational practice (Table I-7-1) (Fedorova et al., 2016). This number decreases very quickly in northern or northeastern BNE and only 2 frontal zone per 3 years were recorded by the same method in the center of BNE (Maceio, Al).

This number of frontal zones in the BNE center increase significantly (up to 26 events per 3 years) if to change the methodology and to use the variation of the equivalent potential temperature as the identification method.

Seasonal variations were weak; frontal zones in the southern BNE were observed all year round with the maximum value (7-6 per month) in autumn (April, May) and spring (October) and with the minimum (2) in December (Table I-7-1). Low seasonal variation was observed in all cities in the center of the BNE (Salvador-BA, Aracaju-SE, Maceio-Al).

Table I-7-1 – Seasonal variation in the number of frontal zones in the BNE for three years (2004-2006).
Number shows quantity of frontal zones in coastal regions of the BNE by the method used in synoptic operational practices in different locations of the BNE from the south (Caravelas, BA) to the BNE center (Maceio, Al). ***Bold*** *numbers show the quantity of frontal zones in the BNE center, by the method used variation in the equivalent potential temperature*
Source: CPTEC, Fedorova et al., 2016.

	Maceio Al	Aracaju SE	Salvador BA	Ilheus BA	Caravelas BA
Lat	9.6	10.9	12.9	14.8	17.6
Lon	35.6	37.0	38.5	39.0	39.3
Jan	0/**2**	0	1	2	4
Feb	0/**2**	0	1	2	3
Mar	0/3	0	1	1	3
Apr	0/**2**	0	2	4	7
May	1/3	2	3	7	7
Jun	0/**2**	1	2	3	5
Jul	0/**2**	0	3	4	5
Aug	0/**2**	0	1	3	4
Sep	0/1	0	1	3	5
Oct	0/3	1	2	5	6
Nov	0/**2**	0	1	3	5
Dec	1/**2**	1	1	2	2
Year	2/**26**	5	18	39	56

4. Modification of the vertical structure of a cold front during its displacement from the subtropical to tropical latitudes

All frontal zones reached the BNE pass from the south. With this displacement, a modification of their vertical structure occurred (Sinhori,

2015). These modifications can be seen by radiosonde data and the vertical profile of some parameters, constructed using numerical models. One example is presented on (Fig. I-7-6), where the modification of the vertical structure of a cold front is visible by radiosonde data during its displacement for 4 days (01-04 November 2003) from the latitude 30°S (Porto Alegre), through Rio de Janeiro (20°S) to latitude 10°S (Petrolina). Vertical profiles of temperature (T) and dew point (Td) at the latitude 30°S show the stable layer 850-650 hPa with the frontal inversion (850-750 hPa) and with the very high humidity from the surface up to 600 hPa (Figure I-7-6a). Instability indices corresponded to K=20 and Lifted index (LI) =10 showed no instability (determination of the instability indices is presented in Fedorova et al. (2017). Usually, values of the K index over 24 indicate a likelihood of showers and negative values of -2 to -6 of LI index are rather common in regions with showers (Djuric 1994; Fedorova, 2008a). Thus, this vertical profile shows *Ns* (Nimbostratus) clouds from the surface up to 600 hPa; these clouds create intense and ongoing precipitations.

In the same frontal zone at a latitude 20°S no more frontal inversion was observed and the moister layer split into two layers 1000-800 hPa and 650-700 hPa (Figure I-7-6b). Subsequently, the entire clouds layer was divided into two parts with *St* (stratus) and *As* (altostratus) clouds. Finally, only a small amount of precipitation from *St* clouds reaches the surface. The low negative LI index (-1.2) also shows weak precipitation. However, K index was equal to 35, and values of the K index over 30 are usually associated with severe thunderstorms with heavy precipitations (Djuric, 1994; Fedorova, 2008a).

At latitude 30°S in the same frontal zone, there were a stable layer in the middle troposphere (about 500 hPa), high humidity only near the surface, and a wetter (T-Td~ 5°C) layer up to 550 hPa (Figure I-7-6c). This vertical profile, together with an LI index of 0.2, indicates a whole layer of clouds with weak instability. Only K index =34 indicates the possibility of heavy precipitation in some regions.

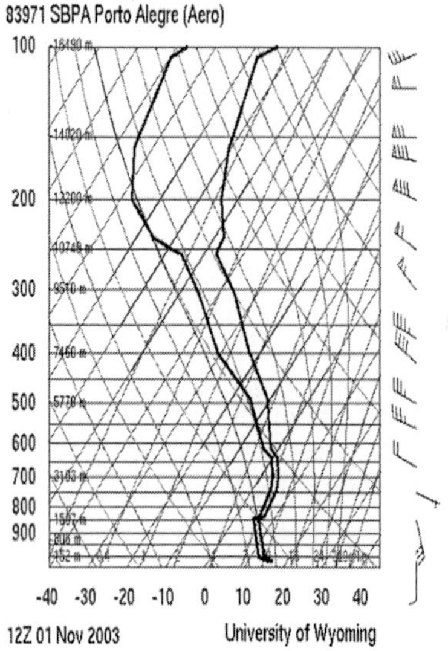

a)

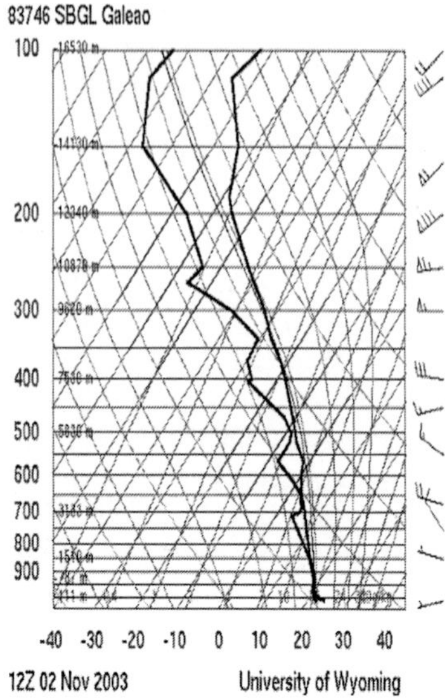

b)

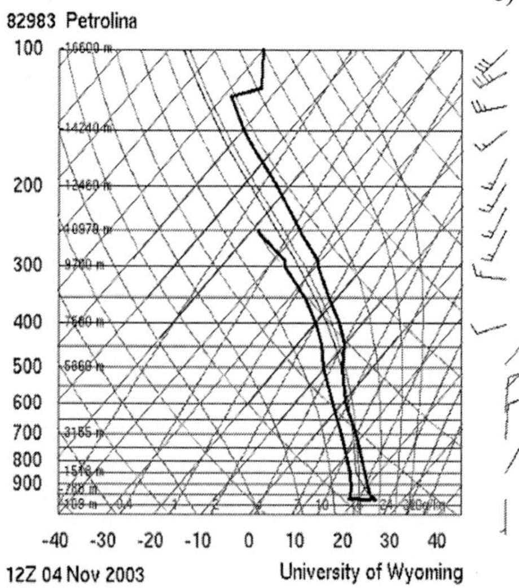

c)

Figure I-7-6 – Modification of the vertical structure of a cold front during its displacement from Porto Alegre (a, ~30ºS, 01 November 2003, 12 UTC), pass through Rio de Janeiro (b, ~20ºS, 02 November 2003, 12 UTC) to Petrolina (c, ~10ºS, 04 November 2003, 12 UTC).
Source: University of Wyoming

Vertical sections of the equivalent potential temperature (θ_e) clearly show the modification of the vertical structure of a cold front during its displacement from subtropical to tropical latitudes (Figure I-7-7). At latitude 30°S, the most intensive θ_e gradients were observed at the low levels up to 800 hPa. Above this level, the inclination of the band of the largest θ_e gradient over cold air corresponds to the inclination of the frontal zone (Figure I-7-7a). When the front displacement to the latitude 23°S was occurred, the frontal regeneration (Sinhori, 2015) and increase in these θ_e gradients were detected. At this latitude, very intense gradients (up to 7K/100km) were observed in the deepest layer, up to 500 hPa (Figure I-7-7b). In the tropical region at the latitudes 12°S and 10°S, θ_e gradients weakened at the low levels and persist at the middle levels (800-500 hPa, 4K/100km)) (Figure I-7-7c and d).

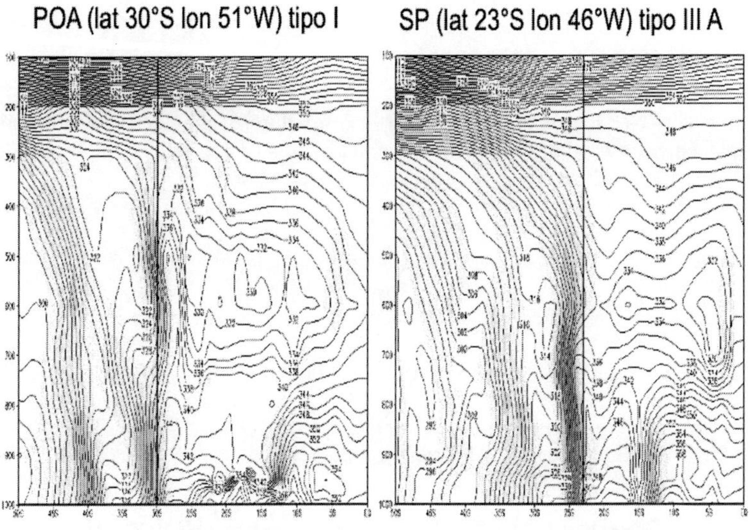

a) b)

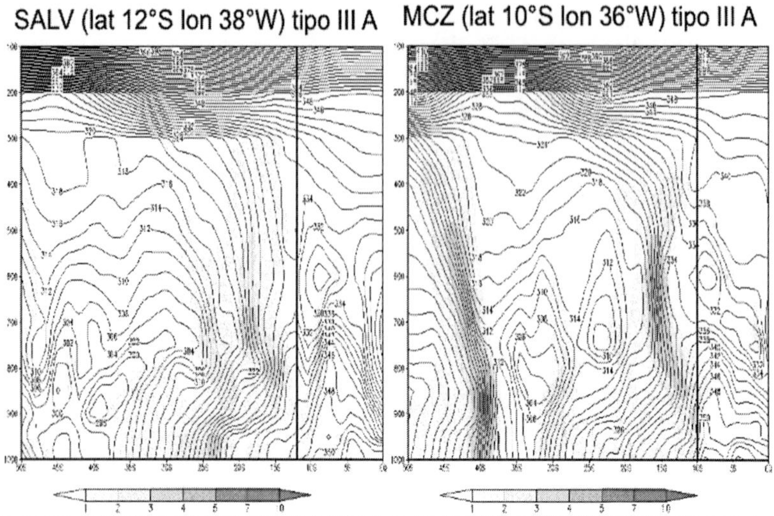

c) d)

Figure I-7-7 – Modification of the vertical structure of a cold front during its displacement from the subtropical to tropical latitudes by vertical section of the equivalent potential temperature (θ_e) on 19-26 May 2002:

a) frontal zone in Porto Alegre, vertical section along longitude 51°W, 19 Maio 2002, 06 UTC

b) frontal zone in San Paulo, vertical section along longitude 46°W, 21May 2002, 12 UTC

c) frontal zone in Salvador, vertical section along longitude 38°W, 24 May 2002, 00 UTC

d) frontal zone in Maceio, vertical section along longitude 36°W, 26 May 2002, 00 UTC.

The vertical black line shows the latitude of the city.

Source: Sinhori, 2015; ECMWF reanalyze data

5. Synoptic conditions of frontal zone displacement from subtropical to tropical latitudes

Frontal zone of a baroclinic cyclone from the first day of formation displaced to the BNE on average in 10 days (Gemiacki, 2005; Fedorova 2008c; Cruz, 2008; Fedorova & Pontes da Silva, 2020). Types of cloudiness of the baroclinic cyclones with a frontal zone, which usually reached the BNE, will be described below by the satellite images.

5. 1 Typical cloudiness of a baroclinic cyclone at the beginning of their displacement

A typical cloud spiral of baroclinic cyclones located in the South Atlantic between 20-40°S is shown in Figure I-7-8 (Sinhori, 2015). The cloud band of the cold front extends along the trough from the ocean to the continent (Figure I-7-8a).

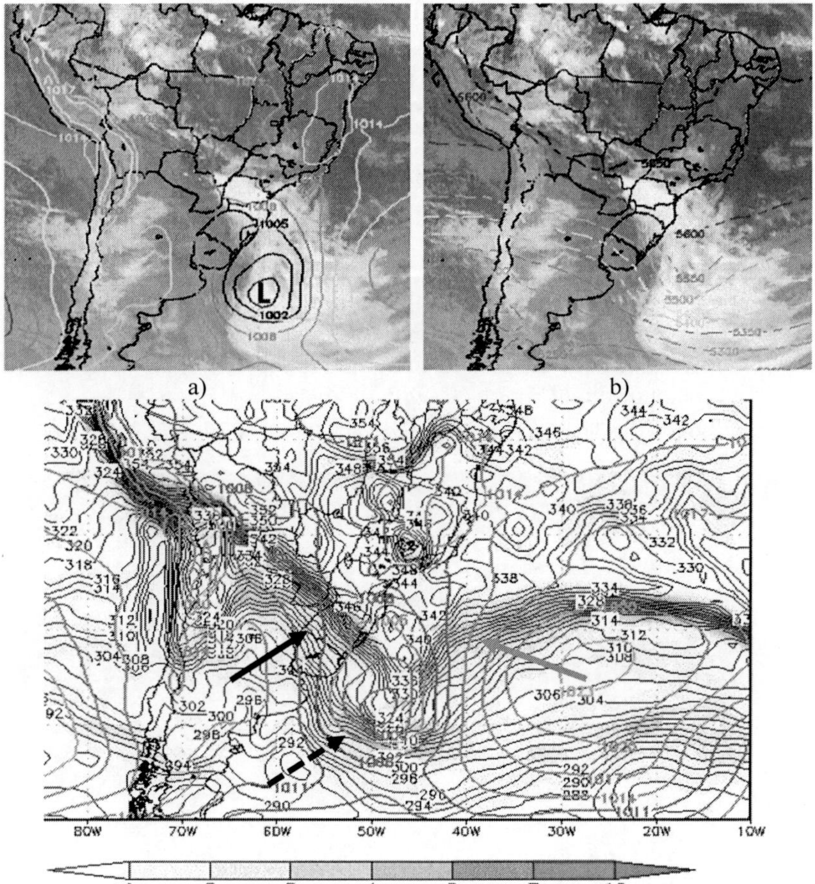

a) b)

c)

Figure I-7-8 – Typical location of baroclinic cyclone with frontal zones, which will influence on the BNE:
Infrared satellite imagery together with the pressure at 1000 hPa (a) and thickness 1000/500 hPa (b) maps. Black arrow shows the axis of the warm wave.

c) *Map of equivalent potential temperature* (θ_e) *at 925 hPa, the gradients of* θ_e *(K/100 km) are show in gray scale. The arrows show the cold (black arrow), warm (grey arrow) and occludes (dashed arrow) fronts.*
Source: Sinhori, 2015; ECMWF reanalyze data

The warm wave on the maps of the thickness 1000/500 hPa is associated with the cloud spiral (Figure I-7-8b). The well-defined edge behind of the cold front clouds shows the cold air invasion in the same figure. High gradients of the equivalent potential temperature (θ_e) correspond to the cold, warm and occluded front's location (Figure I-7-8c). A slight lag of the θ_e gradients to the SW from the trough position indicates the inclination of the cold frond surface above the cold air.

5. 2 Types of cloudiness of baroclinic cyclone and cold front in the beginning of the displacement

Three types of the cloudiness at the cold front periphery over continental region are following (Signorini, 2001; Sinhori, 2015):

- convection clouds at the periphery of cloud spiral (Figure I-7-9a);
- cirrus clouds of the jet stream at the high levels on the periphery of the cloud spiral (Figure I-7-9b);
- cloudless area on the periphery of the cloud spiral (Figure I-7-9c).

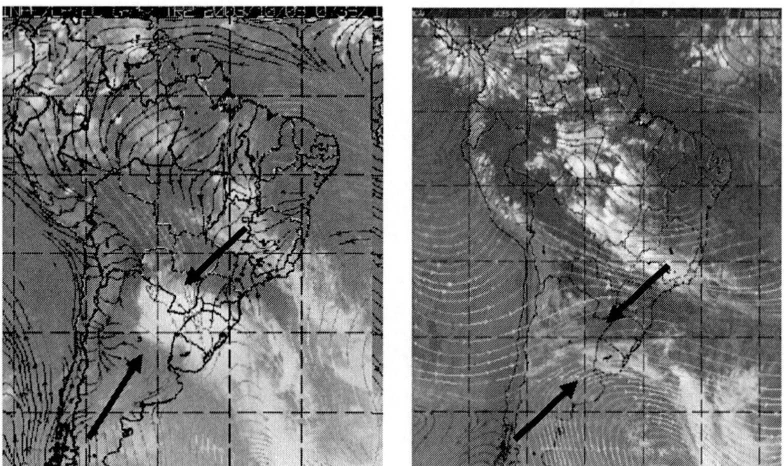

a) b)

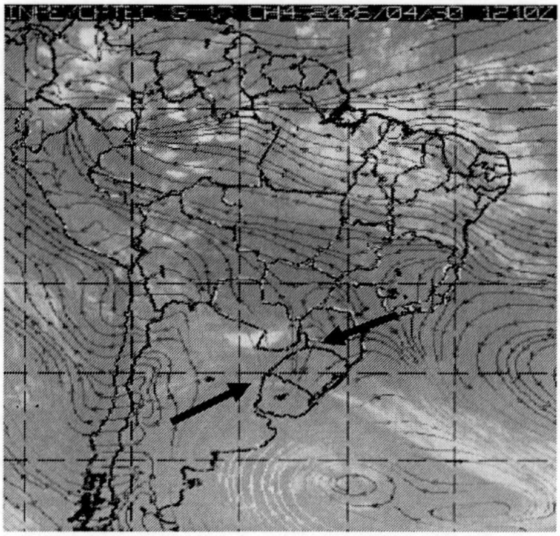

c)

Figure I-7-9 – Types of cloudiness on the frontal periphery:
a) convective clouds; b) cirrus (*Ci*) clouds, associated with jet stream at the high levels; c) cloudless area
Infrared satellite image together with the streamline's maps at 925 hPa (a. and c.) and 200 hPa (b).
Black arrows show the cloudiness on the frontal periphery.
Source: Sinhori, 2015; ECMWF reanalyze data

5.3 Frontogenesis

The processes of cyclogenesis and frontogenesis took place during the whole process of the cold front advance towards the BNE. First between 20°S and 40°S, the process of frontogenesis was observed, and then between 20°S and 30°S, cyclogenesis was detected.

An indicator or characteristic of frontogenesis is an increase in temperature gradients over time (Bluestein, 1993; Djuric, 1994; Satyamurti & Mattos, 1989; Shultz et al., 1998; Shultz, 2005; Fedorova, 1999; Fedorova, 2008b). This gradient can be increased due to cold air advection behind the cold front and warm air advection towards the vanguard side (Chapter I-9, Fedorova & Bakst, 1999; Saldanha et al., 2010). The most intense frontogenetic process occurs with simultaneous advection of warm and cold air.

Equivalent potential temperature (θ_e) and its advection maps most accurately show cold and warm air waves and their advection. Comparison of these maps with temperature advection maps can be seen in Figure I-7-10. The temperature advection map shows cold and warm advection near the center of the cyclone and the absence of temperature variation in the

BNE and the neighboring ocean (Figure I-7-10a). The equivalent potential temperature maps show a high cold air gradient south of frontal zone; this location of the frontal zone is associated with the inclination of the frontal zone over cold air (Figure I-7-10b). A warm wave is also visible at the leading edge of the frontal zone. Maps of θ_e advection (Advθ_e) are presented in Figure I-7-10c.

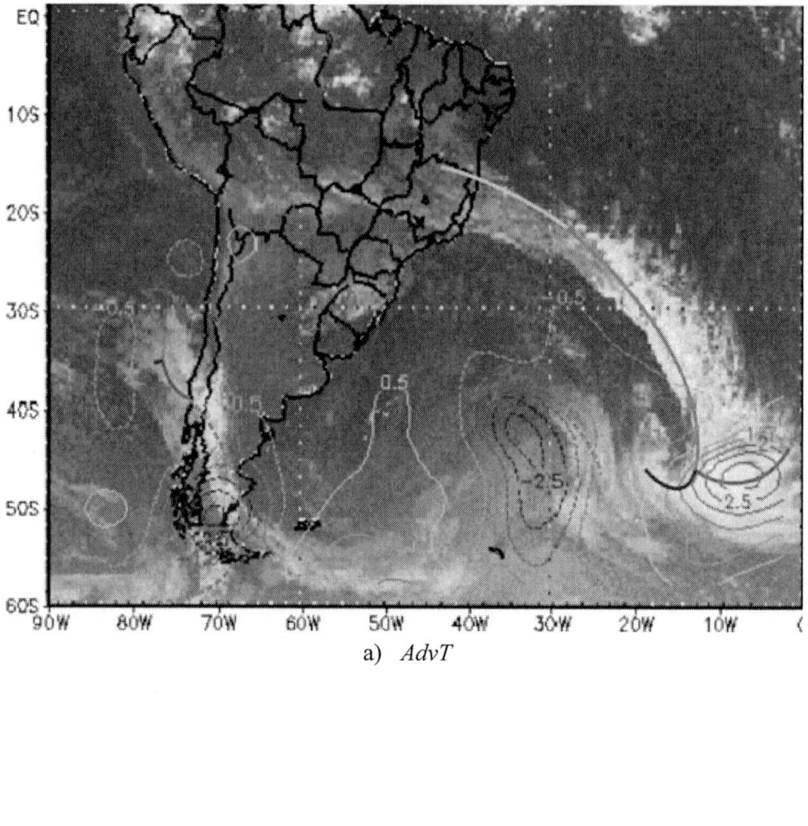

a) *AdvT*

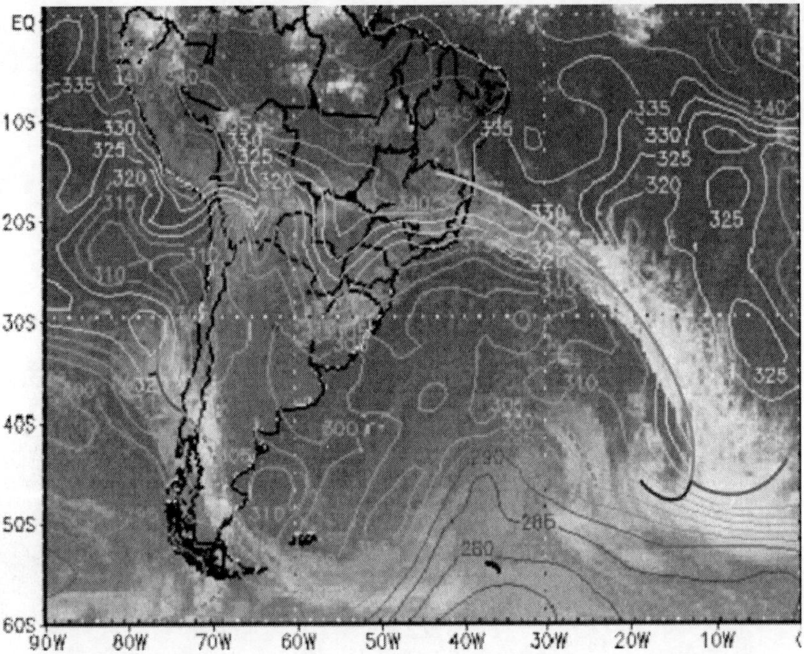
EQ
10S
20S
30S
40S
50S
60S
90W
80W
70W
60W
50W
40W
30W
20W
10W
335
330
325
320
315
310
340
300
290
285
280

b) θ_e

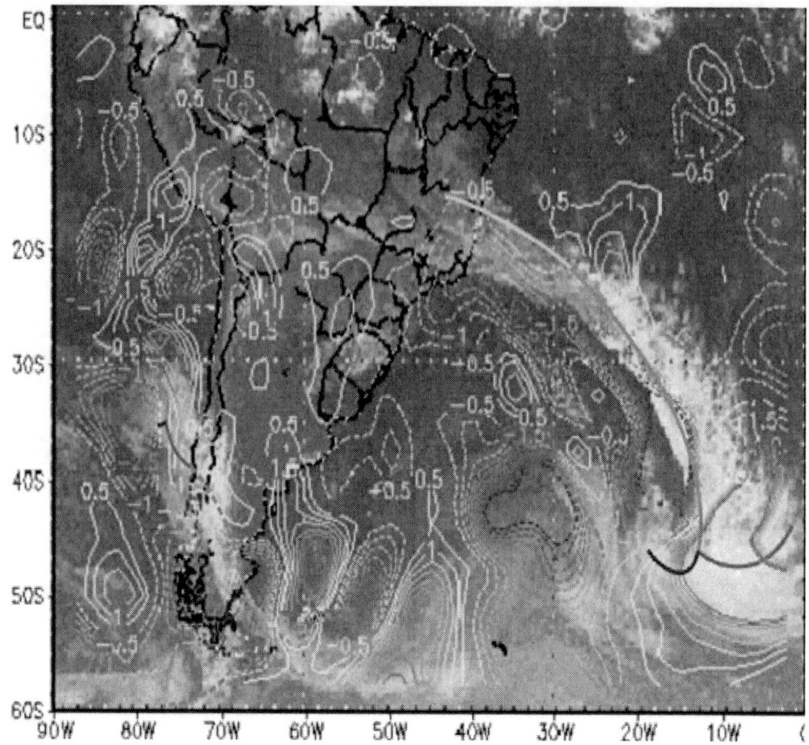

c) $Adv\theta_e$

Figure I-7-10 – Identification of cold and warm air and their advection in the BNE and the neighboring ocean using the next variables:
a) temperature advection (AdvT)
b) equivalent potential temperature (θ_e)
c) advection of θ_e ($Adv\theta_e$)
Frontal zones are shown by full lines: occluded front - black line; warm, cold and periphery of the cold front - grey color from darker to lighter, respectively.
Source: Cruz, 2008

5. 4 Cyclogenesis

The process of cyclogenesis is described in many books (for example, Gan & Rao, 1991, 1994; Sinclair 1995; Fedorova, 2001; Reboita et al., 2010). During the movement of the cold front to the BNE, cyclogenesis processes often occur. The process of cyclogenesis at the cold front of a baroclinic cyclone between 20-30°S is quite typical (Figure I-7-11).

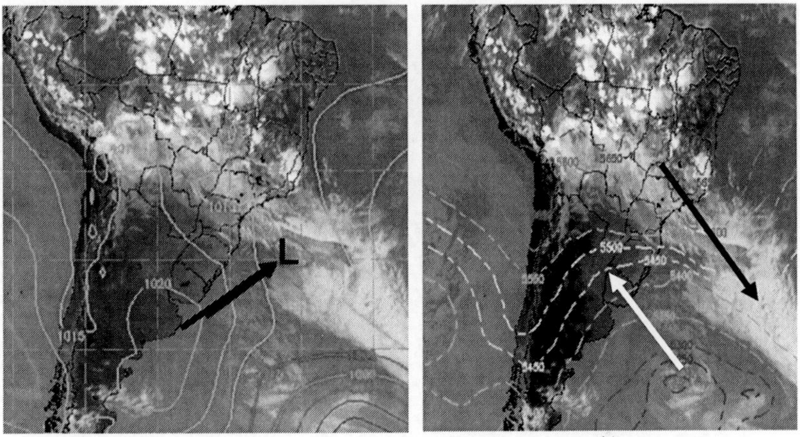

a) b)

Figure I-7-11 – Cyclogeneses process on the cold front of baroclinic cyclone: *Infrared satellite imagery together with the pressure at 1000 hPa (a) and thickness 1000/500 hPa (b) maps. Black arrow shows closed isobar or cyclone center (a) and the axis of the warm wave (b), white arrow shows the axis of the cold wave (b).* Source: Sinhori, 2015; ECMWF reanalyze data

Satellite image shows a cold front cloud band and first closed isobar or cyclone center formation (Figure I-7-11a). The 1000/500 hPa thickness maps show the typical position of a warm wave in the forefront of a new cyclone and a cold wave behind the same cyclone (Figure I-7-11b).

5.5 Coupling of Polar and Subtropical jet streams

The air circulation at the high levels (200 hPa, Figure I-7-12), associated with the penetration of the frontal zone into the tropical region (Cruz, 2008; Fedorova & Pontes da Silva, 2020), show a typical configuration with:

1) bifurcation of the air current in the South Pacific Ocean and
2) coupling of the Polar Jet Stream (PJS) with Subtropical Jet Steam (SJS).

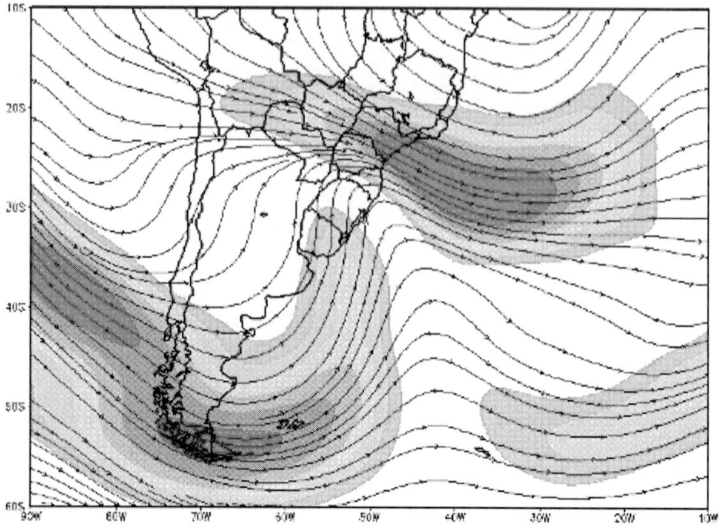

Figure I-7-12 – Coupling PJS with STJS at the latitudes 20-30° and bifurcation of air current over the Pacific Ocean.
Source: Cruz, 2008; Fedorova & Pontes da Silva, 2020

Such a configuration of air currents creates conditions for intensification of the meridional circulation, displacement of the baroclinic anticyclone behind the frontal zone to the NE, and, as a result, more intensive intrusion of cold air.

5.6 Bolivian High displacement to the east

During the entire process of the frontal zone movement towards the BNE, the Bolivian High (BH is anticyclonic circulation at the high level) was displaced to the east (Gemiacki, 2005, Cruz, 2008).

The example in Figure I-7-13 shows the BH location at 200 hPa, and the area with anticyclonic circulation is associated with maximum thickness values of 200-500 hPa. This means that the BH is a warm barotropic anticyclone. The Subtropical Jet Stream passed along the southern periphery of the BH within the whole process. During the entire process of frontal zone moving towards the BNE, the BH was displaced to the east, from 45°W to 28°W.

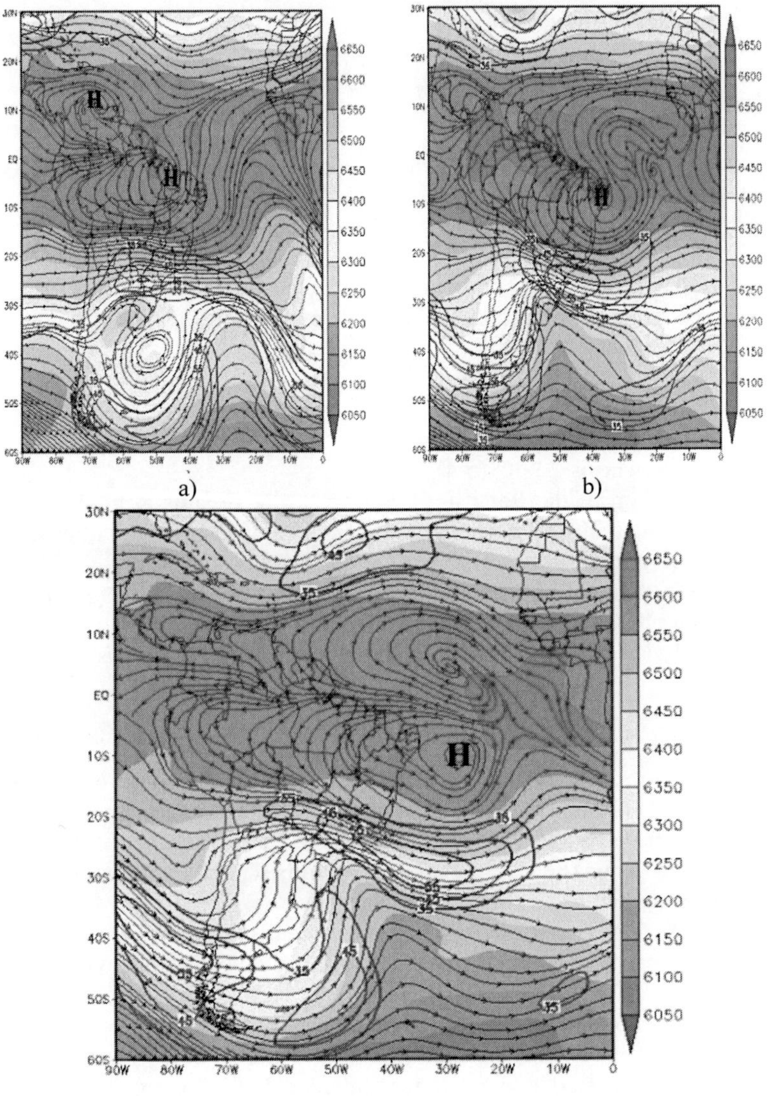

a) b) c)

Figure I-7-13 – Bolivian High displacement to the east during the movement of the frontal zone towards the BNE.
a) 27 April 2005, 12 UTC
b) 29 April 2005, 18 UTC
c) 30 April 2005, 12 UTC
Streamlines at 200 hPa - black lines with arrows

Thickness 200-500 hPa - grey color
Wind velocity above 30/s or Subtropical Jet Stream - black lines with numbers.
H- Bolivian High center at 200 hPa
Source: Cruz, 2008; NCEP/NCAR

Summarizing the main processes for a cold front to reach the BNE, the following can be noted:

1) Baroclinic cyclone with three types of cold front over the continental region (5.1 and 5.2).
2) Intensive cold advection at the low levels (up to 850 hPa), which can be seen as secondary cold front, cyclonic vortex in the cold air (Chapter I-9).
3) Warm advection towards the frontal side of a cold front (1; 5.1; Chapter I-9).
4) Frontogenesis (5.3).
5) Cyclogenesis 20-30° (5.4).
6) Coupling of the Polar and Subtropical jet streams at high levels at latitudes 20-30° over the continental region and bifurcation of air current over the Pacific (5.5).
7) Displacement of the baroclinic anticyclone behind the frontal zone to the NE (5.5).
8) Bolivian High or High at the high levels moved to the east (5.6).

5.7 Frontal zone fracture

Different types of clouds were observed in the separated part of the band. When the cloud band of the frontal zone reaches the continental region of the BNE, the periphery of the cold front separates from the center of the baroclinic cyclone (Fedorova et al., 2016). This separation is very common, and the separated part of the cloud band affects BNE weather. Diagram of the evolution of the baroclinic low in tropical/subtropical regions in the Atlantic Ocean and frontal zone appearance in the BNE are shown in Figure I-7-14. The difference of this diagram from the classical diagrams of the Norwegian model and the Shapiro-Keyser model (Shultz et al., 1998) manifested itself at the dissipation stage (Figure I-7-14, stage V). The separation mentioned above was accompanied by the formation of positive values of $Adv\theta_e$ near the cyclonic center, between this center and the cold front.

Different types of clouds were observed in the separated part of the band. Clouds at the low levels (stratus and stratocumulus) and intensive convection with cumulonimbus development were predominant in 43% and

57% of events, respectively (Fedorova et al., 2016). Conditions for the development of convective or stratus clouds are described in 6.1.

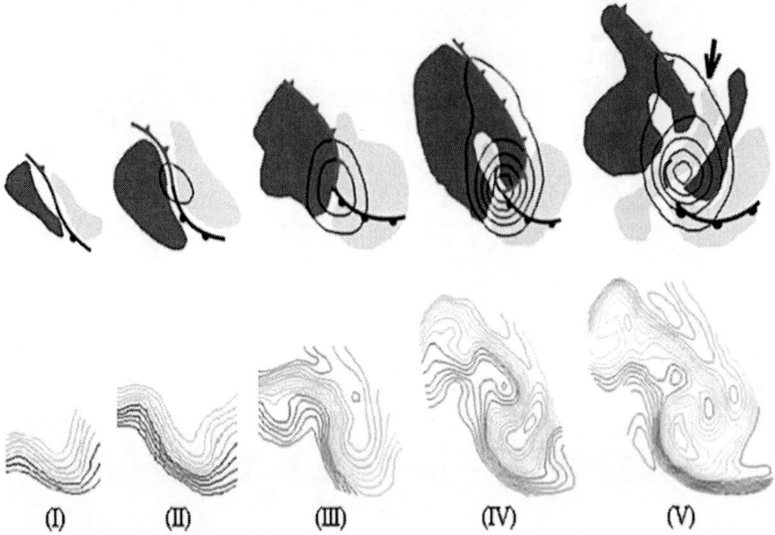

Figure I-7-14 – Diagram of the evolution of the baroclinic low in tropical/subtropical regions in the Atlantic Ocean of the Southern Hemisphere and frontal zone appearance in the BNE.
The figures above show isobars (black line), Advθ_e maps (positive gray; negative gray dark) and frontal zones.
The figures below show (θ_e) maps.
Stages of baroclinic cyclone:
I before low formation, II incipient low with one closed isobar, III young low, IV maximum development stage, V dissipation stage with a frontal fracture.
The arrow shows the position of separation between the cyclonic center and the frontal zone
Source: Cruz, 2008; Fedorova et al., 2016

Example 1

An example of the cold front periphery separation from baroclinic cyclone center can be seen on infrared satellite imagery together with the map of *Adv*θ_e at the 850 hPa during 1 and 2 May 2005 (Figure I-7-15). Positive values of *Adv*θ_e (arrow in the figure) point to the area where the cloud band was divided.

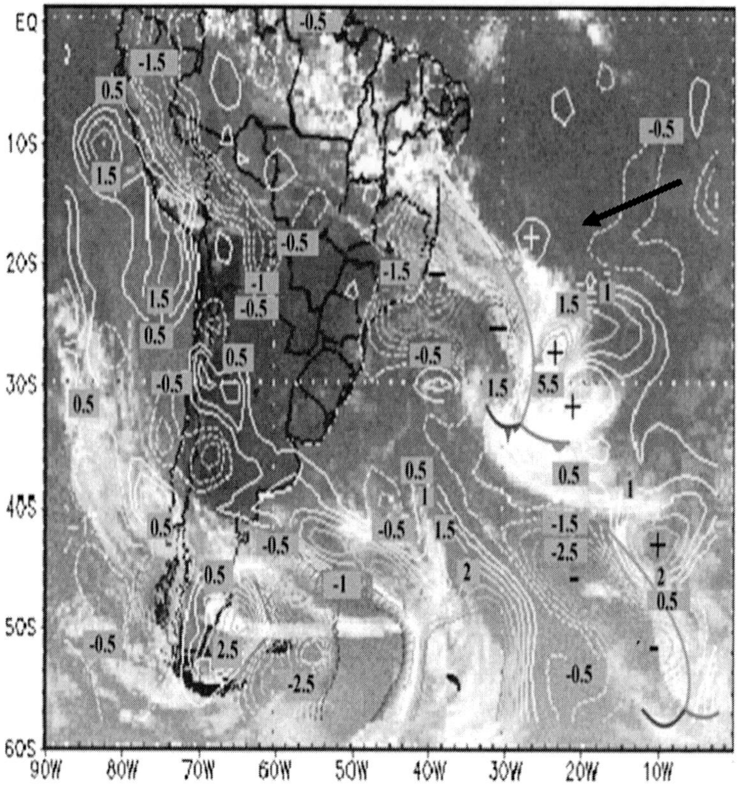

a)

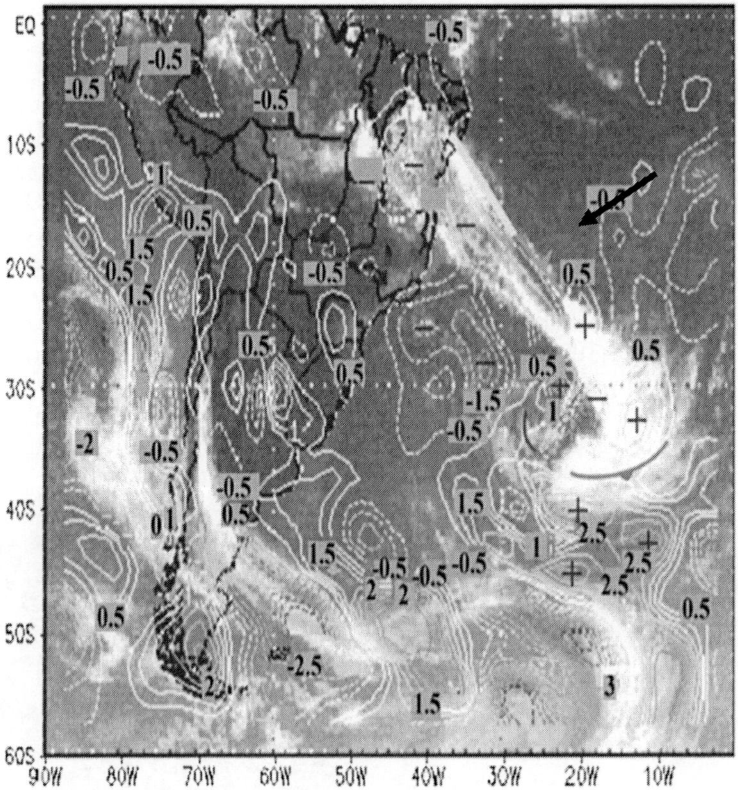

b)

Figure I-7-15 – Separation of a periphery of the cold front from baroclinic cyclone center during 1(a) and 2 (b) May 2005, 18 UTC by infrared satellite imagery together with $Adv\theta_e$ map (Ks^{-1}) at 850 hPa

The arrow shows the position of positive values of $Adv\theta_e$ and cloudiness separation between the cyclonic center and the frontal zone.

Source: Fedorova et al., 2016

Example 2

Frontal zone displacement towards the BNE from 10 to 15 November 2017 is seen in the sequence of the surface pressure maps and IR satellite images in Figure XXX. At the first moment (10 November 2017), the baroclinic cyclone with the frontal zones can be seen near the coast of continental region between 20-40°S (L1, Figure I-7-16a and b). Also, the family of baroclinic cyclones L2 and L3 cross the South Atlantic and reach the south of the BNE. This family of cyclones L2 and L3 on the next day displaced east and was weakening (Figure I-7-16c and d).

The cyclone L1 develops in the following days (10-15 November), but the cyclone center did not move significantly in longitude and was located between 35-40°S and slowly displaced east (from 55 to 35°W) on 11-13 November (Figure I-7-16c, d, e, f, g, h). At the same time, a cold front was moving north, reaching south of the BNE on 12 November. On the following days (12 and 13 November), the cold front was associated with the baroclinic trough of cyclone L1 and was located over the Ocean up to the continental region. The cloudiness of this front penetrates the continental region along a trough. It is very important to note that this trough extends from the equatorial region. The separation of clouds at the periphery of the cold front from the baroclinic cyclone occurred on 15 November (Figure I-7-16i and j). The trough of the main cold front displaced to the south, and at the same time the trough from the equatorial region intensified.

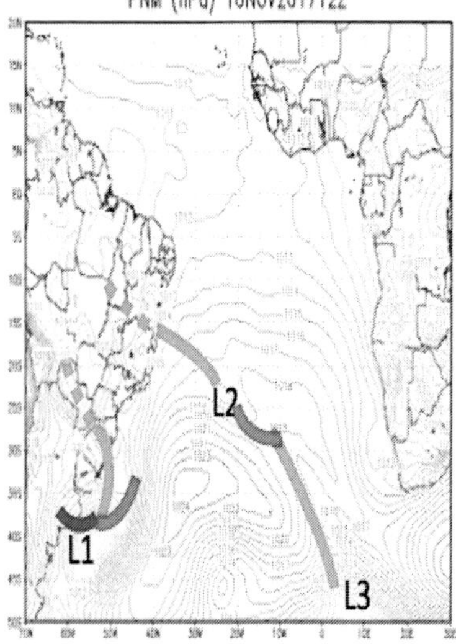

a)

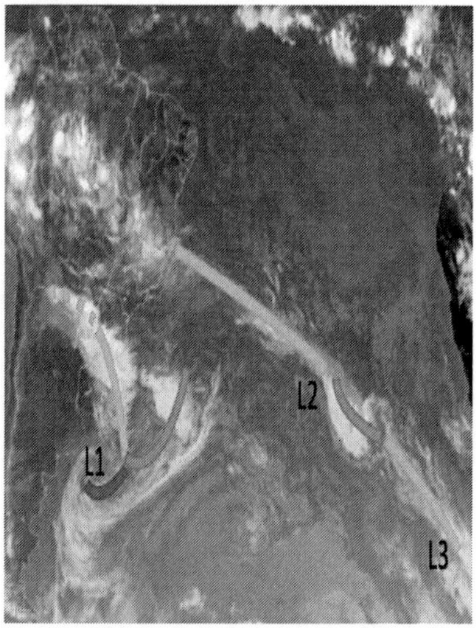

b)

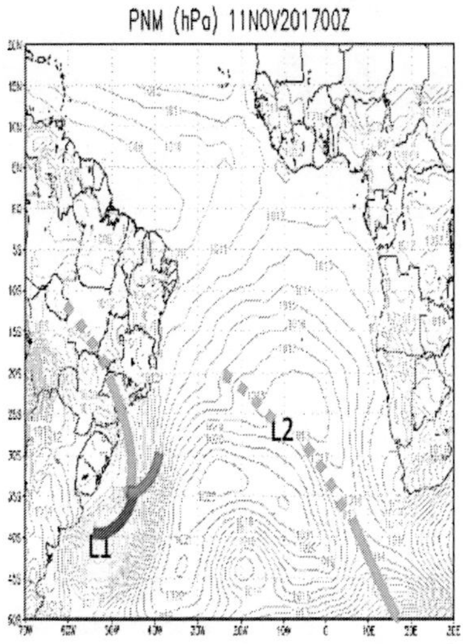

c)

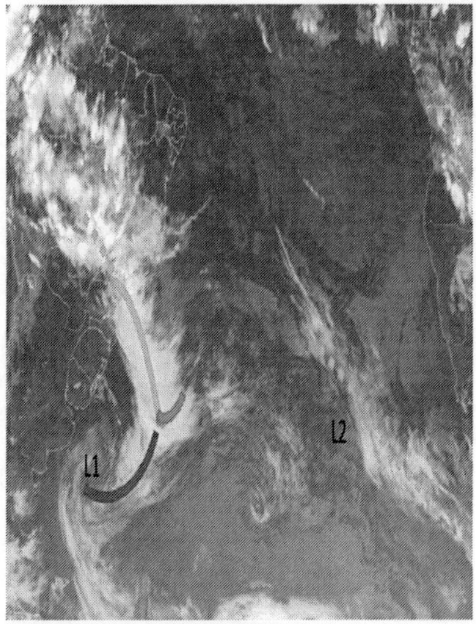

d)

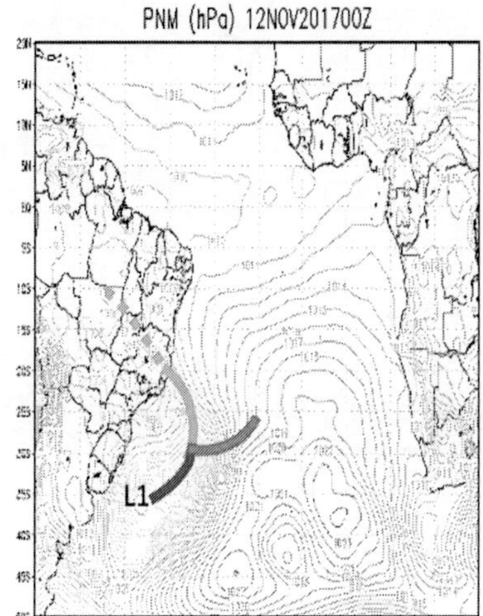

e)

f)

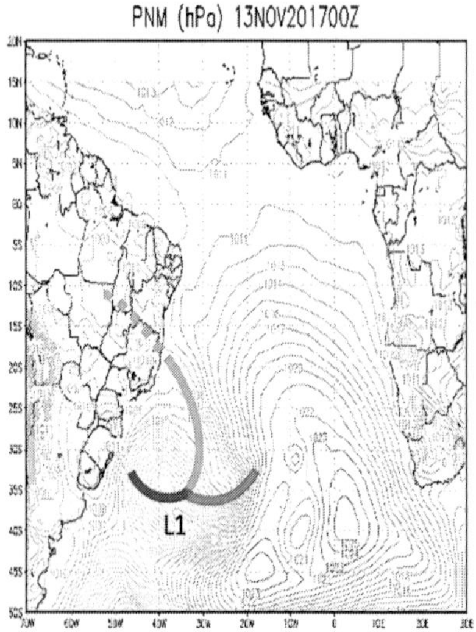

g)

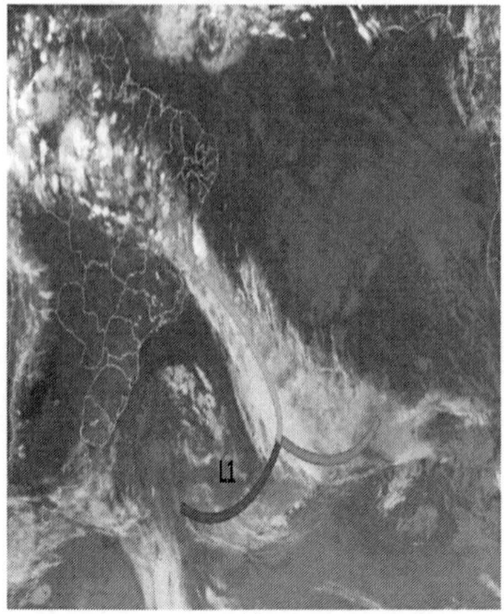

h)

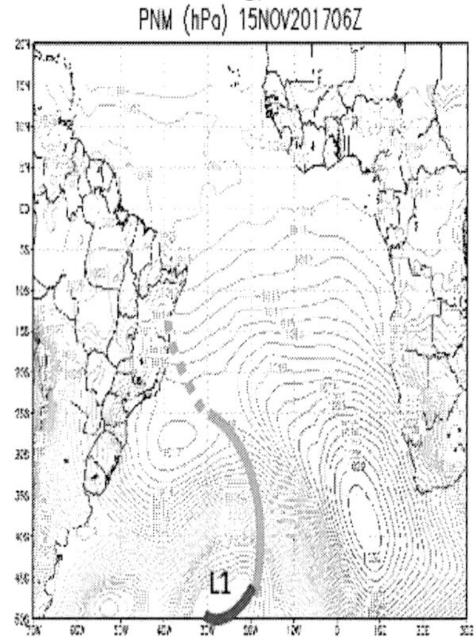

i)

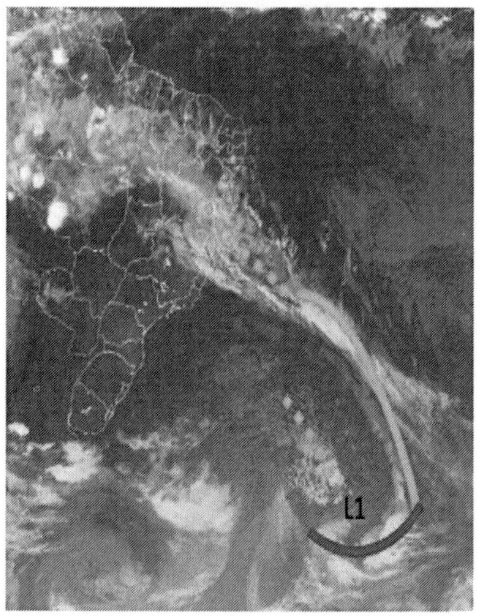

j)

Figure I-7-16 – Frontal zone displacement to the BNE from10 to 15 November 2017:
a), b) 10 November 2017, 12 UTC
c), d) 11 November 2017, 00 UTC
e), f) 12 November 2017, 00 UTC
g), h) 13 November 2017, 00 UTC
i), j) 15 November 2017, 06 UTC

a), c), e), g), i) - pressure maps at the surface level (black line)
b), d), f), h), j) - IR satellite imagery
L1, L2 and L3 - Baroclinic cyclones with frontal zones.
Frontal zones are shown by full lines: occluded front -black line; warm and cold fronts - gray colors from darker to light, respectively,
Dashed lines - periphery of the cold front and secondary front.
Source: CPTEC/INPE

6. Precipitations at the frontal zone periphery

As a rule, the frontal periphery passes over the continent, which creates precipitation and adverse phenomenon (Pontes da Silva et al., 2011). More intense precipitation falling over a large area is observed in the following situations:

1) convection development at the frontal periphery;
2) connection of the frontal extremity with the ITCZ (Muanza Gomes, 2018; Lyra et al., 2019; Muanza Gomes et al., 2021);
3) MCC development (Chapter II-6).

6.1 Convection at the frontal periphery in the BNE

Usually, the frontal extremity passes through the BNE, however, in some situations, the development of convection is observed, and in other situations, stratiform clouds or cloudlessness occurs (Figure I-7-17) (Siqueira & Machado 2004; Veber, 2011; Veber et al., 2020). These situations can be distinguished using synoptic analysis at low, middle and high levels.

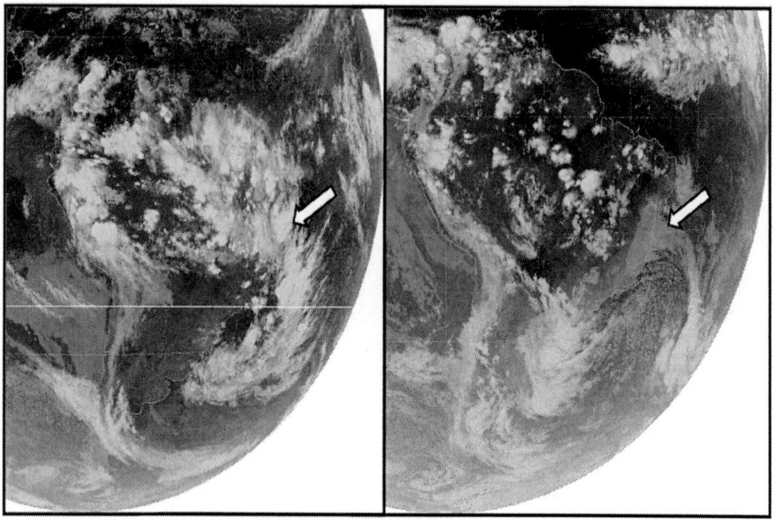

a) b)

Figure I-7-17 – Cold front in the BNE on IR satellite images in two situations: a) with the development of convective cloudiness at the frontal periphery; and b) stratiform cloudiness at the frontal periphery.
White arrow pointed at the frontal periphery.
Source: GOES - NCDC/NOAA, Veber, 2011; Veber et al., 2020

Convection development at the frontal periphery

Typical conditions for the convection development are two situations *at the low levels* (Veber, 2011; Veber et al., 2020). The first shows the cyclone location near the continent between 20-30°S and the frontal zone of this baroclinic cyclone observed over the BNE (Figure I-7-18a). In the second situation, the cyclone is located far from the continent with a center between 40-50°S, the usual situation with cyclones (Figure I-7-18b). It is important to note that the frontal zone, starting from the baroclinic Low, passed along the trough between *H1* and *L* up to 30°S, and north of this latitude, the frontal extremity was observed between the currents from *H1* and *H2*.

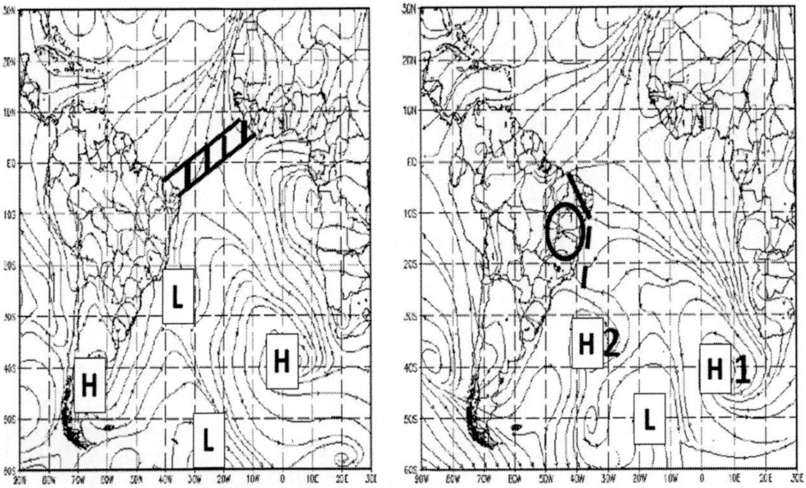

a) b)

Figure I-7-18 – Conditions for the convection development at the frontal periphery at low levels. *Streamlines at low levels (925 hPa);*
dashed line - the axis trough, WDTW; continuous line - crista; double line - ITCZ, circle - confluence of streamlines
Source: Veber, 2011; Veber et al., 2020

Currents from *H1* in tropical regions are the Trade Winds – TW (Chapter I-3). Near the coastal region, above the ocean these TW created a wave, Wave disturbance on the Trade Wind – WDTW. The presence of WDTW is a very important factor for the development of convection. In this typical situation, the confluence of the TW currents over the continent is observed.

At the middle levels, for the convection development, a typical situation is the confluence of air currents from L and H over the BNE with the center over the ocean (Figure I-7-19a). *At the high levels*, the trough or UTCV affects the BNE region (Figure I-7-19b). In these situations, a divergence of air currents is observed over the BNE.

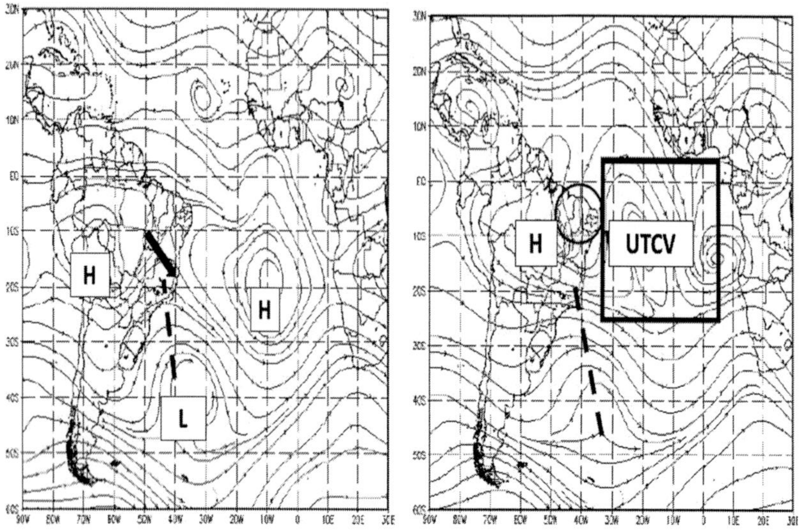

a) 500 hPa b) 200 hPa

Figure I-7-19 – Conditions for the convection development at the frontal periphery at the middle (a, 500 hPa) and high (b, 200 hPa) levels.
Streamlines at the middle (500 hPa) and high (200 hPa) levels;
Black arrow pointed at the confluence region between L and H over ocean.
UTCV- Upper Tropospheric Cyclonic Vortex; dashed line - trough's axis;
black arrow- confluence of the air currents; circle -difluence of the air currents.
Source: Veber, 2011; Veber et al., 2020

Stratiform clouds or cloudlessness at the frontal periphery.

Stratiform clouds or cloudlessness at the frontal periphery were observed when centers of all baroclinic cyclones and anticyclones were located above the ocean at the extratropical latitudes (south of 30°S) (Veber, 2011; Veber et al., 2020). In these situations, the anticyclonic circulation prevailed over the BNE *at the low levels* (Figure I-7-20a). These currents are formed by the High centered in the subtropical Atlantic.

At the middle and high levels (500 hPa and above), the main synoptic system over the continent of South America was the anticyclonic circulation (Figure I-7-20b and I-7-21). This High was centered at 20°S at 500 hPa level and at 10°S at 200 hPa. The baroclinic trough in these situations presented meridional position over the Atlantic at 500 hPa level and frequently at 200 hPa. Moreover, the trough can influence the BNE at 200 hPa (Figure I-7-21a).

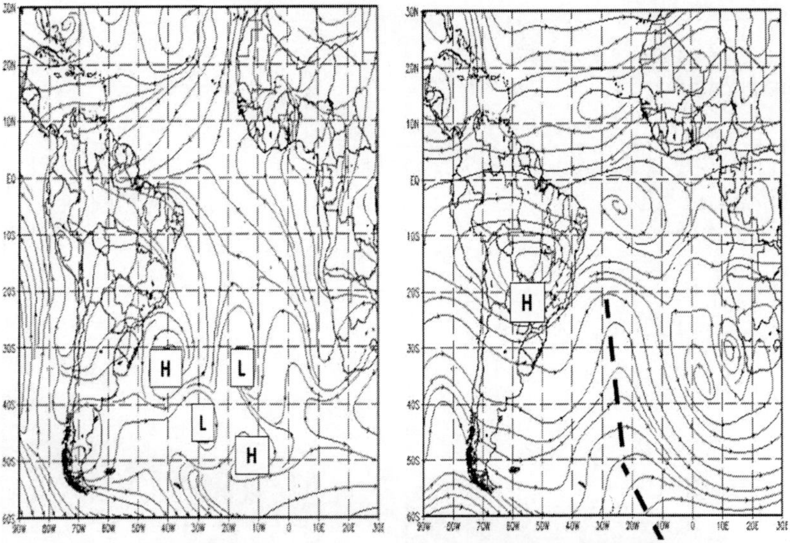

a) 925 hPa b) 500 hPa

Figure I-7-20 – Conditions for the Stratiform clouds or cloudlessness formation at the frontal periphery at low (a, 925 hPa) and middle (b, 500 hPa) levels.
Streamlines at the middle (925 hPa) and high (500 hPa) levels;
dashed line - trough's axis;
Source: Veber, 2011; Veber et al., 2020

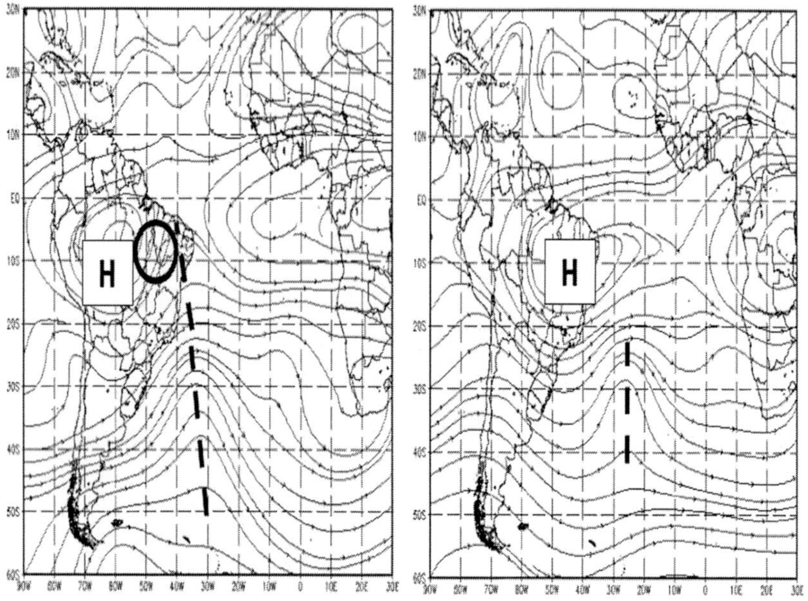

a) b)

Figure I-7-21 – Conditions for the Stratiform clouds or cloudlessness formation at the frontal periphery at the high (200 hPa) levels.
Streamlines at high (200 hPa) levels; dashed line - trough's axis; circle -difluence of the air currents.
Source: Veber, 2011; Veber et al., 2020

6.2 Connection of the frontal periphery with the ITCZ

The formation of precipitation occurred in the region of the connection of the frontal periphery with the ITCZ (Muanza Gomes, 2018; Muanza Gomes et al., 2021). Therefore, the main problem for precipitation forecasting is to determine whether this connection will be formed or not.

To answer this question, one can use synoptic models for these situations (Figure I-7-22). The fundamental difference between these models is the BNEJS location. The connection of the frontal periphery with ITCZ did not occur when the BNEJS passed together with trough at a 200 hPa, reaching approximately 5°S and closer to the ITCZ (Figure I-7-22b). When the BNEJS is located in the south, at 15-20°S, the frontal periphery is connected to the ITCZ (Figure I-7-22a). In this situation, an anticyclonic circulation is observed above the frontal periphery at 200 hPa level.

Frequencies of synoptic situation with the connection or not of the frontal periphery with ITCZ are presented in Table I-7-2. For 10 years, 29 frontal zones were observed in the BNE. Spring was a season with an increased number (13 events) of the frontal zones, but a connection with ITCZ was not formed. In summer, only two frontal zones were reveled and both of them had no connection with the ITCZ.

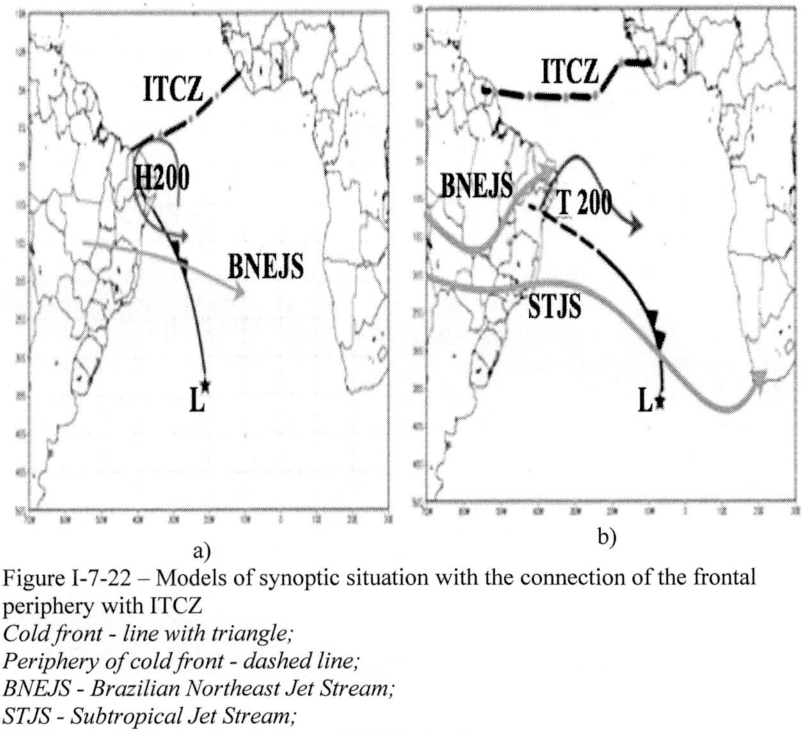

a) b)

Figure I-7-22 – Models of synoptic situation with the connection of the frontal periphery with ITCZ
Cold front - line with triangle;
Periphery of cold front - dashed line;
BNEJS - Brazilian Northeast Jet Stream;
STJS - Subtropical Jet Stream;
H200 - anticyclonic circulation at 200 hPa level;
T 200 - trough at 200 hPa level.
Source: Muanza Gomes, 2018; Muanza Gomes et al., 2021

Table I-7-2 – Seasonal and monthly frequencies of the frontal zone in the BNE for 10 years (2008-2017) with or without connection to the ITCZ
C*- connection;* ***NC****-without connection*
Source: Muanza Gomes, 2018; Muanza Gomes et al., 2021.

	D	J	F	M	A	M	J	J	A	S	O	N	Year
C	-	-	-	-	2	3	1	-	-	-	-	-	6
NC	1	1	-	-	1	1	2	1	3	6	4	3	23
Total	1	1	-	-	3	4	3	1	3	6	4	3	29
	summer			autumn			winter			spring			
C	-			5			1			-			6
NC	2			2			6			13			23
	2			7			7			13			29

Example of precipitation at the frontal periphery with ITCZ

The connection of the frontal periphery with the ITCZ was detected within 4 days 8-11 April 2010. The connection was visible on IR satellite images throughout the entire period (Figure I-7-23). Moreover, this connection is clearly visible on the relative vorticity (Figure I-7-24) and $Adv\theta_e$ (Figure I-7-25) maps. In this regard, a vast area of precipitation was formed, and the maximum amount of precipitation exceeds 350 mm/24h on 10 April 2010 (Figure I-7-26).

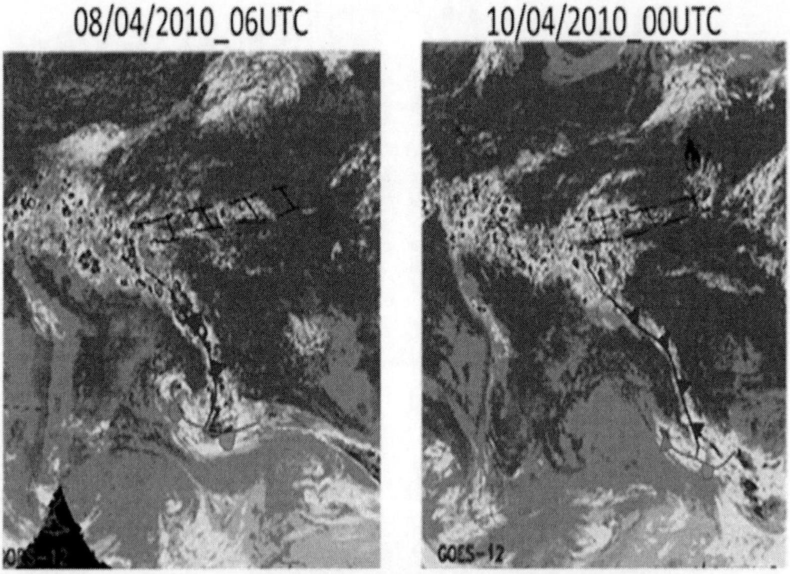

a) b)

c)

Figure I-7-23 – Connection of the frontal periphery with ITCZ on 8 (a), 10 (b) and 11 (c) April 2010: IR satellite images with frontal zone (*line with triangle*) and ZCIT (double *line*) locations.
Source: Muanza Gomes, 2018; Muanza Gomes et al., 2021.

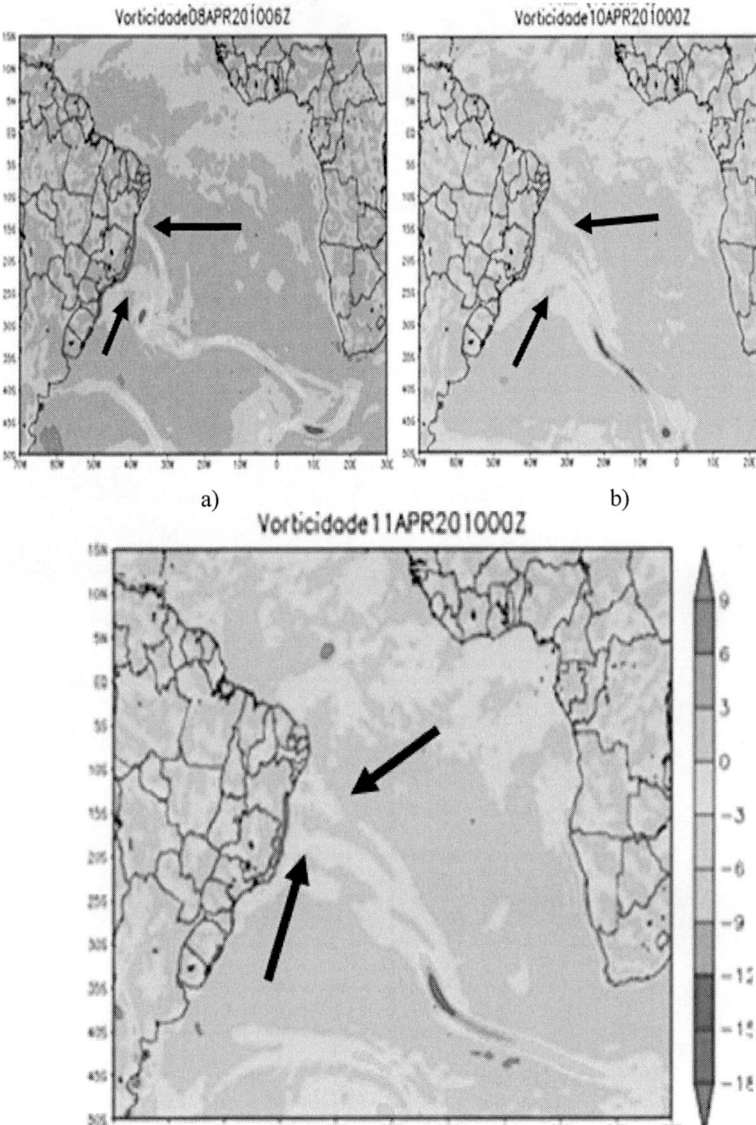

c)

Figure I-7-24 – Connection of the frontal periphery with ITCZ on 8 (a), 10 (b) and 11 (c) April 2010: maps of relative vorticity (s^{-1}). *Light grey: between 0 and -3 s^{-1} and arrow pointed in this region at the frontal periphery.*
Source: Muanza Gomes, 2018; Muanza Gomes et al., 2021

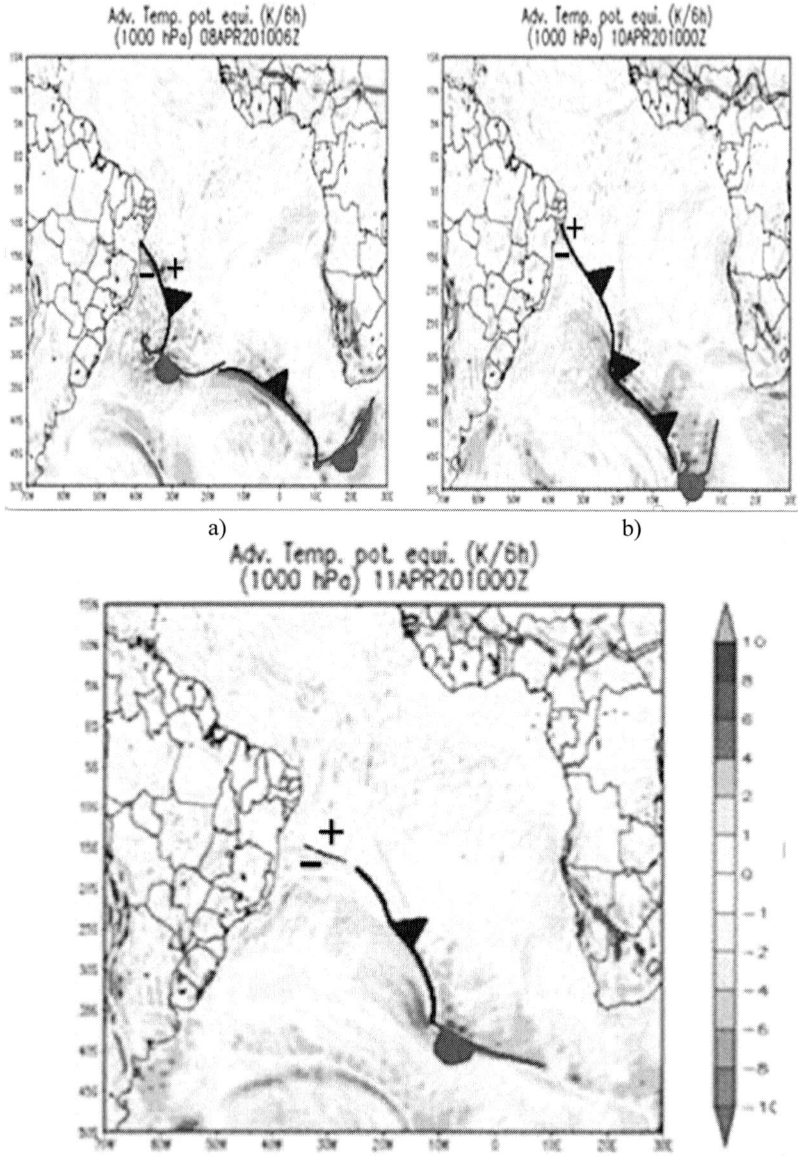

Figure I-7-25 – Connection of the frontal periphery with ITCZ on 8 (a), 10 (b) and 11 (c) April 2010: maps of Advθ_e (K).
Source: Muanza Gomes, 2018; Muanza Gomes et al., 2021

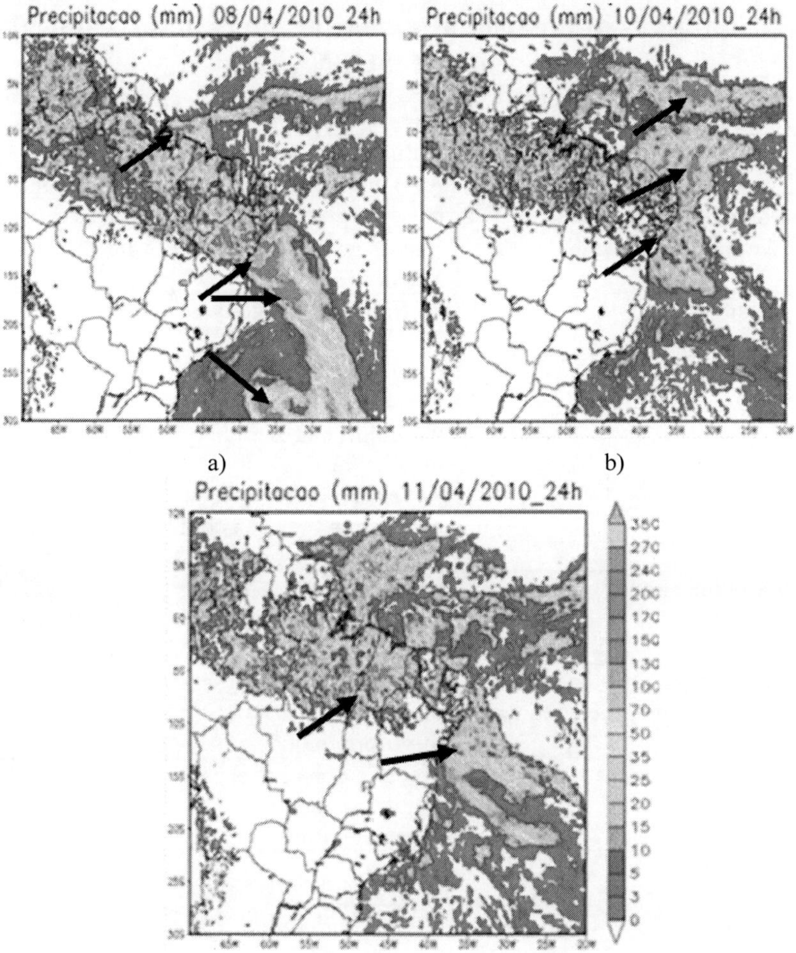

c)

Figure I-7-26 – Connection of the frontal periphery with ITCZ on 8 (a), 10 (b) and 11 (c) April 2010: maps of precipitation (mm/24h). *(Arrow pointed at the area with precipitation more than 200 mm/24h)*
Source: Muanza Gomes, 2018; Muanza Gomes et al., 2021

7. Summary of the frontal zone characters in the tropical zone of the BNE

Table I-7-3 – Principal characteristics of the frontal zone in the BNE

Characteristic	**Value or description**
Levels of more intensive gradients	At low levels weak temperature and pressure gradients were observed Intense gradients of θ_e and Advθ_e at 900-500 hPa
Frequencies	~8 frontal zones per year were observed
Seasonal variation	*Cold front zones during year* *Summer and spring:* connection between the cold front periphery and ITCZ was not observed *Autumn*: high frequency of the connection between cold front periphery and ITCZ was observed
Baroclinic zone	weak baroclinic zone at the low levels and more intensive one at the middle levels
Cloudiness at the frontal periphery over the continent	Convection cloudiness, Cb, MCC or Stratiform clouds
Precipitation and phenomena at the periphery of the cold front	Precipitation can reach high intensity Light fog, mist, and fog Thunderstorms.

8. References

Albuquerque Iolly Sousa. "Associação entre as correntes de jato e zonas frontais no Nordeste brasileiro." Thesis, Universidade Federal de Alagoas, Maceió, 2022.

Bluestein H. B. *Observations and Theory of Weather Systems. Vol. II. Synoptic–Dynamic Meteorology in Midlatitudes*, Oxford University Press, 594p, 1993.

Bolton D. "*The computation of equivalent potential temperature.*" Monthly Weather Review 108: 1046–1053, (1980).

Cavalcanti I.F.A, Kousky V.E. *"Frentes frias sobre o Brasil."* In: Cavalcanti IFA, Ferreira N., Silva MGA, Silva Dias MAF. Tempo e Clima no Brasil. Sao Paulo: Oficina de Textos, 135–147, (2009).

Cruz Cesar Duarte. *Determinação sinótica dos fatores que favorecem as influências frontais sobre o estado de Alagoas,* MSc. Thesis, Federal University of Alagoas, Maceio, Brazil, 117p., MET-UFAL-MS-052, 2008.

Djuric Dusan, *Weather Analysis*. New Jersey: Prentice Hall, 304p, ISBN: 0-13-501149-3, 1994.

Fedorova Natalia. *Meteorologia Sinótica. Volume 1.* Pelotas: Ed. Universitária/UFPel, 260p, ISBN: 85-7192-108-3, 1999.

Fedorova N., Bakst L. "*Identificação da advecção de ar quente através da analise de dados de satélites.*" Revista Brasileira de Geofísica 17, n. 1., 67-78, (1999).

Fedorova N, Carvalho MH. "*Synoptic processes during La- Nina and El-Nino years. Part II: Frontal Zones.*" Revista Brasileira de Meteorologia 15-2: 57–72, (2000).

Fedorova Natalia. *Meteorologia Sinótica Volume 2.* Pelotas: Ed. Universitária/UFPel, 242p, ISBN: 85-7192-145-7, 2001.

Fedorova Natalia. *Sinótica I: Fenômenos atmosféricos, dados sinóticos e massas de ar.* Material didático: sinopses, figuras, equações Maceió: Ed. Universitária/ UFAL, 175 p, ISBN: 978-85-7177-419-3, 2008a.

Fedorova Natalia. *Sinótica III: Frentes, correntes de jato, ciclones e anticiclones.* Material didático: sinopses, figuras, equações Maceió: Ed. Universitária/ UFAL, 192 p, ISBN:978-85-7177-421-6, 2008b.

Fedorova N. Sinótica IV: Sistemas e processos sinóticos atuantes na américa do sul. Material didático: sinopses, figuras, equações Maceió: Ed. Universitária/ UFAL, 192 p. ISBN: 978-85-7177-422-3, 2008c.

Fedorova N., Levit, V., Cruz C.D. *"On Frontal Zone Analysis in the Tropical Region of the Northeast Brazil*", Pure and Applied Geophysics 173, 1403-1421, DOI 10.1007/s00024-015-1166-y. (2016).

Fedorova Natalia, Pontes Da Silva Bruce Francisco, Levit Vladimir. *Análise e previsão prática das nuvens pelo diagrama termodinâmico SKEW-T/LOG-P,* 227p. ISBN: 978-85-923360-1-1, 2017.

Fedorova, N, Pontes Da Silva B.F. "*Capítulo 1 - Zonas frontais no Nordeste Brasileiro.*" In: Meteorologia em tópicos, v. 6, 9 – 65, ISBN 978-65-991641-0-1, 2020

Gan MA, Rao VB. *"Surface cyclogenesis over South America."* Monthly Weather Review 119, 1293–1302, (1991).

Gan MA, Rao VB. *"The influence of the Andes Cordillera on Transient Disturbances.*" Monthly Weather Review 122, 1141–1157, (1994).

Gemiacki L. "*Atuacao de Sistemas Frontais na Estac ̧a ̃o Seca do Nordeste do Brasil. (Influence of the frontal systems in the dry season in the*

Northeast of Brazil." 101p., MSc. Thesis, Federal University of Alagoas, Maceio, Brazil. (2005). MET-UFAL-MS-035.

Kousky VE. *"Frontal Influences on Northeast Brazil.*" Monthly Weather Review, 107, 1140–1153, (1979).

Lyra M.J.A., Cavalcante L.C.V., Levit, V., Fedorova N. "*Connection between frontal extremity and intertropical convergence zone on the northeast Brazil."* Anuário do Instituto de Geociências – UFRJ, 42 - 1, 413-424, DOI: http://dx.doi.org/10.11137/2019_1_413_424, (2019).

Muanza Gomes A., Fedorova N., Levit, V. *"Ligação da extremidade frontal com zona de convergência intertropical (ZCIT) e análise de fenômenos associados."* Nordheast Geosciences Journal 7, No 1, (2021).

Muanza Gomes A. *"Padrões de circulação atmosférico associados á ligação da extremidade frontal com a ZCIT"*, MSc. Thesis, Federal University of Alagoas, Maceio, Brazil, (2018).

Oliveira AS. *"Interaction between the South American frontal systems and the Amazonian convection."* MSc. Thesis, Instituto Nacional de Pesquisas Espaciais. Sao Jose dos Campos, Brazil, INPE-4008-TDL/239, (1986).

Parmenter FC. "*A Southern Hemisphere cold-front passage near the Equator.*" Bulletin of the American Meteorology Society 57, 1435–1440, (1976).

Petterssen, Sverre. *Weather analysis and Forecasting.* v. 1, New York, Toronto, London: McGraw-Hill, 428., p1956.

Pontes da Silva BF, Fedorova N, Levit V, Peresetsky A. *"Synoptic systems associated with the heavy precipitations in the Alagoas State",* Revista Brasileira de Meteorologia 26-3, 295–310, (2011).

Reboita MS, Da Rocha RP, Ambrizzi T, Sugahara S. *"South Atlantic Ocean Cyclogenesis Climatology Simulated by Regional Climate Model (RegCM3).*" Climate Dynamics 35 (7-8): 1331–1347, (2010).

Saldanha C. B., Fedorova N, Carvalho, M.H. *"Estudo de um caso de frente fria secundária sobre o sul do Brasil",* Revista Brasileira de Meteorologia, 25, n.1, 32 – 45, (2010).

Satyamurti P, Mattos LF. *"Climatological lower tropospheric frontogenesis in the midlatitudes due to horizontal deformation and divergence."* Monthly Weather Review 117-6, 1355–1364, (1989).

Satyamurty P, Nobre CA, Silva Dias PL. *"South America."* In Meteorology of Southern Hemisphere, Karoly DJ, Vincent DG (eds). American Meteorological Society: Boston, MA; 119–139, (1998).

Siqueira JR, Machado LAT. *"Influence of the Frontal Systems on the Day-to-Day Convection Variability over South America."* Journal of Climate 17-9, 1754–1766, (2004).

Simmonds I, Keay K, Bye JAT. "*Identification and Climatology of Southern Hemisphere Mobile Fronts in a Modern Reanalysis.*" Journal of Climate 25, 1945–1962, (2012).

Sinclair MR. "*A Climatology of Cyclogenesis for the Southern Hemisphere.*", Monthly Weather Review 123, 1601–1619, (1995).

Signorini E. Synoptic processes associated with intensification of the end of the cold front in southern Brazil. MSc. Thesis, Federal University of Pelotas, Pelotas, Brazil, (2001).

Sinhori Natalia Parisotto "*Estrutura vertical das frentes frias durante a passagem do sul até o nordeste brasileiro, entre 2000 e 2012*", MSc. Thesis, Federal University of Alagoas, Maceio, Brazil, (2015).

Smith RK, Reeder M J, Tapper NJ, Christie DR. "*Central Australian Cold Fronts*", Monthly Weather Review 123, 16–38, (1995).

Shultz DM, Keyser D, Bosart, LF. "*The effect of large- scale flow on low-level frontal structure and evolution in midlatitude cyclones.*" Monthly Weather Review 126-7, 1767–1791, (1998).

Schultz DM. "*A Review of Cold Fronts with Prefrontal Troughs and Wind Shifts*", Monthly Weather Review 133, 2449–2472, (2005). Veber Maicon Eirolico. "*Desenvolvimento de atividades convectivas sobre a região nordeste do brasil, organizada pela extremidade frontal*", 129p, MSc. Thesis, Federal University of Alagoas, Maceio, Brazil: MET-UFAL-MS-090, 2011

Veber ME, Fedorova N., Levit, V. "*Desenvolvimento de atividades convectivas sobre a região nordeste do brasil, organizada pela extremidade frontal.*", Revista Brasileira de Meteorologia 35, n. Especial, 995-1003, DOI: http://dx.doi.org/10.1590/0102-77863550071, (2020).

CHAPTER I.8

IMPACT OF A NORTHERN HEMISPHERE TROPICAL CYCLONE ON THE BNE

NATALIA FEDOROVA
LUCAS CARVALHO VIEIRA CAVALCANTE
AND VLADIMIR LEVIT

1. Identification

Synoptic scale systems of the Southern Hemisphere (SH), which influence the weather formation in the BNE, have been described in previous chapters. At the same time, the synoptic scale systems of the Northern Hemisphere (NH) also influence the formation of weather in the BNE by creating air currents passing through the equator.

This effect has already been discussed in the displacement and intensification of the frontal zone (Chapter I-7) and the UTCV formation (Chapter I-4).

This chapter will provide information on the impact of tropical depression and tropical cyclone of the NH on fog and mist formation in the BNE. A detailed description of the process of formation and forecasting of these adverse phenomena is shown in Chapter II-5. The mechanism of the influence of the NH on the BNE is presented in this chapter, since a similar process can also affect other weather phenomena.

Tropical cyclones (TCs) are systems that have great destructive power in their area, in addition, they create air currents across the equator and affect the weather at a distance.

TCs have been classified according to their sustained wind speed (Vasquez, 2000) as follows:

Tropical disturbance (TD). In this stage, winds are less than 17 m/s with open circulation (no closed isobars).

Tropical depression (TDep). Winds are less than 17 m/s, but there is a closed circulation (closed isobars).
Tropical storm (TS). Winds are equal to or greater than 17 m/s but less than 33 m/s with a closed circulation. The storm is usually assigned a name.
Hurricane (H). Winds are equal to or greater than 33 m/s.
Major hurricane (MH). Typhoon with 1 min of sustained winds of 65 m/s or greater (MH was not observed during the study period).

Tropical cyclones have been studied for many years due to the complexity of their formation and destructive power. At the same time, only a few studies of the air circulation between the hemispheres created by TCs have been carried out to date (Fedorova & Levit, 2016; Cavalcante 2018, Fedorova et al., 2020; Cavalcante et al., 2020).

2. Location, duration, and season

All tropical cyclones (TC) formed over the North Atlantic Ocean are reported by National Hurricane Center's Tropical Cyclone Reports https://www.nhc.noaa.gov/data/ter/. Based on its data set, it is possible to trace the TC trajectory from genesis to dissipation and determine its coordinate for each stage of development (Table I-8-1).

Table I-8-1 – Information about synoptic systems in the tropical region of the Northern Atlantic and adverse phenomena by METAR data in the BNE. *Adverse phenomena: RA (rain), DZ (drizzle), TS (thunderstorm), FG (fog), BR (mist), HZ (haze) and FU (smoke).*
METAR complementary Information: PR (partial), BC (patches), - (light intensity) and + (heavy intensity).
Synoptic systems: TD (tropical disturbance), TDep (tropical depression), TS (tropical storm), H (Hurricane) and D (dissipation).
Source: Kimberlain, 2013; Cangialosi, 2015; Cavalcante, 2018, Fedorova et al, 2020.

Name	**Days (Duration)**	**Synoptic systems**		**Adverse phenomena** *(per day)*
		Maximum stage	Stage over BNE	
		2013		
Chantal	06-10/07 (**5**)	TS	TS	RA BR; HZ; -DZ BR; -TSRA; FG.

Dorian	22/07-03/08 **(13)**	TS	TS	BCFG; TS; BR; -DZ BR; TS RA; FG; HZ
Erin	15-20/08 (**6**)	TS	TS-D	PR FG; TS RA; -DZ BR; +RA DZ
		2014		
Bertha	29/07-06/08 **(8)**	H	*TD*	-RA BR; DZ; -TS RA; FG
Edouard	10-23/09 **(14)**	H	*TD* -H	-RA BR; -DZ BR
		2015		
Danny	17-24/08 **(8)**	H	*TD* -H	FG; -DZ BR; RA BR
Erika	24-28/08 (**5**)	TS	TS	BCFG; BR
Fred	30/08-06/09 **(8)**	H	TS- *TDep*	-TSRA; HZ; FU; -RA BR
Grace	05-09/09 **(5)**	TS	TS-D	-TSRA; -RA BR
Ida	15-28/09 **(14)**	TS	*TD* -D	BC FG; FU; HZ

The indirect influence of the Atlantic North TC on the BNE weather conditions for three years 2013-2015 during the TC season (from July to September) was analyzed (Cavalcante 2018, Fedorova et al., 2020). NH TCs lasted from 5 to 14 days and only 1-3 days had any influence on the BNE weather when passing between 35 and 50° W. TC passing between 35 and 50° W in the Northern Hemisphere had two stages: a tropical storm (TS) and a tropical depression (TDep) (Table I-8-1).

3. Vertical circulation

The circulation trough the equator is formed by two currents in each hemisphere (Fedorova & Levit, 2016; Cavalcante 2018, Fedorova et al., 2020, Cavalcante et al., 2020). These currents are formed by the following synoptic scale systems:

1) In the northern hemisphere, cyclonic circulation creates an air convergence at low levels in tropical storm (TS) or tropical depression (TDep), and air divergence at high or middle levels (this height depends on the TC intensity).
2) In the southern hemisphere, the trough of an extratropical baroclinic cyclone at low and middle levels creates a current towards the BNE, and anticyclonic circulation associated with this process at middle

and high levels. WDTW near the BNE coast can also be observed at low levels. The role of WDTV in a fog formation in the BNE will be discussed in detail in Chapter II-5.

Consequently, an important sinking mechanism in the BNE from the confluence of currents at high levels from the NH and SH was described in Fedorova & Levit (2016), Cavalcante (2018), Fedorova et al. (2020), Cavalcante et al. (2020). Air flow at high levels from the Northern Hemisphere formed by the merger of two currents: (1) from TC (TS or TDep) and (2) from the High located over the equatorial and southern Atlantic. The air flows from the SH were created by a frontal trough associated with STJSSH (subtropical jet stream of the Southern Hemisphere).

In addition, a convergence zone at middle levels is formed between the currents from a tropical cyclone of the NH and anticyclonic circulation in the SH. This anticyclonic circulation dominated most of the BNE. Middle tropospheric cyclonic vortices (MTCV) (Chapter I-5) were observed in half the events, but do not change the main influence of the air flow from TC on the BNE weather. The locations of all these synoptic scale systems are illustrated in Figure I-8-1.

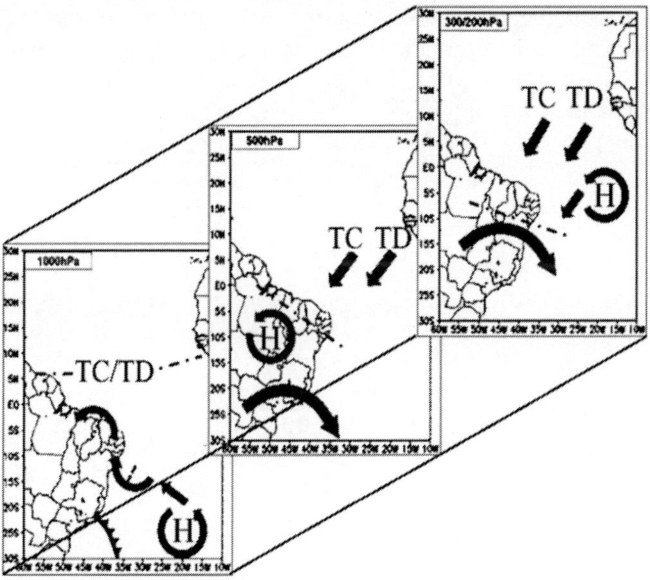

Figure I-8-1 – Synoptic systems observed more frequently at all levels that make up the circulation.
Acronyms: TC (Tropical Cyclone), TD (Tropical Disturbance or tropical depression) and H, High/Anticyclone

Dashed line –Trough (WDTW), Line dot dash – Convergence
Source: Cavalcante 2018, Fedorova et al., 2020.

4. Relationship with adverse phenomena

The effect of high-level air current from tropical depression and tropical cyclone (for example Danny-15) of the NH on the BNE weather was studied for cases of fog, mist, and stratus clouds at various points in the BNE (Fedorova & Levit, 2016; Cavalcante 2018; Fedorova et al., 2020; Cavalcante et al., 2020).

It was concluded that the synoptic scale circulation between the two hemispheres, between the tropical cyclone in the NH and the frontal zone in the SH, leads to the sinking and, therefore, the conditions favorable for the formation of low visibility.

An adverse phenomena formation during all TC events for three years 2013 – 2015 from July to September was analyzed (Cavalcante, 2018; Fedorova et al., 2020). Table 1 shows adverse phenomena found in each TC event.

Sinking over the BNE formed the humidity accumulation at low levels and, as a result, fog, mist, and weak rain formation (Figure I-8-2). Therefore, fog, mist, and weak rain in the BNE were adverse phenomena, which were most often associated with TC events. These phenomena were observed mainly near the coast of the BNE. Fog days were also accompanied by light rain.

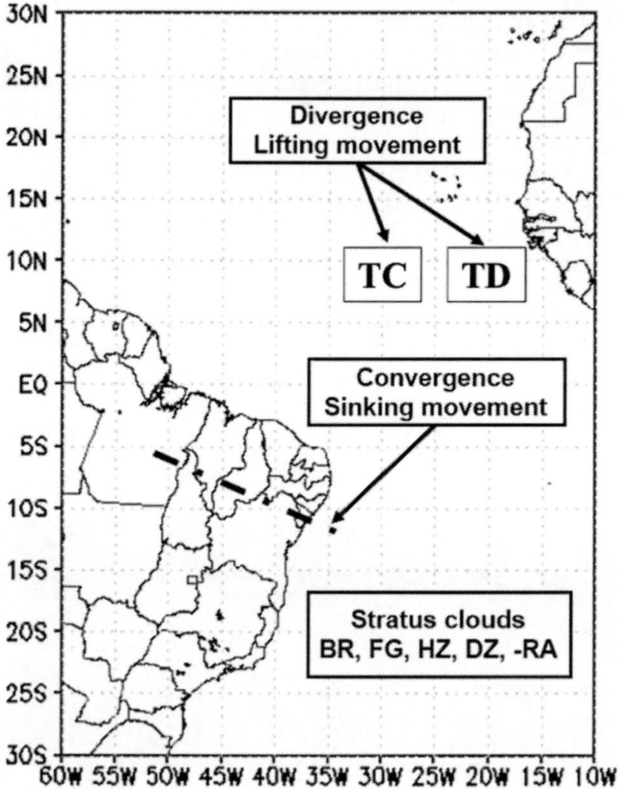

Figure I-8-2 – Schematic representation of TC and TD locations, main vertical movements, and adverse phenomena.
Dash-dotted line – Convergence.
Acronyms: TC (Tropical Cyclone), TD (Tropical Disturbance or tropical depression).
RA (rain), DZ (drizzle), FG (fog), BR (mist), HZ (haze).
Source: adapted from Fedorova et al., 2020.

5. Example 1, 17-18 September 2015

The first example shows the mist event on 17 and 18 September 2015 in Alagoas. Tropical storm IDA passed the longitude of the BNE these days. According to the surface stations of the Maceio Airport (SBMO), the mist with rain occurred at 12 UTC on 17 September. Also, the mist was identified on SYNOP maps on 18 September at 00 and 12 UTC.

According to the National Hurricane Center (NHC) report (Cangialosi, 2015), Tropical Storm IDA was originated from a tropical wave that moved off the west coast of Africa on 13 September. The wave moved westward until a well-defined low-pressure area formed around 12 UTC on 15 September, passing through the NEB longitude interval at 00 UTC on 17 September. The track of TC IDA is shown in Figure I-8-3.

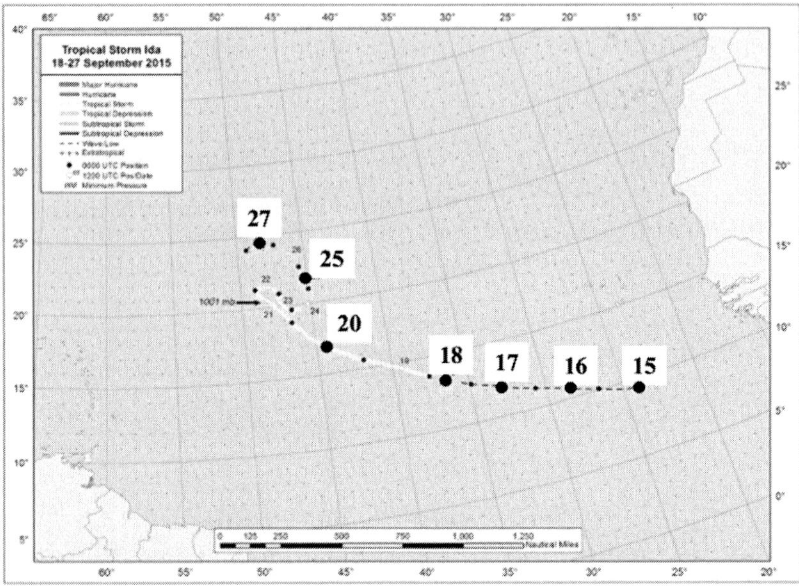

Figure I-8-3- Track for Tropical Storm Ida, 18-27 September 2015. Source: Cangialosi, 2015.

On the streamline maps on September 16, 18 UTC, it is possible to see the cyclonic circulation associated with the tropical storm IDA (1000hPa Figure I-8-4a) and the divergence of the air flow at middle levels (500 hPa, Figure I-8-4b). This divergence generates a southward current across the equator. Another air flow from the High of the SH has a northerly direction. These two currents converge over the central region of the BNE (Alagoas State). Subsequently, MTCV and High were formed at middle levels near the BNE on 17 September, 00 UTC (Figure I-8-4c). On the same day 12 UTC, the cyclonic circulation of the NH reached 300 hPa (Figure I-8-4d). This circulation showed a high intensity on 18 September, 12 UTC (Figure I-8-4e). A wave disturbance at the trade winds (WDTW) was formed over the coastal region (Figure I-8-4f) and created humidity convergence at low levels (Figure I-8-4g).

On the vertical section of the vertical wind speed along 35°W the air lifting is visible in the entire troposphere at 10°N in the IDA region and sinking at 9.6°S in the upper and middle layer (200-800 hPa) (Figure I-8-4k). It is important to note the presence of lifting in the WDTW at 9.6°S at low levels (1000-800 hPa).

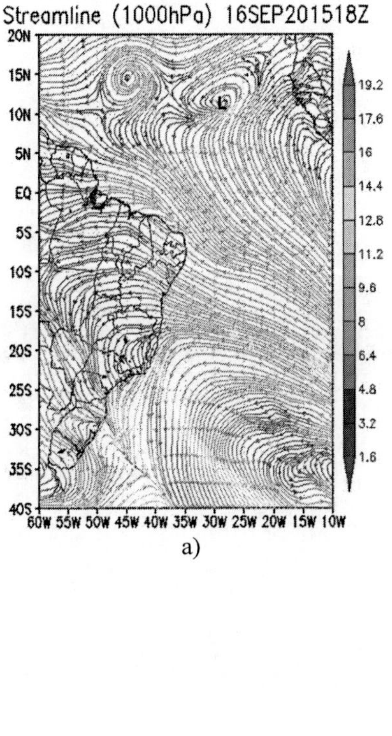

a)

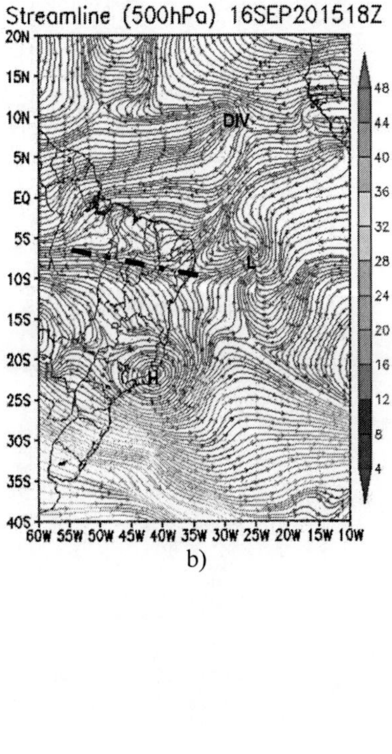

b)

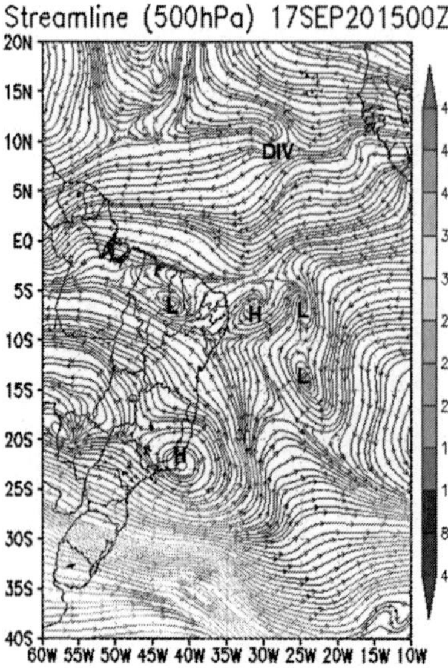
Streamline (500hPa) 17SEP201500Z
DIV
H
L
H
L
L
H
20N 15N 10N 5N EQ 5S 10S 15S 20S 25S 30S 35S 40S
60W 55W 50W 45W 40W 35W 30W 25W 20W 15W 10W
48 44 40 36 32 28 24 20 16 12 8 4

c)

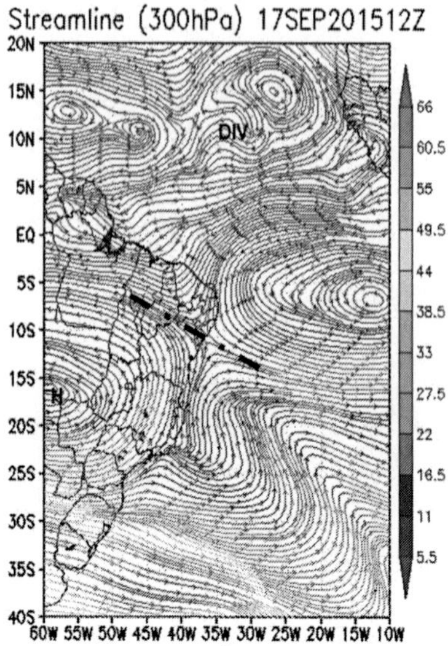
Streamline (300hPa) 17SEP201512Z
DIV
H
20N 15N 10N 5N EQ 5S 10S 15S 20S 25S 30S 35S 40S
60W 55W 50W 45W 40W 35W 30W 25W 20W 15W 10W
66 60.5 55 49.5 44 38.5 33 27.5 22 16.5 11 5.5

d)

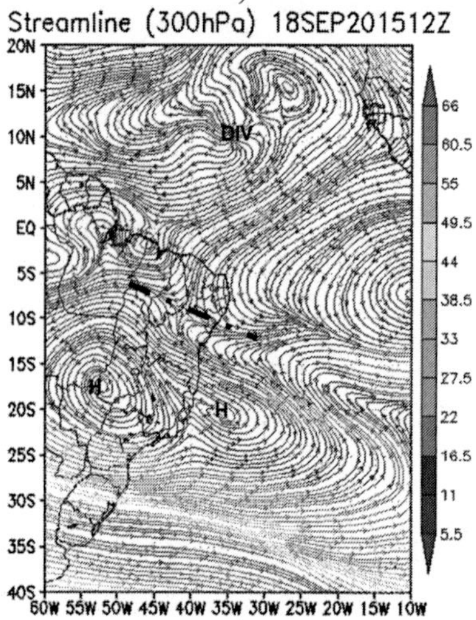
Streamline (300hPa) 18SEP201512Z
DIV
H
H
20N 15N 10N 5N EQ 5S 10S 15S 20S 25S 30S 35S 40S
60W 55W 50W 45W 40W 35W 30W 25W 20W 15W 10W
66 60.5 55 49.5 44 38.5 33 27.5 22 16.5 11 5.5

e)

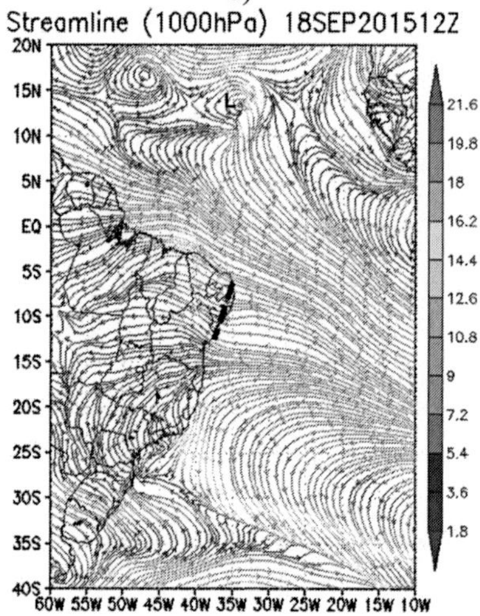
Streamline (1000hPa) 18SEP201512Z
L
20N 15N 10N 5N EQ 5S 10S 15S 20S 25S 30S 35S 40S
60W 55W 50W 45W 40W 35W 30W 25W 20W 15W 10W
21.6 19.8 18 16.2 14.4 12.6 10.8 9 7.2 5.4 3.6 1.8

f)

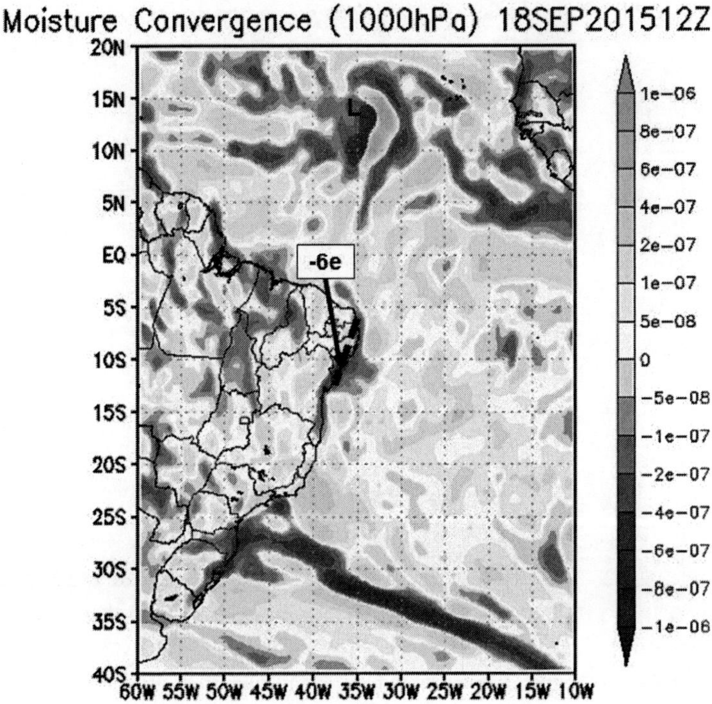
Moisture Convergence (1000hPa) 18SEP201512Z
20N
15N
10N
5N
EQ
5S
10S
15S
20S
25S
30S
35S
40S
60W 55W 50W 45W 40W 35W 30W 25W 20W 15W 10W
-6e
1e-06
8e-07
6e-07
4e-07
2e-07
1e-07
5e-08
0
-5e-08
-1e-07
-2e-07
-4e-07
-6e-07
-8e-07
-1e-06

g)

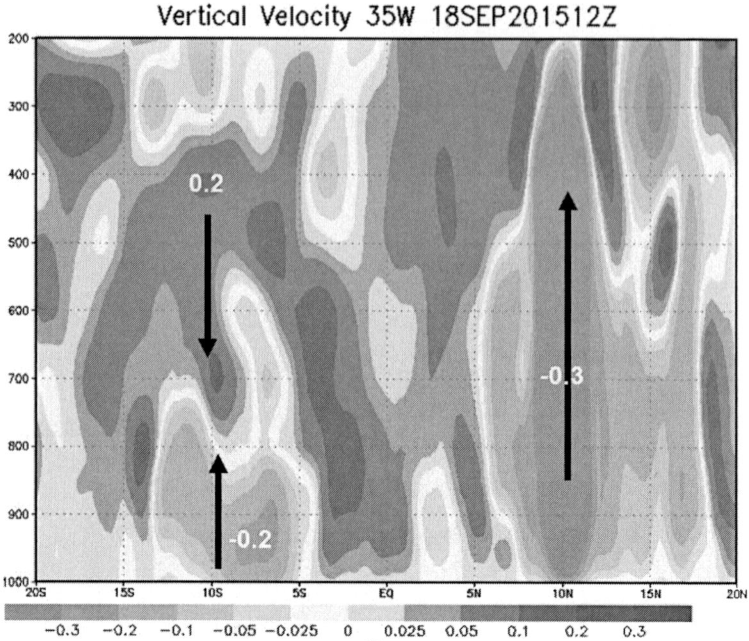

k)

Figure I-8-4 – Circulation between the two hemispheres on 16-18 September 2015.
Streamline ($m\ s^{-1}$) maps:
September 16, 18 UTC, at 1000 hPa (a), 500 hPa (b);
September 17, 00 UTC, at 500 hPa (c), September 17, 12 UTC, at 300 hPa (d);
September 18, 12 UTC, at 500 hPa (e), September 18, 12 UTC, at 1000 hPa (f);
Humidity Convergence map (g, mm s^{-1}): September 18, 12 UTC, at 1000 hPa (g);
Vertical section of vertical speed (k, Pa s^{-1}) along 35°W; September 18, 12 UTC
L, Low/cyclone; and H, High/Anticyclone; DIV, divergence.
Dashed line – Trough, Line dot dash – Convergence
Source: Model ERA interim, 0.5x0.5°.

6. Example 2, June 11, 2010

The second example shows the fog event resulting from the circulation between the currents of both hemispheres on June 11, 2010 in Alagoas. According to Maceio Airport surface stations (SBMO), the fog event occurred between 04 UTC and 07 UTC.

The Low and tropical disturbance at 5°N with airflow convergence at 1000 hPa level was identified from streamline maps and satellite images (Figure I-8-5a and g). These two cyclonic circulations can be seen in the same region at 850 hPa level too (Figure I-8-5b). Intense convection with large vertical development of clouds is visible on the infrared satellite image in the regions of these cyclonic circulations. These systems created a diffluent air flow at 500 hPa (Figure I-8-5c) and 400 hPa (Figure I-8-5d) levels.

In the southern hemisphere, WDTW close to the BNE coast and a circulation associated with an extratropical baroclinic cyclone at 35°S were observed (Figure I-8-5a). The trough of this cyclone is very intense at middle and high levels (500 and 400 hPa) (Figure I-8-5c, d). A weak anticyclonic circulation at 500 hPa level at this trough extremity was observed over the Alagoas State (Figure I-8-5c).

Air currents from this tropical NH disturbance and from SH trough at 400 hPa level created intense confluence over Alagoas. Vertical section of the vertical wind speed along 36°W show the air lifting in the entire troposphere at 5°N in the tropical disturbance and sinking in the 200-700 hPa layer at 9.6°S (Figure I-8-5f). The lifting in the WDTW of SH area was in the 700-1000 layer at 9.6°S. This mean, that the circulation between hemispheres creates subsidence and hence conditions for humidity accumulation at low levels. This increase in humidity is also contributed by the air flow confluence in the WDTV (Figure I-8-5e). As a result, fog formed in the region of high humidity in the WDTW area (Figure I-8-5e).

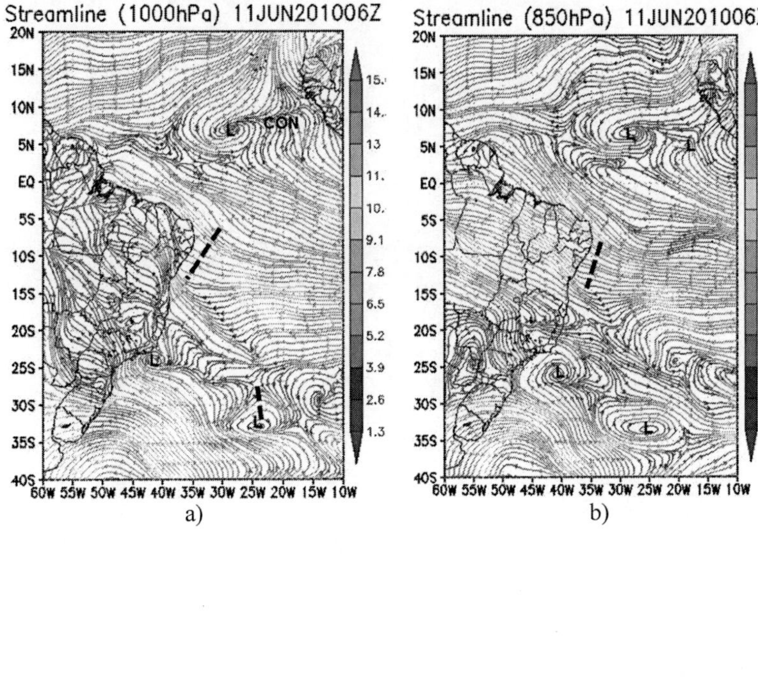

a) b)

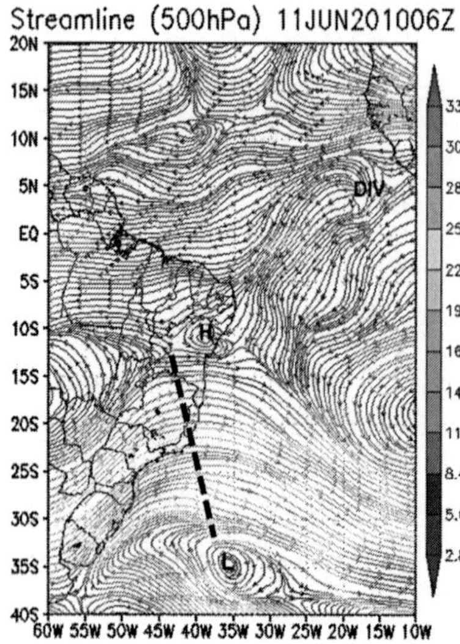
Streamline (500hPa) 11JUN201006Z
20N
15N
10N
5N
EQ
5S
10S
15S
20S
25S
30S
35S
40S
60W 55W 50W 45W 40W 35W 30W 25W 20W 15W 10W
DIV
H
L
33.6
30.8
28
25.2
22.4
19.6
16.8
14
11.2
8.4
5.6
2.8

c)

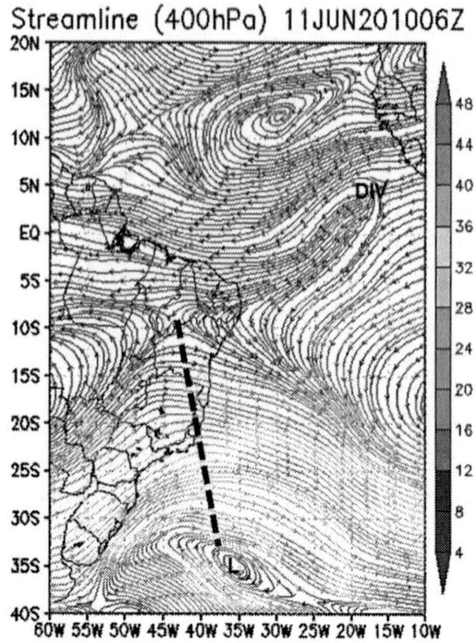
Streamline (400hPa) 11JUN201006Z
20N
15N
10N
5N
EQ
5S
10S
15S
20S
25S
30S
35S
40S
60W 55W 50W 45W 40W 35W 30W 25W 20W 15W 10W
DIV
L
48
44
40
36
32
28
24
20
16
12
8
4

d)

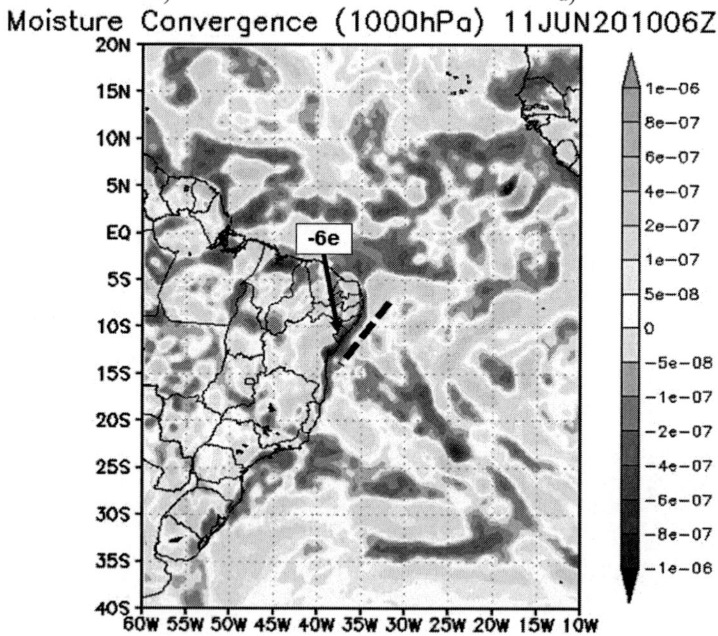
Moisture Convergence (1000hPa) 11JUN201006Z
20N
15N
10N
5N
EQ
5S
10S
15S
20S
25S
30S
35S
40S
60W 55W 50W 45W 40W 35W 30W 25W 20W 15W 10W
-6e
1e-06
8e-07
6e-07
4e-07
2e-07
1e-07
5e-08
0
-5e-08
-1e-07
-2e-07
-4e-07
-6e-07
-8e-07
-1e-06

e)

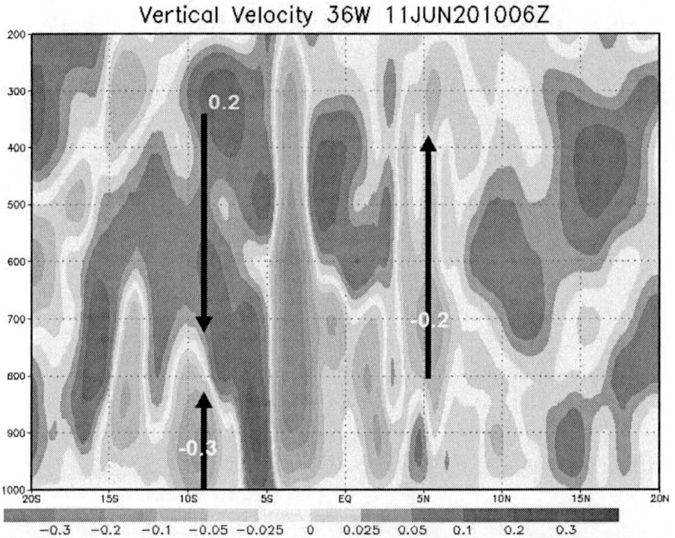

f)

g)

Figure I-8-5 – Circulation between the two hemispheres on 11 July 2010, 06 UTC. *Streamline ($m\ s^{-1}$) maps:* at 1000 hPa (a), 850 hPa (b); 500 hPa (c), 400 hPa (d).

Humidity Convergence map (g, mm s^{-1}) at 1000 hPa (e).
Vertical section of vertical velocity (k, Pa s^{-1}) along 36°W (f).
Satellite imagery in IR channel (g), GOES 13 + METEOSAT 9
L - Low/cyclone; H - High/Anticyclone; DIV - divergence, Dashed line - Trough
Source: a), b), c), d), e), f) Model ERA interim, 0.5x0.5°;
g) GOES 13+METEOSAT 9

7. References

Cangialosi, J. P. *Tropical Cyclone Report (AL102015): Tropical Storm Ida; National Hurricane Center. 7 November 2015.* Available online: https://www.nhc.noaa.gov/data/tcr/AL102015_Ida.pdf (accessed on 28 November 2022). (2015).

Cavalcante Lucas Carvalho Vieira "*Análise das ligações dos ciclones tropicais do hemisfério norte e fenômenos adversos no NEB.*" M. Sc. Teses, Universidade Federal de Alagoas, Maceió, 2018.

Cavalcante L.C.V; Fedorova N.; Levit V., *"Influência dos distúrbios tropicais na formação de fenômenos adversos de atmosfera estável no nordeste brasileiro"* Revista Brasileira de Meteorologia 35, DOI: http://dx.doi.org/10.1590/0102-77863550002, (2020)

Fedorova Natalia; Levit Vladimir. "*Fog in the tropical region. Fog formation in the tropical region of the northeast of Brazil.*" LAP LAMBERT Academic Publishing, Saarbrücken, Germany. 82 f, 2016.

Fedorova, N., Levit, V., Cavalcante, L. C. V. *"Impact of Tropical Cyclones in the Nothern Atlantic on adverse phenomena formation in northeastern Brazil.*" In: LUPO, A., R. Current Topics in Tropical Cyclone Research. IntechOpen, London, United Kingdom, (2020).

Kimberlain, T.B. *Tropical Cyclone Report: Tropical Storm Ida. Miami: National Hurricane Center.* Disponível em https://www.nhc.noaa.gov/data/tcr/ AL032013_Chantal.pdf, acessado em 02 mai. 2020, 2013.

Vasquesz, T. *Weather Forecasting Handbook*, Garland, Texas: Weather Graphic Technologies, 98p, 2000.

CHAPTER I.9

SYNOPTIC SYSTEMS ANALYSIS USING SATELLITE DATA

NATALIA FEDOROVA AND VLADIMIR LEVIT

Information on the use of the satellite images for the synoptic analyses and forecasting is described below. The main methods for analyzing of cloud types and synoptic and mesoscale systems were described in many books, for example Conway (1997), Vetlov & Veltischev (1982), Kidder et al. (1995), and Satyamurty et al. (1998).

Synoptic systems located in subtropical latitudes directly or indirectly influence the process of weather formation in the tropical region and especially in the BNE. Therefore, an interpretation of typical cloudiness from satellite data over all South America is presented below.

1. Cold air advection

Satellite imagery provide addition information for identification of cold air advection at low levels (up to 850 hPa) (Reeder and Smith, 1998; Gemiacki, 2005; Fedorova, 2008a, 2008b, 2008c; Brazil Saldanha et. al., 2010; Fedorova et al., 2016). Cold air advection is visible on IR satellite imagery as a cyclonic curvature of the cloud band. Also, *secondary cold front* or *cyclonic vortex in the cold air* (section 1.1) and *instant occlusion* (section 1.2) are associated with a cold air invasion.

A typical situation with cold air advection can be seen in figure I-9-1. The most pronounced curvature is associated with more intensive cold advection.

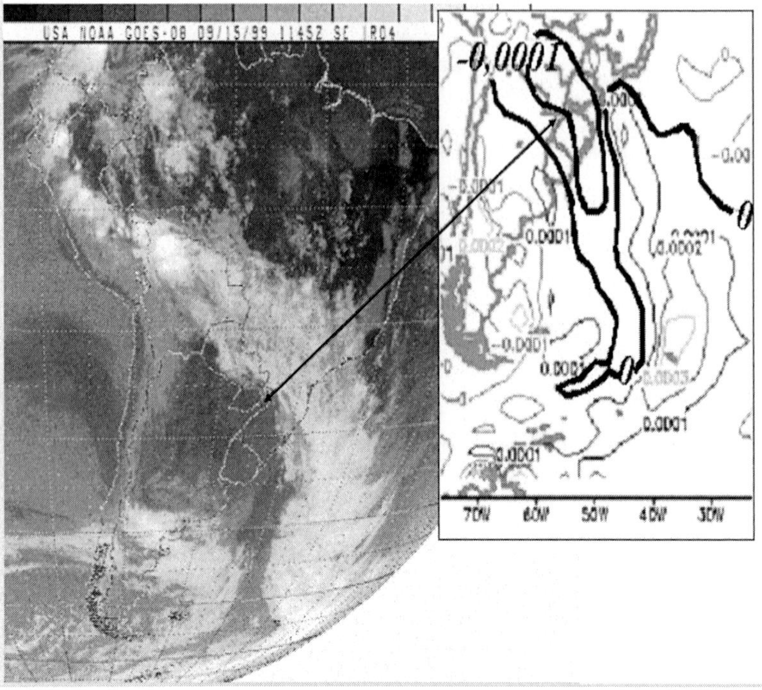

a) b)

Figure I-9-1 – Cold air advection on the infrared satellite imagery (a) and temperature advection at 850 hPa (b) on 15 September 1999, 11:45 UTC (a) and 12:00 UTC (b). Source: INPE/CPTEC, Fedorova, 2008b.

1.1 Secondary frontal zone and vortex in the cold air

On satellite images, a cold air wave is represented as a *line* or a *curved line* or *mesovortex* with intense convection. These convective systems are located behind the cold front, into the region of cold air. Therefore, a *line* or *curved line* of convection clouds is called a *secondary cold front.* When a vortex forms on a curved convective line, the term *cold air vortex* is used.

A secondary cold front formed in the cold air, which occurred, as a rule, two days after the start of the cold air intrusion. After half a day or a day, this front developed into a mesovortex, but this process did not always occur, and the intrusion of cold air was associated only with a secondary cold front.

An example of a secondary cold front in the form of a cold air vortex is shown in the satellite image in Figure I-9-2. The cloudiness periphery of the main cold front reached the south of the BNE. Behind this front, inside the region with convective cells, a mesovortex is visible on satellite imagery.

The structure of the main and secondary cold fronts is represented by maps of different variables in Figure I-9-3. The main cold front identified by the baroclinic cyclone trough (with center located 25°S, 35°W), confluence of streamlines at low levels (Figure I-9-3a), negative numbers of relative vorticity (Figure I-9-3b), intense gradients of thickness 700-925 hPa (Figure I-9-3a) and high volumes of precipitable water (Figure I-9-3c).

The continuation of the cloud band to the south of the BNE or periphery of the cold front is visible by: 1) the negative relative vorticity numbers; 2) the confluence of the streamlines 3) the temperature advection at 925 hPa and 4) the precipitable water shows the periphery of the main cold front.

A secondary cold front is associated with: 1) a weak trough visible on pressure and streamline maps at 925 hPa, 2) negative relative vorticity numbers along the secondary front, 3) cold air advection behind the front, and 4) dry air after the passage of the secondary front.

Figure I-9-2 – Secondary frontal zone (*mesovortex*) on the IR satellite image on 3 November 2003, 15 UTC
Main cold front – dark grey full line,
Periphery of the cold front – light grey full line;
Secondary cold front – dashed line
Source: Gemiacki, 2005; Fedorova 2008c.

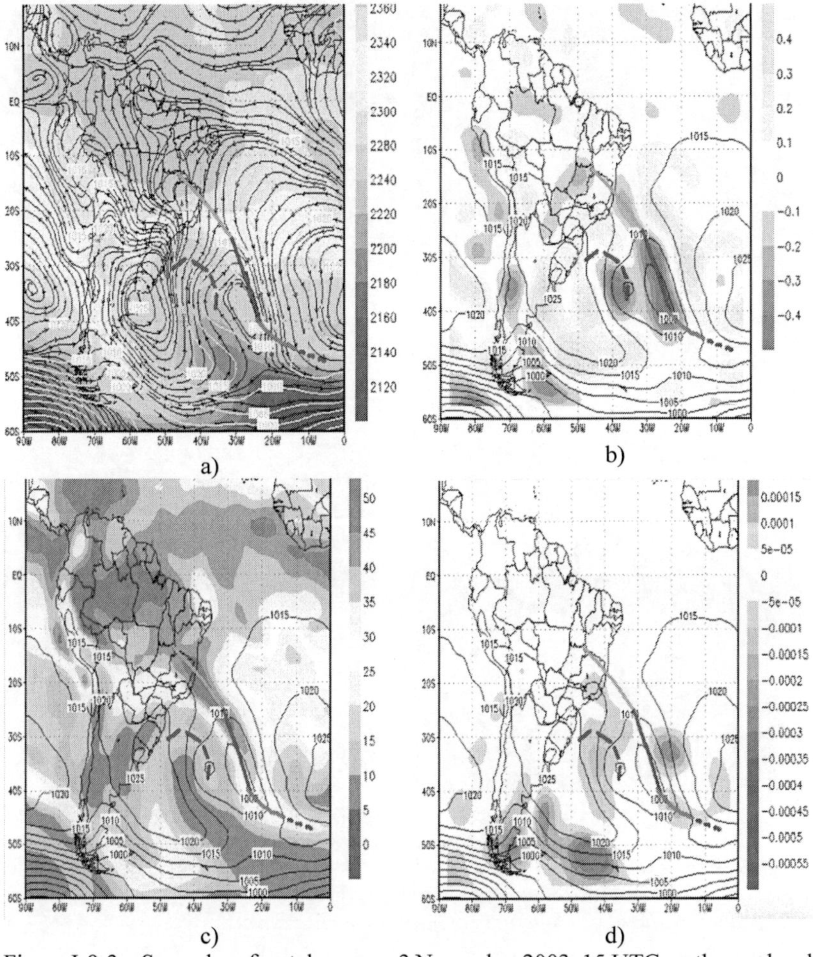

Figure I-9-3 – Secondary frontal zone on 3 November 2003, 15 UTC on the sea level pressure maps with: a) streamlines at 925 hPa (*black line with arrow*) and thickness 700-925 hPa (*grey colors*), b) relative vorticity at 925 hPa, c) precipitable water, d) temperature advection at 925 hPa
Source: Gemiacki, 2005; Fedorova 2008c.

The following example shows two cold air waves or secondary frontal zones or mesovortices in the South Atlantic on IR satellite imagery from 23 September 2022, 12 UTC (Figure I-9-4). More than two vortexes in cold air have not yet been observed on satellite images.

Figure I-9-4 – Two waves of cold air or secondary frontal zones or mesovortices on IR satellite imagery on 23 September 2022, 12 UTC
Black arrows show the cloudiness of mesovortices.
White arrows show the direction of entry of cold air waves into the mesovortex.
Source: Living Earth;

1.2 Instant occlusion

If the cold air intrusion continues, then the cold air vortex joins with the main cold front. This process is called "*instant occlusion*" because the form of cloudiness on satellite images is similar to the cloudiness of a baroclinic cyclone at the stage of maximum development (Zillman and Price, 1972; Reeder, Smith, 1998; Fedorova, 2008c). This process took place one- or two-days after the secondary front formation.

An example of "*instant occlusion*" can be seen in Figures I-9-5 and I-9-6. This process is more clearly visible on satellite imagery in the form of a cloudiness vortex. This vortex is associated with: 1) secondary nuclei of the negative relative vorticity numbers; 2) cold advection along the occlusion; and 3) dry air behind the instant occlusion.

Figure I-9-5 – Instant occlusion on IR satellite image on 4 November 2003, 18 UTC
Main cold front – dark grey full line,
Periphery of the cold front – light grey full line; cloudiness
Secondary cold front – dashed line
Source: Gemiacki, 2005, Fedorova 2008c.

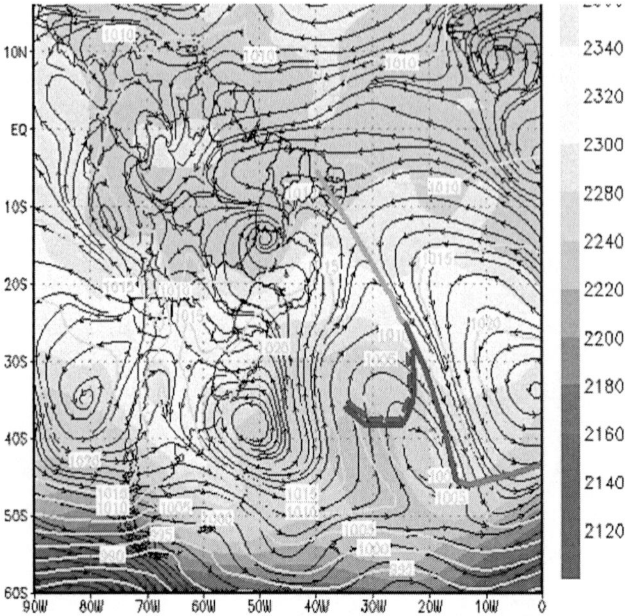
10N
EQ
10S
20S
30S
40S
50S
60S
90W 80W 70W 60W 50W 40W 30W 20W 10W 0
2340
2320
2300
2280
2240
2220
2200
2180
2160
2140
2120

a)

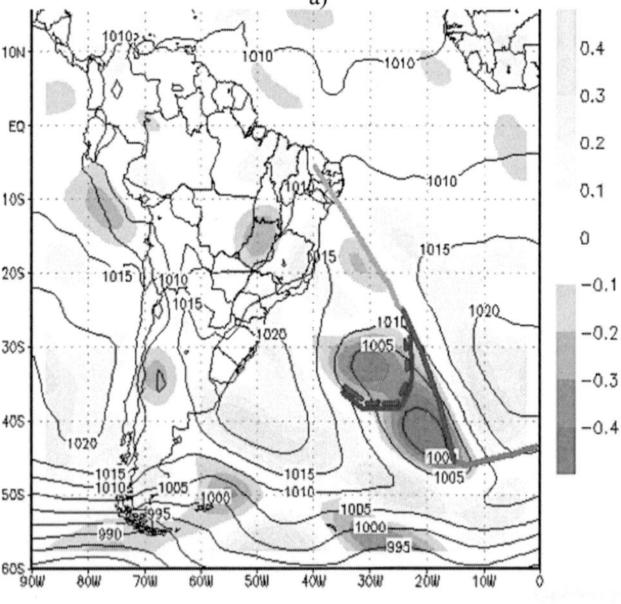
10N
EQ
10S
20S
30S
40S
50S
60S
90W 80W 70W 60W 50W 40W 30W 20W 10W 0
1010
1015
1020
1005
1000
995
990
0.4
0.3
0.2
0.1
0
−0.1
−0.2
−0.3
−0.4

b)

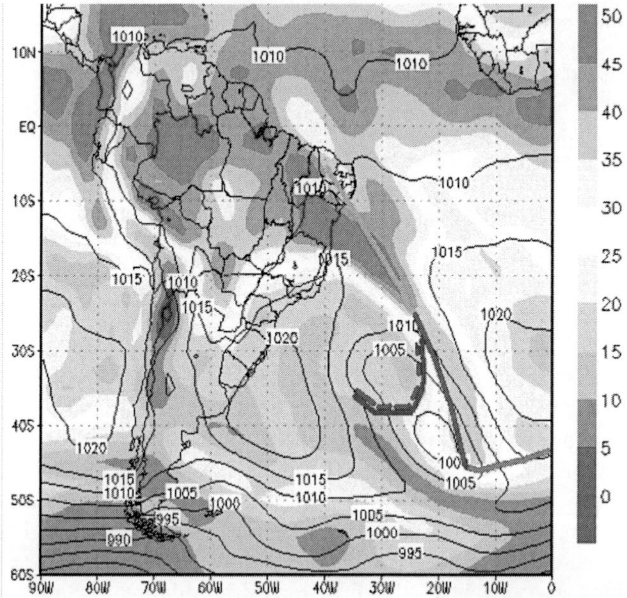

c)

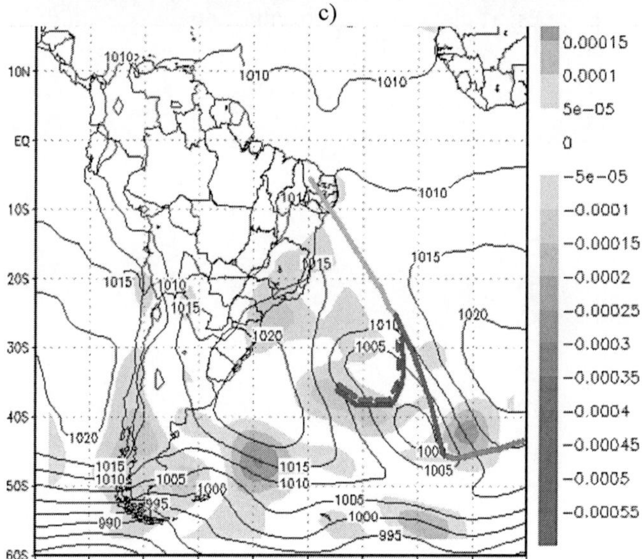

d)

Figure I-9-6 – Instant occlusion on 4 November 2003, 18 UTC on the sea level pressure maps with:

a) streamlines at 925 hPa (*black line with arrow*) and thickness 700-925 hPa (*grey colors*)
b) relative vorticity at 925 hPa
c) precipitable water
d) temperature advection at 925 hPa
Source: Gemiacki, 2005, Fedorova 2008c.

2. Warm air advection

Warm air advection is visible on satellite images as a large region with *Ci* clouds and anticyclonic curvature (Fedorova & Bakst, 1998g, 1999a; Fedorova, 2008a, 2008b, 2008c; Carvalho et al., 2011). This intensification in curvature is associated with an intensification in warm advection and vice versa, the destruction of these areas with Ci and the weaking of curvature show a decrease in warm advection.

As an example (Figure I-9-7a), IR satellite images in northern Argentina and southern Brazil show a large area with Ci clouds of anticyclonic curvature. The hot core (up to 13°C), (Figure I-9-7b) and the highest values of warm advection (up to 4°C), (Figure I-9-7c) were located north of this cloud. This curvature shows the direction of this advection. Satellite images with various types of cloudiness with anticyclonic curvature, indicating warm advection, are shown in Figure I-9-8.

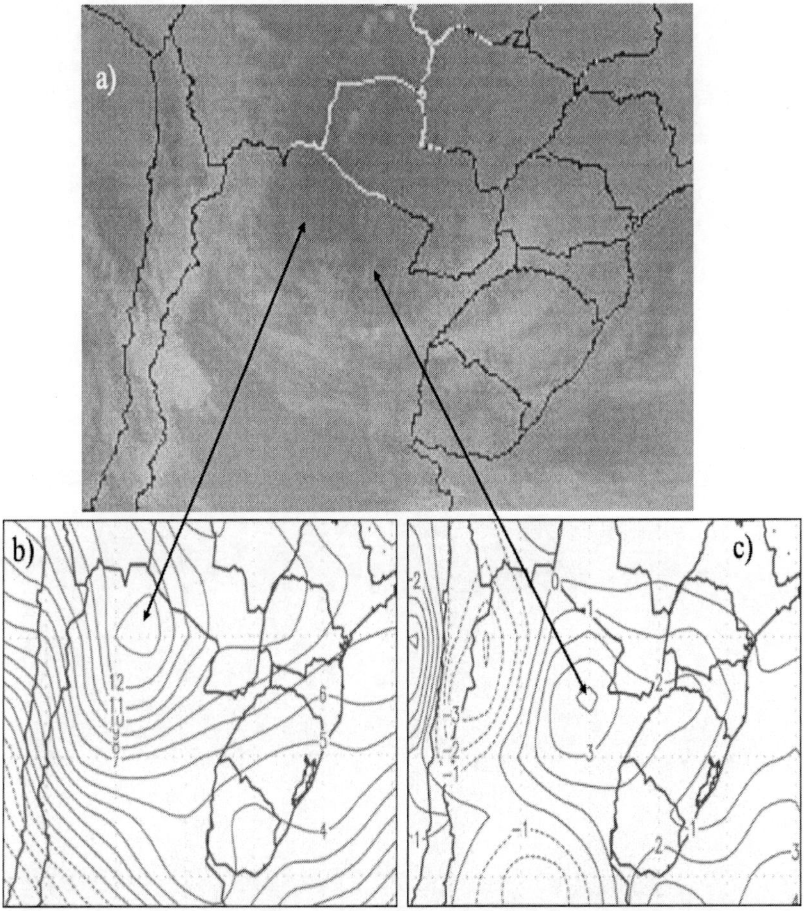

Figure I-9-7 – Warm air advection in the infrared satellite imagery (a) and in the temperature maps at 700 hPa (b) and temperature advection at 700 hPa (c) on 4 August 1996

Source: INPE/CPTEC, Fedorova, 2008b.

Black arrows show the hot nucleus (b) and the highest values of warm advection (c) on maps and satellite imagery (a).

Source: INPE/CPTEC, Fedorova et al., 1998g, 1999a.

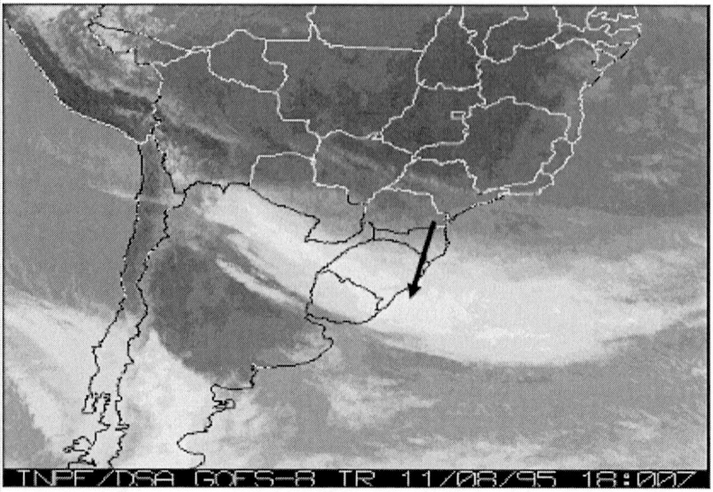

a)

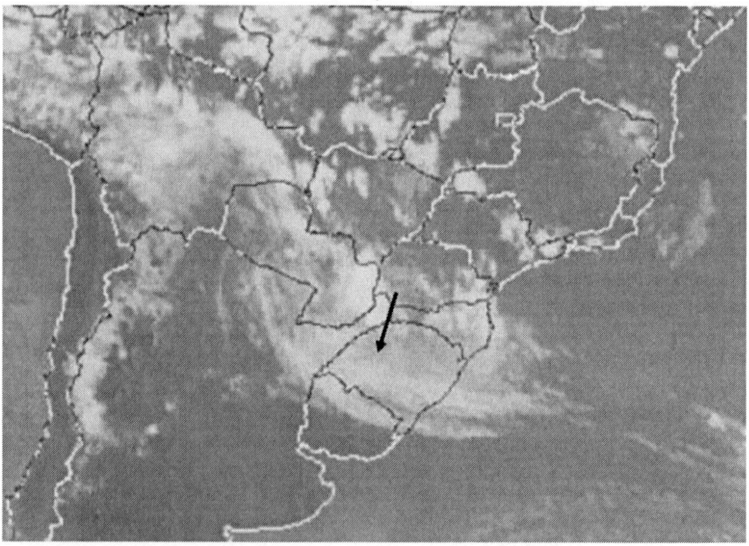

b)

Figure I-9-8 – Warm air advection in the infrared satellite imagery on:) 11 August 1995, 18 :00 UTC; b) 28 January 1996, 21:00UTC.
Black arrows show the direction of the warm advection.
Source: INPE/CPTEC, Fedorova et al., 1998g, 1999a; Fedorova, 2008a.

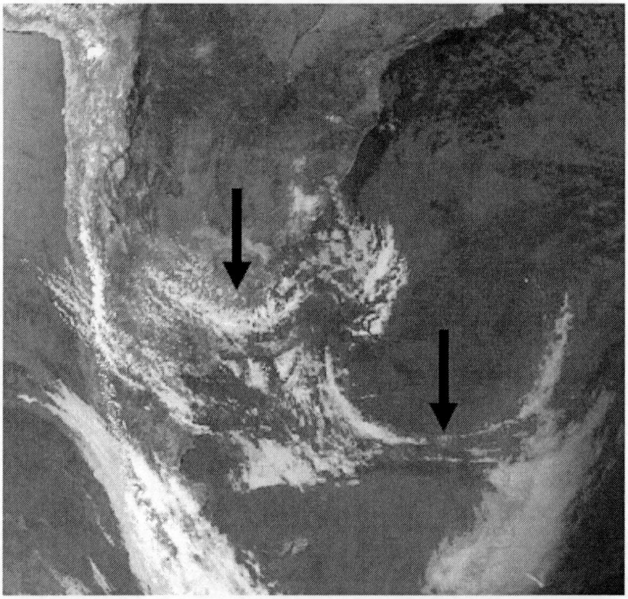

a)

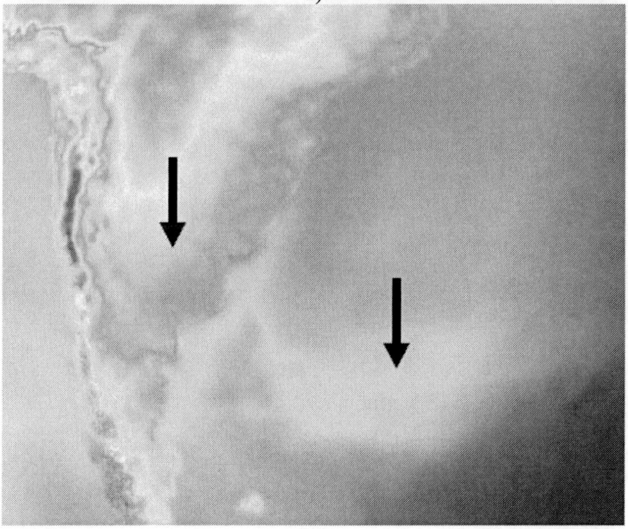

b)

Figure I-9-9 – Warm air waves on satellite image (a) and on a surface temperature map (b) on 31 August 2022.
Arrows show waves of warm air.
Source: Living Earth.

The following example shows warm air waves in an infrared satellite image and a surface temperature map on 31 August 2022 (Figure I-9-9). Anticyclonic cloudiness curvature in the south of South America and in the South Atlantic is similar to two warm waves in these regions.

3. Cyclogenesis and frontogenesis processes

The formation of baroclinic cyclones or cyclogenetic process is observed around the east coast of Southern America, more frequently between 40-20°S (Gan & Rao, 1991, 1994). The frontal zones of baroclinic cyclones, which form in the 30-20°S band, can reach the NEB (Sinhory, 2015; Fedorova et al., 2016; Chapter 7). Therefore, it is very important to identify the cyclone formation with its frontal zones and then monitor the frontal displacement.

Satellite images provide a lot of information to identify the cyclogenesis process. It is important to note that specific characteristics signs in cloudiness appear before the formation of a cyclone on the surface (before the formation the first closed isobar). This means that the forecast of the formation of a cyclone with a frontal zone can be developed with a greater antecedence.

Four types of cloudiness for a typical cyclogenetic region in southern Brazil are presented in figure I-9-10 (Fedorova et al., 1998a, 1998c, 1998d, 1998e, 1998f, 1999). The predominance of warm advection during cyclogenesis process is shown in Figures I-9-10a and I-9-10b. The cyclogenesis cloudiness at the periphery of the frontal zone is shown in Figures I-9-10c and I-9-10d. The formation of a fast cyclone around MCC cloudiness occurred in a situation where satellite images show a relationship between frontal periphery cloudiness, MCC, and large convective cloudiness in the northern part of South America (Figure I-9-10d).

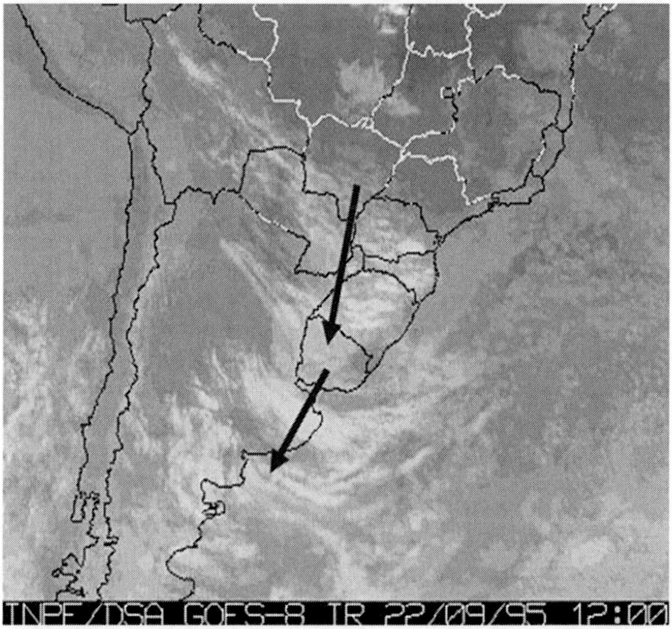

a)

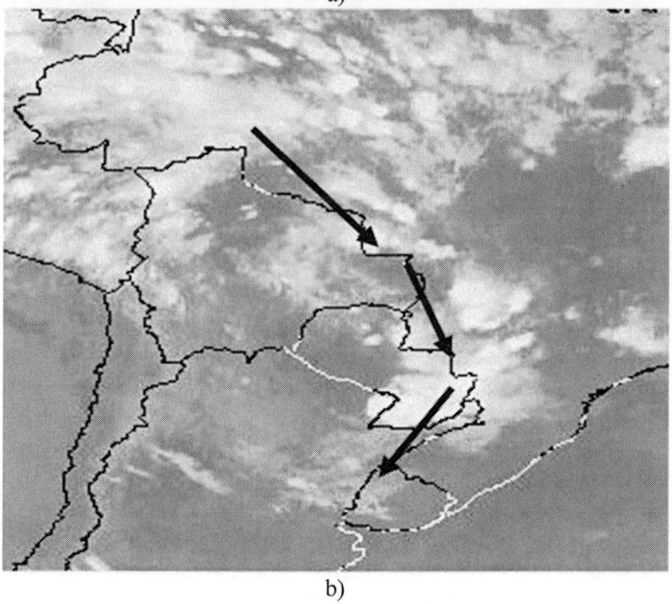

b)

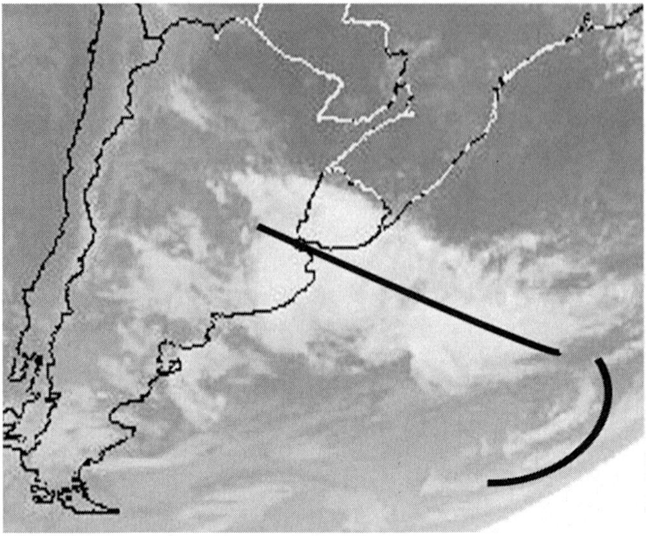

c)

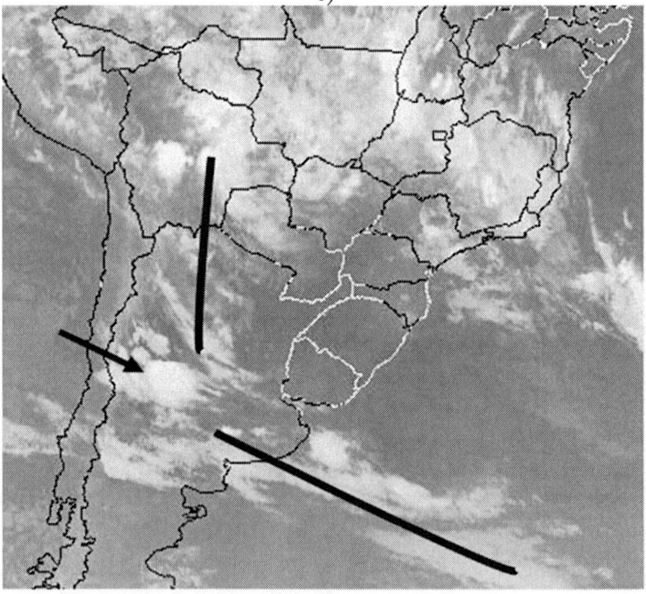

d)

Figure I-9-10 – Four types of typical cyclogenetic cloudiness:
a) and b) situations with predominant warm advection (black arrows),
c) and d) at the frontal zone periphery *black lines),*

d) CCM location (black arrows) and connection between the frontal periphery, MCC and large convective cloudiness (black lines)
Source: INPE – CPTEC, Fedorova et al., 1999; *Fedorova, 2008b*

The formation of a baroclinic cyclone with a frontal zone, the direction and speed of its displacement will determine the part of the fontal zone that reaches the BNE.

4. The direction of baroclinic cyclone vortex displacement by satellite data

The characteristics of cloudiness on satellite images to determine the direction of baroclinic cyclone displacement are presented below (Fedorova & Bakst, 1999b).

Cloudiness characteristics on satellite images showing the *main direction* of displacement of the baroclinic cyclonic vortex and frontal zones are following:

1. The more intense curvature of the rear side of the cold front cloudiness band indicates the direction of cold air intrusion (Figure I-9-11).

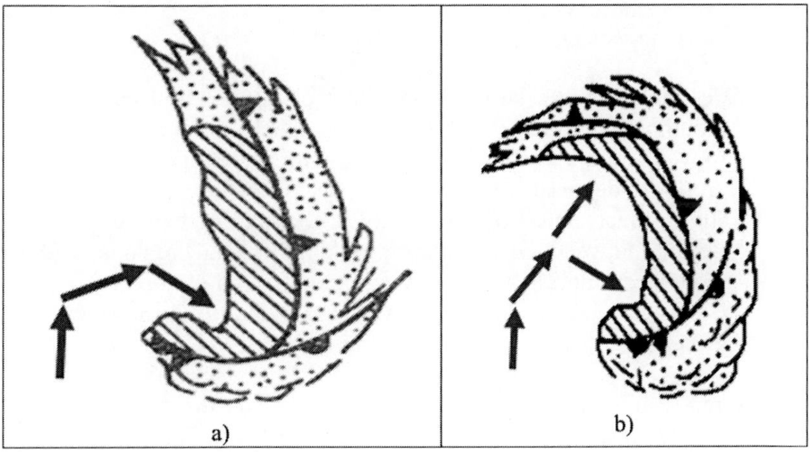

Figure I-9-11 – Regions with cold air intrusion into the rear side of the cold front cloudiness band: a) in one direction and b) in two directions.
The shaded area shows the heavy precipitation zone
Source: Fedorova, 1980; Ferorova & Bakst, 1994.

2. Large areas of Cirrus (Ci) clouds show regions with warm air advection (Figure I-9-12).

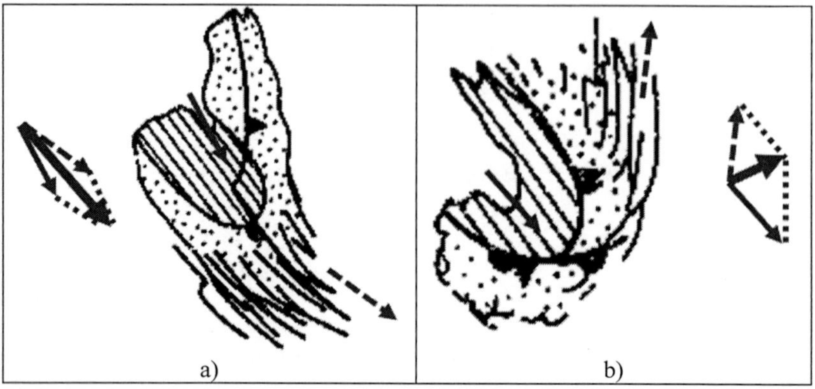

Figure I-9-12 – Vectors for the direction identification of the cyclonic vortex displacement to:
a) SW and b) NW
- Direction of a large area of Ci clouds (dashed arrow);
- Vector between a large area without cloudiness and more intense cloudiness (black narrow arrow)
- Vector sum (wide arrow)
The shaded area shows the heavy precipitation zone

3. Vector between a large area without cloudiness and more intense cloudiness (Figure I-9-12). This rule is like the synoptic rule: cyclone moves along a vector from an area with increasing pressure to an area with decreasing pressure.
4. The displacement direction vector is the sum of two vectors: 1) a vector between more intense cloudiness and regions without cloudiness and 2) a vector indicating the direction of a large area of Cirrus cloudiness. The sum of vectors (wide arrow) shows the total direction of the vortex displacement (Figure I-9-12).

Examples of using satellite images to determine the displacement of baroclinic cyclones are presented in Sections 9, 10, 11 of this Chapter.

5. Displacement velocity of baroclinic cyclones and frontal zones by satellite data

The displacement velocity of baroclinic cyclonic vortex depends on: 1) stages of cyclone (Fedorova, 2008b), 2) direction of displacement of the cyclonic vortex and 3) region of this vortex (Fedorova & Bakst, 1993, 1994, 1997b, 1999b; Bakst & Fedorova, 1994). The acceleration (increase or decrease of velocity) is calculated using actual speed of the vortex. This actual speed can be determined from the displacement of various parts of the vortex, such as: 1) the frontal and back sides of the frontal cloudiness band, 2) the center of vortex and 3) the equatorial and polar edges of the cloudiness.

Table I-9-1 shows standard values of acceleration (%) for five stages of cyclone vortex development, for various displacement trajectories and for its most important parts. Acceleration is observed in the wave and young stages of baroclinic cyclone vortex. The maximum acceleration reached 20% of the velocity in the center and at the polar edge of the vortex, which had a zonal displacement from west to east. Deceleration is observed at the stages of maximum development, occluded and old vortexes in the frontal and back edges of the cloudiness band on all displacement trajectories.

An example (Figure I-9-13) shows the automatic calculation of the velocity of the rear side of a cold front from IR satellite images (Fedorova et al., 1998b). Based on the location of one point on the rear side of the cold front on two images (circles of the starting and final points in Figure I-9-13) with an interval of 6 hours, the real velocity of this frontal side was calculated. Using this velocity and the information from the Table I-9-1, a 12-hour and 24-hour velocity forecast (V-FACT and V-PROGN) can be calculated.

This forecasted velocity allowed the forecasted location of this starting point to be calculated. The forecasted location of the starting point is represented by forecasted coordinates (longitude and latitude) for 12 h and 24 hours.

This example shows the forecast of the backside of the cold front and therefore it is possible to predict when one air mass in front of the frontal zone will be replaced by another air mass behind this front.

Table I-9-1 – Velocity acceleration (%) for the five stages of cyclone development, for the various displacement trajectories and for its more important parts.

Trajectories	Stage	Part of cyclone	Acceleration %
Component from the pole	Wave, young cyclone	Center, equatorial edge	+15
Component from the equator	Wave, young cyclone	Center, polar edge	+15
Zonal	Wave	Center, polar edge	+20
Zonal	Young cyclone	Center, equatorial edge	+10
Everyone else	Wave, Wave, young cyclone	All parts not mentioned above	-10
All	Maximal development	front edge back edge	-10 -20
All	Occluded	front edge back edge	-20 -30
All	Old	front edge back edge	-30 -20

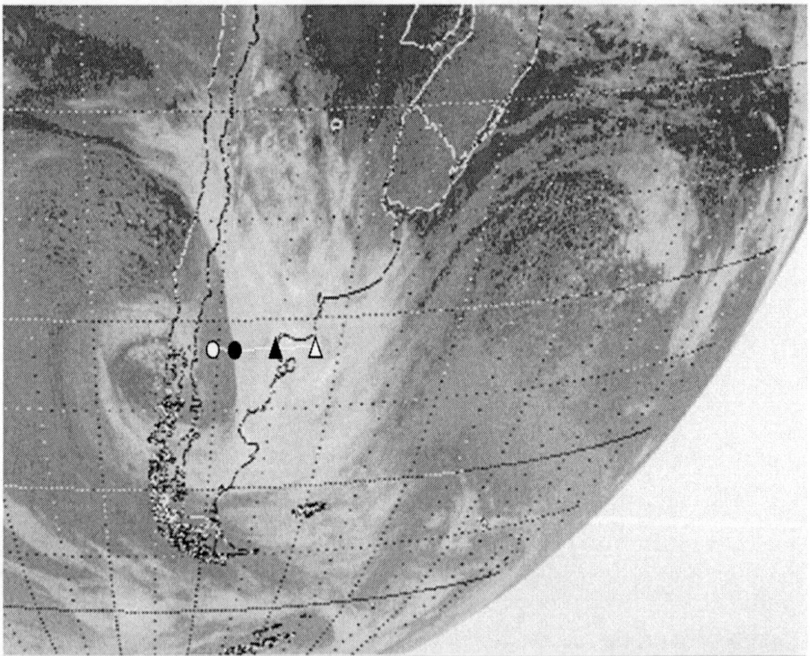

V_FACT	V-PROGN	D12	S12	V-FACT	V-PROGN	D24	S24
43.1	34.5	-65.4	-41.4	34.5	24.2	-61.9	-41.4

Figure I-9-13 – Automatic calculation of the velocity of the rear side of the cold front from IR satellite imagery.
The circles show starting (white circle) and final (black circle) location of the rear side of the cold front with an interval of 6 hours.
The triangles show the forecasted location of the rear side of the cold front. The black and white triangle show forecasts for 12 h and 24 hours, respectively.
Table: V-FACT – actual velocity calculated from satellite imagery, V-PROGN – forecasted velocity calculated from satellite imagery using the information in Table I-9-1.
D12, S12, D24, S24 show forecasted coordinates (D-longitude; S-latitude) for 12 h and 24 hours.
Source: Fedorova et al., 1998b.

6. Cloudiness evolution on satellite images

The evolution of cloudiness can be traced by comparing satellite images at different times. The example in Figure I-9-14 shows the weakening of intense cumulonimbus convection in tropical South America. The dissipation of Mesoscale Convective Cloudiness (MCC) is observed in the extratropical region.

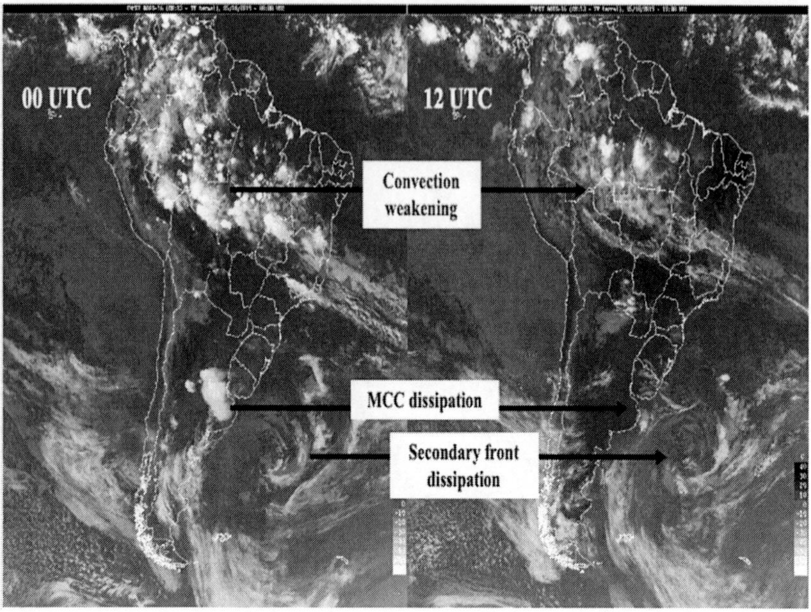

a) b)

Figure I-9-14 – Cloudiness evolution over the South America based on satellite imagery on 25/10/2019, 00h (a) and 12 UTC(b).
Line with triangles shows the cold front.
Source: INMET, GOES-16, Ch:13 – IV.

The same figure shows a cold front in the south of the BNE at 00 UTC and a decrease in cloudiness at 12 UTC. Also in the South Atlantic region, the secondary cold front dissipates due to its coupling with the cloudiness remnants of the main cold front. This cloudiness modification in time indicates the processes of frontogenesis or frontolysis.

The displacement, intensification and weakening of frontal cloudiness bands occurred simultaneously, and satellite IR images present the result of all these processes. The method and technique for automatic identification of the cloudiness evolution, intensification or weakening, from satellite images are described in Fedorova et al., 1998b, 2001. The method of automatic determination of the frontal zone from muitispectral satellite data is based on the identification of the area between the intensification and weakening in the cloudiness of a defined frontal zone located in this region

(Fedorova & Bakst, 1997a). Figure I-9-15 shows an example of the evolution of two cloud groups: 1) Cb and Ns (*Cumulonimbus* and *Nimbostratus*) and 2) other clouds.

This information will make it possible to predict the type of cloudiness and its evolution. The result of such an analysis is the determination of the presence or absence of the frontal band of a baroclinic cyclone, which can affect the weather in the BNE. Consequently, it will be possible to make a conclusion about the characteristic vertical movements and humidity in the BNE and, as a result, about the precipitation formation.

Figure I-9-15 – Automatic identification of the cloudiness evolution using satellite imagery.

Development and intensification are shown by white arrows and weakening by black arrow.

Evolution of Cb and Ns clouds are shown by numbers 2 and 3; evolution of other clouds by numbers 1 and 4.

Source: Fedorova et al., 2001.

7. Location of heavy precipitation in the cyclonic vortex by satellite data

The basic idea of using satellite cloud images for precipitation forecasting was described in (Popova & Fedorova,1975). As a result of many studies, areas of synoptic scale systems on satellite images associated with heavy precipitation were identified (Fedorova, 1976, 1978, 1980, 1981, 1982 a,b,c, 1983, 1985, 1986, 1988a, 1988b). Cyclone vortex areas and frontal zone that may affect the forecast area in the BNE are described in Chapter I-7. Usually, the periphery of cold fronts reaches the BNE.

The main signs of vortex cloudiness on satellite images with a precipitation zone are shown in Figure I-9-16.

The area of heavy precipitation is located in the following parts of the cyclonic baroclinic vortex:

- an area with a more intense curvature at the following stages of the vortex: wave, young and maximum development (Figure I-9-16 a, b and c).
- an area along a cloudiness band with intense intrusion of cold air into the rear side of a cold front in one or two directions in the stages of young and maximum vortex development (Figure I-9-11a and b).
- all areas of the cloudiness spiral at the final stage of vortex development (Figure I-9-16d and e).

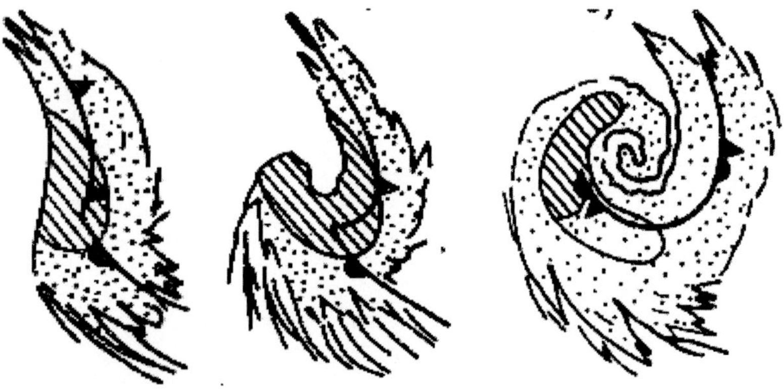

a) b) c)

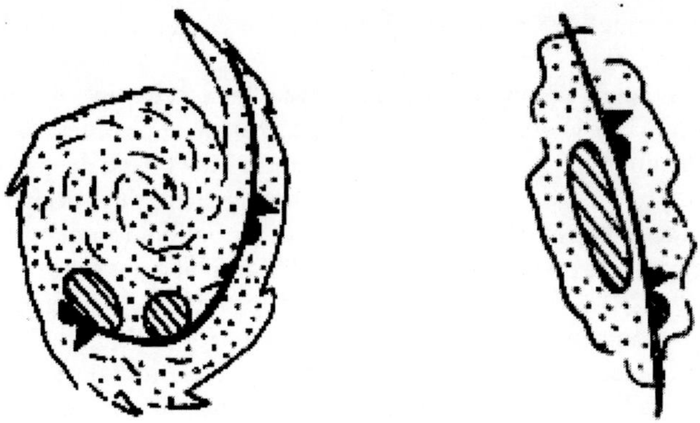

d) e)

Figure I-9-16 – Scheme of the typical location of heavy precipitation (shaded areas) at different stages of the cyclonic vortex.
Source: Fedorova, 1980, 1982 a,b,c.

The result of such a cloudiness analysis is the determination of the presence or absence of the frontal band of a baroclinic cyclone, which can affect the BNE weather. This information will allow the forecaster to infer typical large-scale vertical motions and humidity, and consequently precipitation formation.

8. Summary, practical recommendations and examples

Weather in the BNE is influenced by various systems of extratropical regions. Therefore, the analysis of these systems evolution in the extratropical region is very important for weather forecasting in the tropical region.

First of all, this influence occurs trough the periphery of the frontal zones of extratropical cyclones. Among them, baroclinic cyclones with frontal zones have the greatest influence on the precipitation formation (Chapter I-7). In order to find out whether this effect will occur or not, it is necessary to determine the stage of the cyclone, the direction and velocity of its movement. Three examples showing this effect are presented below.

9. Example 27-28 October 2022

The following example shows the influence of the frontal zone of a baroclinic cyclone in the south of the South Atlantic on the weather in the

BNE, using IR satellite images on 27-28 October 2022 (Figures I-9-17, I-9-18, I-9-19, I-9-20).

Satellite images, commonly used for weather forecasting in the BNE (Figure I-9-17) show one fragment of the frontal zone over the North Atlantic, reaching the central region of the BNE. This image does not show the structure of the baroclinic cyclone that created this frontal zone. Therefore, the process of this frontal zone formation is not presented. Consequently, the use of images to predict this frontal zone is not possible.

Figure (I-9-18) shows the large area and location of the center of the baroclinic cyclone in the extreme south of the South Atlantic, near the Antarctic continent. The cloudiness band of the frontal zone has a meridional direction and reaches the BNE. The cold air entrance also has a meridional direction along the back side of the cloudiness band.

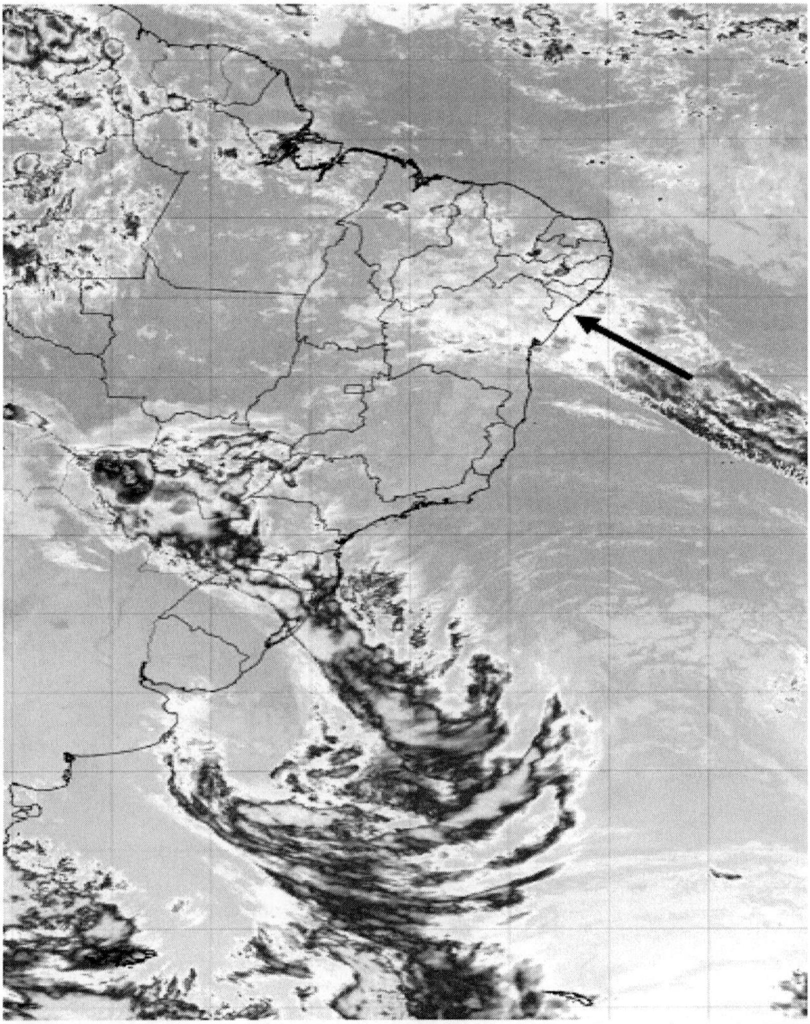

Figure I-9-17 – Influence of the frontal zone (*black arrow*) on the weather in the BNE on IR (Channel 13) satellite imagery on 27 October 2022, 6 UTC
Source: CPTEC/INPE, GOES -16.

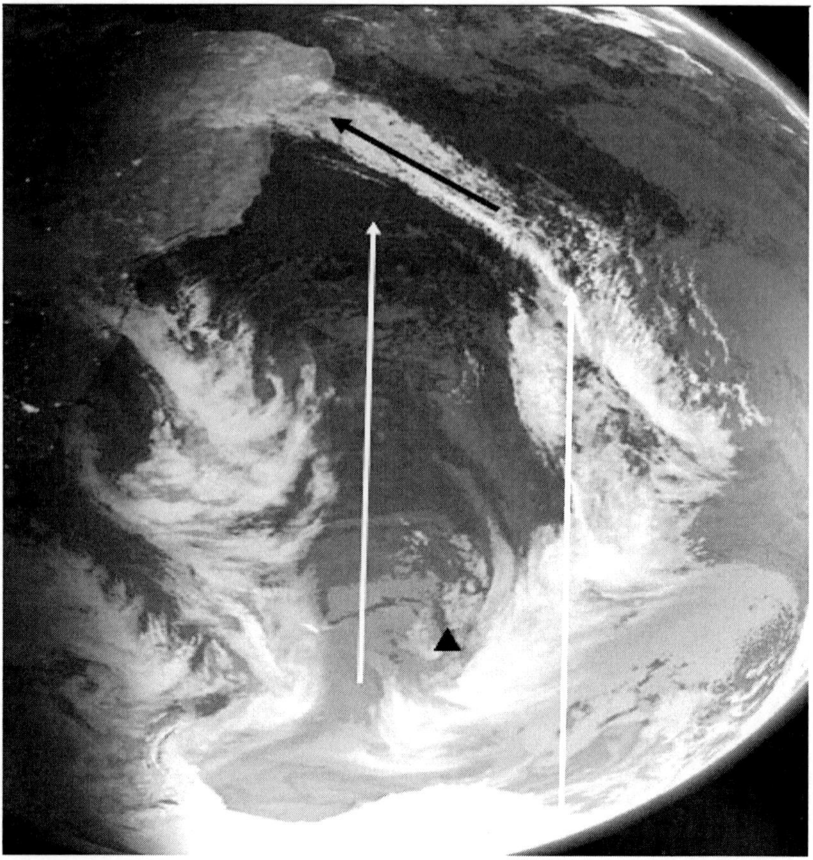

Figure I-9-18– Influence of the frontal zone of baroclinic cyclone in the south of the South Atlantic on the weather in the BNE on satellite imagery on 27 October 2022, 6 UTC

White arrow shows the main direction of the cold air entrance.
Black arrow shows the extremity of the frontal zone in the BNE.
White dot shows the center of the BNE coastal region.
Black triangle shows the center of the baroclinic cyclone in the south of the South Atlantic.
Source: Living Earth.

After 6h (Figure I-9-19), the curvature of the back side of the main frontal zone increases, and small white arrows show the direction of the cold air entry. The next day (Figure I-9-20) this curvature decreases, and the meridional current weakens. A decrease in the supply of cold air to the rear of the frontal zone weakens the temperature gradient in the frontal zone, and therefore leads to frontolysis. Comparing the satellite images in Figures I-9-18, I-9-19 and I-9-20 one can be see a degrade of the cloudiness band near the BNE (around black arrows). This confirms the frontolyzes and, consequently, the weaking of cloudiness and the absences of precipitation. Thus, to precipitation forecasting in the BNE, it is necessary to analyze the sings of cloudiness on satellite images over the central and southern Atlantic.

Figure I-9-19 – Influence of the frontal zone of baroclinic cyclone in the south of the South Atlantic on the weather in the BNE on satellite imagery on 27 October 2022, 12 UTC
White arrows show the main directions of the cold air entry.
Black arrow shows the extremity of the frontal zone at the BNE.
White dot shows the center of the BNE coastal region.
Source: Living Earth.

Figure I-9-20 – Influence of the frontal zone of baroclinic cyclone in the south of the South Atlantic on the weather in the BNE on satellite imagery on 28 October 2022, 12 UTC
White arrows show the main directions of the cold air entry.
Black arrow shows the extremity of the frontal zone.
White dot shows the center of the BNE coastal region.
Source: Living Earth.

10. Example 4-7 November 2022

Heavy precipitation and thunderstorms were observed in the central area of the BNE on 4 and 5 November 2022. The synoptic systems associated with these adverse phenomena are described below (Figures I-9-21– I-9-27).

The frontal zone crosses the South Atlantic from southern Africa to the BNE on 4 November 2022 (Figure I-9-21a). On the frontal periphery over the BNE (inside the circle in the same figure), a large area with intense convection was observed. Temperature anomaly maps confirmed the frontal location across the Atlantic Ocean between the positive anomaly on the frontal side and the negative anomaly on the rear side on 4 and 5 November 2022 (Figure I-9-23 and I-9-24).

a) 4 November 15 UTC

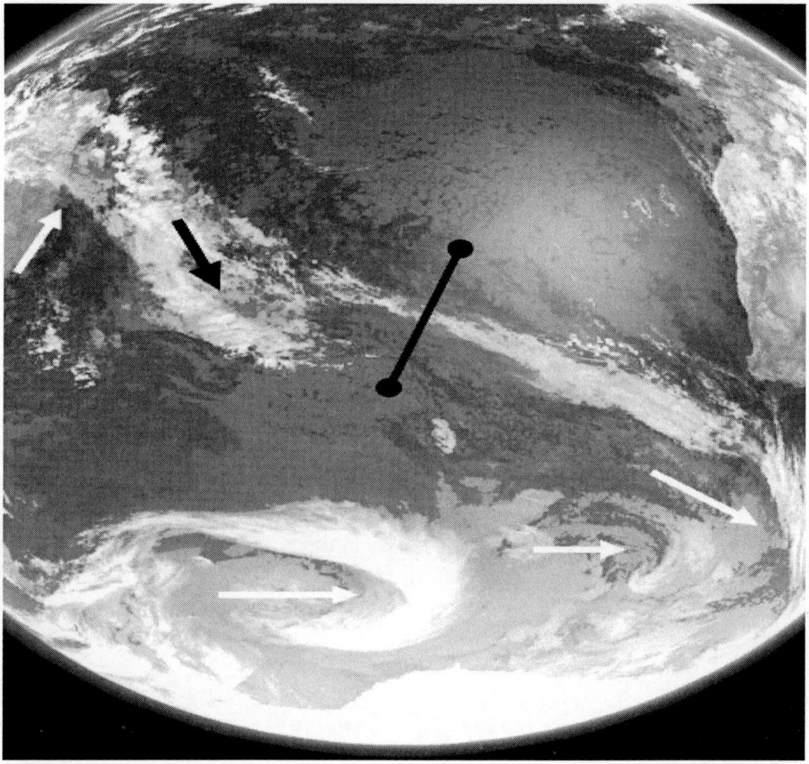

b) 5 November 8 UTC

Figure I-9-21 – Cyclogenesis near the BNE and influence of the frontal zone of baroclinic cyclone in the south of the South Atlantic on the weather in the BNE on the satellite imagery on 4 and 5 November 2022.
White arrows show the main directions of the cold air entry.
Black arrow shows the main directions of the warm air entry.
Black circle in the (a) shows intense convection over the BNE.
Black line with two points shows the region of separation of the cloudiness of the new cyclone from the principal frontal zone.
White dot shows the center of the BNE coastal region.
Source: Living Earth.

In the aforementioned frontal zone, a cyclogenesis process was observed and a new baroclinic cyclone formed near the BNE with a minimum pressure at the center of 1007 hPa, as shown the sea level pressure map on 5 November 2022, 12 UTC (Figure I-9-26). One trough extended to the BNE, and the other to the SSW over the Atlantic. The thickness 500-1000 hPa map also shows a warm wave on the frontal side and a cold wave on the rear side of this young cyclone (Figure I-9-25).

This cyclogenesis process can be identified from satellite imagery. A comparation of the cloudiness band between 4 and 5 November 2022, 6 UTC (Figure I-9-21) shows a significant deformation of cloudiness band. The anticyclonic and cyclonic curvature of the back side of the band indicate warm and cold advection (*black and white arrows, respectively*). A similar curvature can be seen between the positive and negative temperature anomaly (Figure I-9-23b), which confirms this advection. Also, these air currents at high levels (250 hPa, Figure I-9-26) show the subtropical jet stream displacement to the NW during the period from 4 to 5 November 2022 (when comparing Figures I-9-26a and b). It is very important to note that cyclogenesis occurred under the area of change in the curvature of this current, while the cyclone center on the surface was formed between the ridge (anticyclonic curvature) and trough (cyclonic curvature) at a high level. This can be seen on high-level maps with the location of the cyclone center on the surface on 5 November 2022 (Figure I-9-26b). *L* shows the center of a surface cyclone located between the ridge and trough at 250 hPa, so that the ridge is in the forefront of the cyclone and the trough is behind.

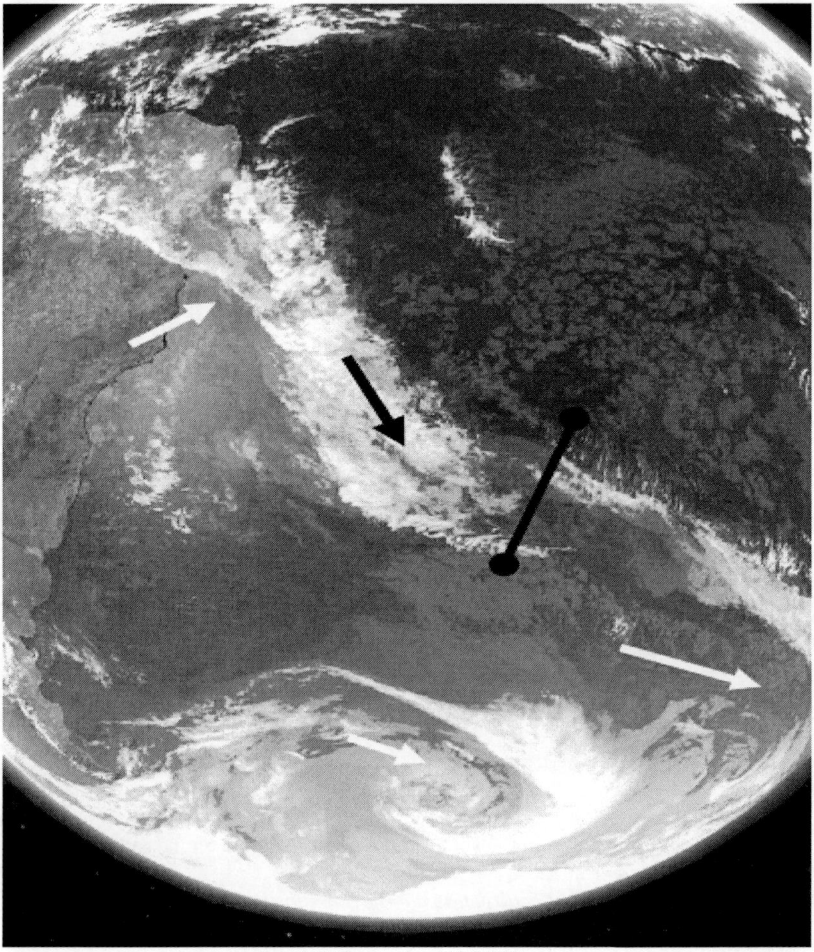

a) 5 November 13 UTC

b) 6 November, 13 UTC

Figure I-9-22 – Weaking of cyclone near the BNE and separation from the frontal zone of baroclinic cyclone in the south of the South Atlantic on the satellite imagery on 5 and 6 November 2022.
White arrows show the main directions of the cold air entry.
Black arrow shows the extremity of the frontal zone.
Black circle in the (a) shows intense convection over the BNE.
Black line with two points shows the region of separation of the cloudiness of the new cyclone from the principal frontal zone.
White dot shows the center of the BNE coastal region.
Source: Living Earth.

The weakening of the cyclone near the BNE and its separation from the frontal zone of the baroclinic cyclone in the south of the South Atlantic can be identified from satellite images on 5 and 6 November 2022. The cyclone weakening can be identified by the decrease in the curvature of the backside of the clouds band in the period from 5 to 6 November 2022 (Figure I-9-

22). A similar process is shown by temperature anomalies maps for this period of time (Figure I-9-24).

The separation of these systems indicates a decrease in the cloudiness band, and it is visible near the *black line with two dots* on satellite images (Figures I-9-21b, I-9-22a). This separation was facilitated by the zonal displacement of cyclones in the South Atlantic Ocean. On satellite images on 5-7 November (Figures I-9-21b and I-9-22a and b) the curvature of the back side of the vortex cloudiness is clearly visible. This indicates an eastward displacement of the baroclinic cyclones. Therefore, the frontal zone is stretching and torn. As a result, the cloudiness near the BNE weakened and adverse phenomena in the form of heavy rains and thunderstorms ceased.

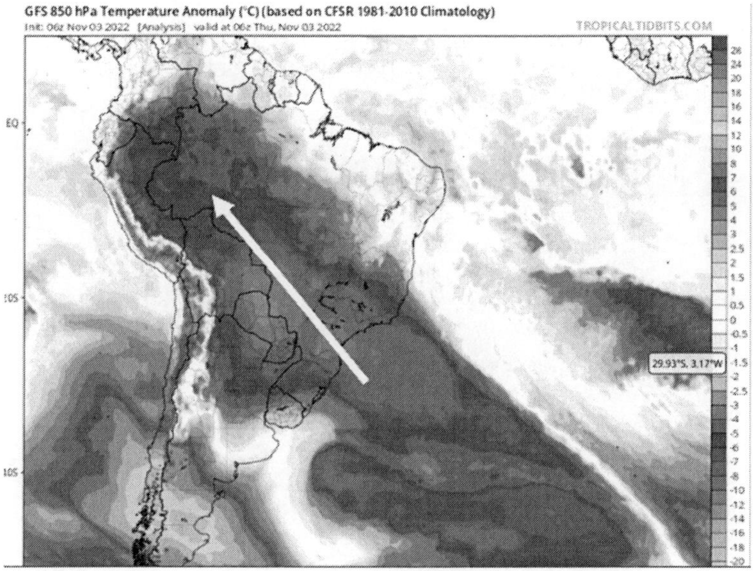

a) 3 November 6 UTC

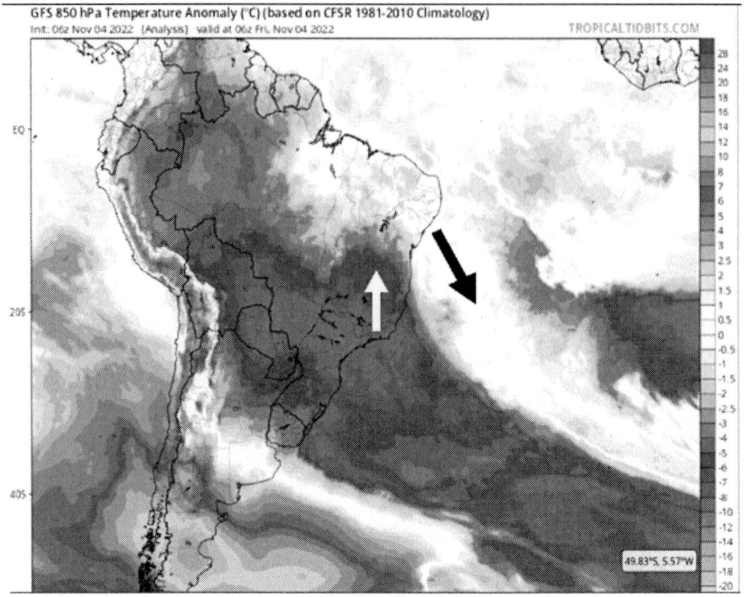

b) 4 November 6 UTC

Figure I-9-23 – Maps of temperature anomalies (°C) based on CFSR 1981-2010 climatology on 3 (a) and 4 (b) November 2022.
Positive temperature anomaly – white area with black arrow.
Negative temperature anomaly – black area with white arrow.
Source: Tropical Tidbits.

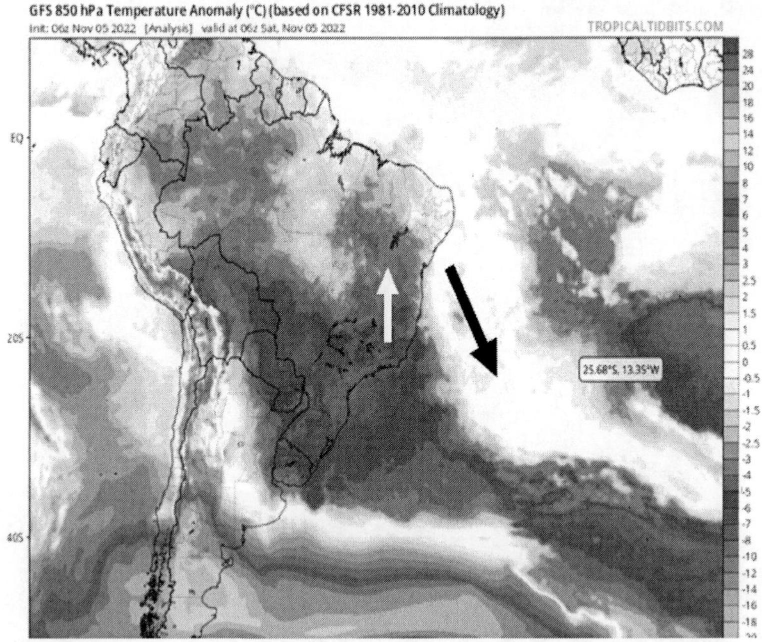

a) 5 November 6 UTC

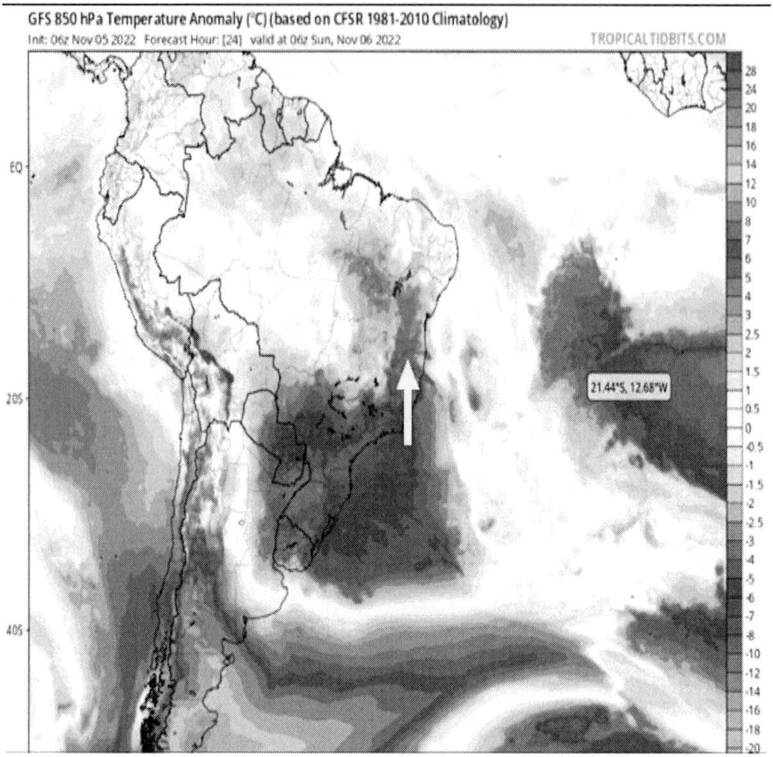

b) 6 November 6 UTC

Figure I-9-24 – Maps of temperature anomalies- (°C) based on CFSR 1981-2010 climatology on 5 (a) and 6 (b) November 2022.
Source: Tropical Tidbits.

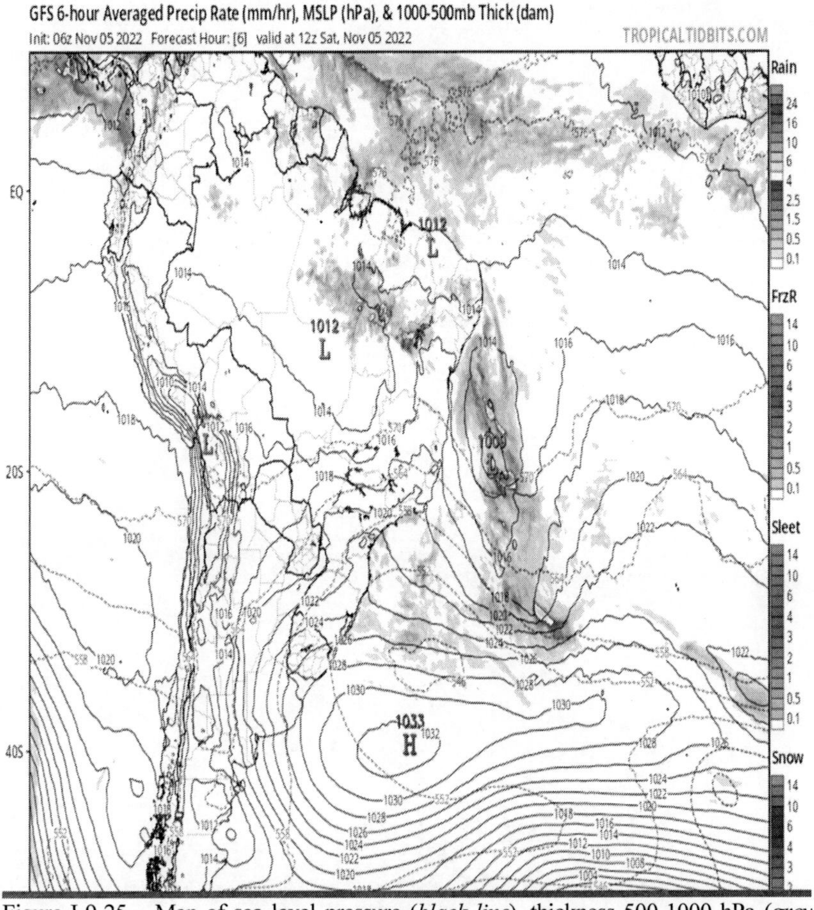

Figure I-9-25 – Map of sea level pressure (*black line*), thickness 500-1000 hPa (*grey dotted line*) and average precipitation (*grey colors, mm/h*) on 5 November 2022, 12 UTC Source: Tropical Tidbits.

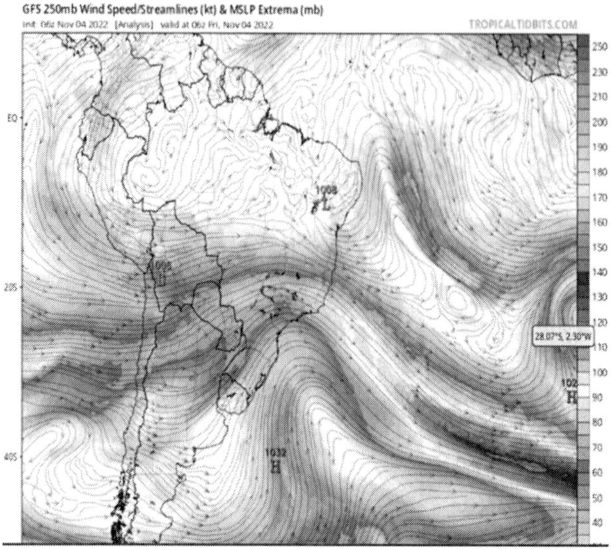

a)

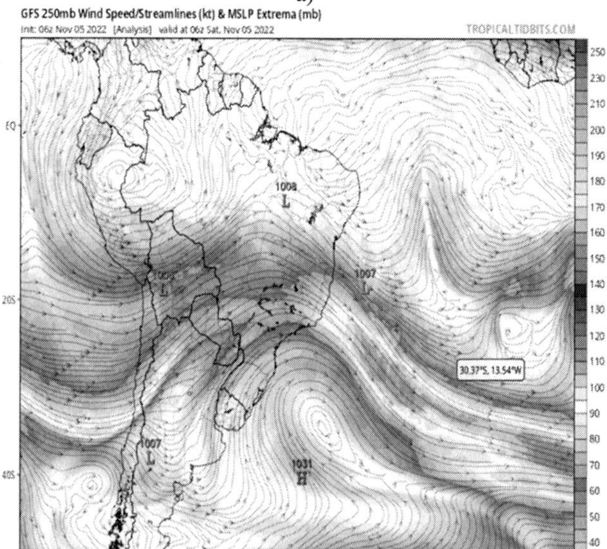

b)

Figure I-9-26 – Maps of streamlines at 250 hPa (*black line with arrow*) and symbols (*L*) of cyclone center at the sea level pressure with number of minimum pressures on 4 (a) and 5 (b) November 2022, 06 UTC
Source: Tropical Tidbits.

11. Example 15-16 November 2022

Frontal zones often passed from the south to 20ºS, as described in Chapter I-7. Therefore, to weather forecasting in the BNE, it is necessary to decide if this front will pass over the BNE or not. Satellite imagery can help make decision. The two examples given above show the structure of baroclinic cyclones in extratropical regions, the frontal zones of which reached the BNE. The example below presents the cloudiness structure on satellite imagery when the frontal zone is not displaced to the BNE.

Satellite images on 15 November 2022 (Figure I-9-27a) show a cloudiness vortex of a baroclinic cyclone centered in the south of the South Atlantic. The black arrow shows the main direction of the cold air inflow into this cyclone to the south. This signal indicates that the cloud vortex is moving south. The next day, 16 November 2022 (Figure I-9-27b), this direction changes little to the southwest. The southward vortex displacement was also confirmed.

White arrows in the same figure show the direction of cold air inflow to the rear side of the frontal zone. It is important to note, that the back border of this cloudiness band is not sharp compared to the similar region in another example on 5 November 13 UTC. The next day (Figure I-9-27b), the entire frontal band was straight. The absence of curvature indicates the absence of cold air advection. This is very important sign that the cold front will not passe in a northeasterly direction and this frontal zone will not affect the weather formation in the BNE.

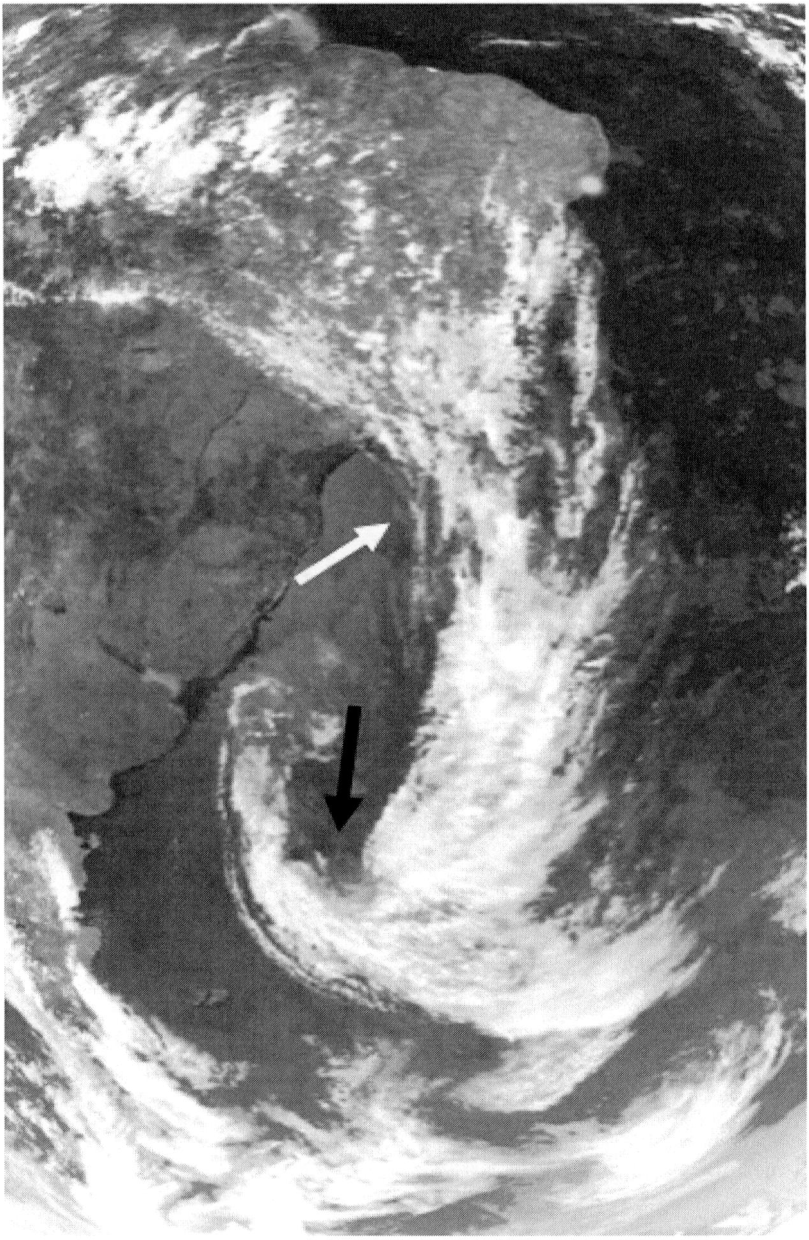

a)

b)

Figure I-9-27 – Forecasting of the passage or not of the frontal zone over the BNE from satellite images on 15 and 16 November 2022.

Black arrows show the main directions of the cold air entry into the baroclinic cyclone in the south of the South Atlantic.

White arrows show the direction of cold air entry beyond the frontal zone (a) and the absence of curvature and, as a result, cold air advection of the back side of the cloudiness band (b).

Source: Living Earth.

12. References

Brazil Saldanha C., Fedorova N., Carvalho M.H. *"Estudo de um caso de frente fria secundária sobre o sul do Brasil",* Revista Brasileira de Meteorologia 25-1, 32 – 45, (2010).

Bakst, L.; Fedorova, N., *"Forecast of displacement velocity of the basic sites of large - scale cyclonic cloud systems and frontal sections from satellite data."* Earth exploration from space 5, 3-8; Moscow, ISSN: 0205-9614, (1994).

Carvalho M.H., Fedorova N. *"Study about warm fronts that occur in southern Brazil."* Revista Brasileira de Meteorologia 26- 2, 257–272, (2011).

Conway Eric D. *An introduction to Satellite Image Interpretation.* London: The Johns Hopkins Press, 242p, 1997.

Fedorova, N, Levit V., Cruz, C.D. *"On frontal zone analysis in the tropical region of the Northeast Brazil"* Pure and Applied Geophysics 173,1403-1421, (2016), doi 10.1007/s00024-015-1166-y.

Fedorova Natalia. *Sinótica II: Conceitos básicos de cinemática e de dinâmica da atmosfera e análise dos campos meteorológicos.* Maceió: Ed. Universitária/ UFAL, 180 p, ISBN: 978-85-7177-420-9, 2008a.

Fedorova Natalia. *Sinótica III: Frentes, correntes de jato, ciclones, anticiclones.* Maceió: Ed. Universitária/ UFAL, 192 p, ISBN: 978-85-7177-421-6, 2008b.

Fedorova Natalia. *Sinótica IV: Sistemas e processos sinóticos atuantes na América do Sul.* Maceió: Ed. Universitária/ UFAL, 180 p, ISBN: 978-85-7177-422-3, 2008c.

Fedorova, N., Fedorov, D., N. J. Ferreira, Carvalho, M. H.*"Mapa da evolução de nebulosidade."* In: Congresso Latinoamericano e Iberico de meteorologia - CLIMET IX, Congresso Argentino de Meteorologia VIII- CONGREMET VIII, Argentina, Buenos Aires, nº 108 (2001)

Fedorova, N., Carvalho, M.H., Fedorov D., *"Formação de vórtices observados no campo de nebulosidade sobre a América do Sul"* Revista Brasileira de Meteorologia 14- 2, 15-27, (1999).

Fedorova, N.; Pinheiro J., Fedorov, D. *"Cyclogenetic cloudiness over South America"*, P1.14B, In: 9 Conference on Satellite Meteorology and Oceanography, Paris, (1998a).

Fedorova, N.; Carvalho, M. H.; Fedorov, D. *"Desenvolvimento das Nuvens e dos Sistemas de Nebulosidade Através dos Dados de Satélite Nos Canais Infravermelho e Vapor D'Água."* In: Congresso Brasileiro de Meteorologia X, CBMET- X, VIII Congresso da FLISMET, Brasilia-DF, (1998b).

Fedorova, N.; Carvalho, M. H.; Fedorov, D. *"Formação de vórtices no campo de nebulosidade sobre a América do Sul. Parte I. Nebulosidade ciclogenética através dos dados de satélite. campos de temperatura na região do vórtice ciclônico."* In: Congresso Brasileiro de Meteorologia X, CBMET- X, VIII Congresso da FLISMET, Brasilia-DF, (1998c)

Fedorova, N.; Carvalho, M. H.; Fedorov, D. *"Formação de vórtices no campo de nebulosidade sobre a América do Sul. Parte II: A corrente de jato, as correntes de ar e a formação do vórtice ciclônico."* In: Congresso Brasileiro de Meteorologia X, CBMET- X, VIII Congresso da FLISMET, Brasilia-DF, (1998d).

Fedorova, N.; Carvalho, M. H.; Fedorov, D. *"Formação de vórtices no campo de nebulosidade sobre a América do Sul. Parte III: Campos de vorticidade relativa e divergência na região do vórtice ciclônico."* In: Congresso Brasileiro de Meteorologia X, CBMET- X, VIII Congresso da FLISMET, Brasilia-DF, (1998e).

Fedorova, N.; Carvalho, M. H.; Fedorov, D. *"Formação de vórtices no campo de nebulosidade sobre a América do Sul. Parte IV. O ciclone em médios e altos níveis."*. In: Congresso Brasileiro de Meteorologia X, CBMET- X, VIII Congresso da FLISMET, Brasilia-DF, (1998f).

Fedorova, N.; Bakst L. *"Identification of warm advection on satellite data."* In: 9 Conference on Satellite Meteorology and Oceanography, Paris. P2.60B, (1998g).

Fedorova, N.; Bakst L. *"Identificação da advecção de ar quente através da análise de dados de satélites."* Revista Brasileira de Geofísica 17-1, 67-78, (1999a).

Fedorova, N.; Bakst L. *"Evolução e velocidade da nebulosidade dos ciclones através de dados de satélites meteorológicos."* Revista Brasileira de Meteorologia 14-1, 67-85, (1999b).

Fedorova, N.; Bakst L. *"The results of automatic frontal section identification from the muitispectral Satellite Data."* In: The International Association of Meteorology and Atmospheric Science & International Association for Physical Science of the Oceans. IAMAS-IAPSO. N° IM2GGG, ISBN 0646 32298 2. Australia, Melbourne, (1997a).

Fedorova, N.; Bakst L. *"Evolution And Displacement Velocity of the Lage-Scale Cyclonic Cloud Systems from Satellite Data",* In: The International Association of Meteorology and Atmospheric Science & International Association for Physical Science of the Oceans. IAMAS-IAPSO. N° IM2HHH, ISBN 0646 32298 2, Australia, Melbourne, (1997b).

Fedorova, N.; Bakst L. *"The testing of the method of the large-scale cloud systems forecasting velocity from satellite data."* In: Informational Collection No 22, 42-46, Meteorological Service of Russia, Moscow, (1994).

Fedorova Natalia, Bakst Leonid. *"Forecast of the speed of the main sections of large-scale cloud systems based on satellite data."* Methodological Letter, Federal Service of Russia for Hydrometeorology and Environmental Monitoring, Hydrometeorological Center of the Russian Federation, Moscow, 3-20, 1993.

Fedorova Natalia. *"Forecast of the precipitation and its probabilities with 12-36h antecedence from satellite data in the cold season."* Methodological Letter. Russia, Hidrometeoisdat: Leningrad, 551.509.324:551.507.362.2 (083.13); F 1805040400-073/069(02)-88, 1988a.

Fedorova Natalia. *"Taking into account the speed of cloud systems moving in the forecast of precipitation by satellite data"*. In: Information letter No2. State committee of USSR in Hydrometeorology. Informational Collection of Natural Resources Centre of Russia, Russia, Hydrometeoisdat, Moscow, 37-49, (1988b).

Fedorova Natalia. *"Optimization of the calculation of precipitation by satellite data"*, Proceedings of the Hydrometeorological Center of the USSR, Moscow, 276, 34-45, Russia, Hydrometeoisdat, Moscow, ISSN: 0371- 7089, (1986).

Fedorova Natalia. *"The results of testing the method of calculation of precipitation from data on cloudiness from satellite data"*. Proceedings of the Hydrometeorological Center of the USSR, Moscow, 268, 95-100 Russia, Hydrometeoisdat, Leningrad, ISSN: 0371 -7089, (1985).

Fedorova Natalia. *"Diagnosis and forecast of rainfall by characteristics of cloud systems obtained from satellite data",* Proceedings of the Hydrometeorological Center of the USSR, Moscow, 245, 87-101, Russia, Hydrometeoisdat, Leningrad, ISSN: 0371- 7089, (1983).

Fedorova Natalia. *"Chapter 8.3 Analysis of the precipitation more than 10 mm/12h in summer over the Europe by satellite data"*. In: Guide to the use of satellite data in analysis and weather forecasting. Leningrad: Hidrometeoisdat, 201-204, (1982a).

Fedorova Natalia. *"Chapter 8.4 Analysis of the precipitation more than 3 mm/12h in winter over the Europe by satellite data."* In: Guide to the use of satellite data in analysis and weather forecasting. Leningrad: Hidrometeoisdat, 204-207, (1982b).

Fedorova Natalia. *"Chapter 8.5 Recommendations for the forecast of the location of heavy precipitation zones (more than 10 mm/12h in summer, more than 3 mm/12h in winter"*. In: Guide to the use of satellite data in analysis and weather forecasting. Leningrad: Hidrometeoisdat, 207-210, (1982c).

Fedorova Natalia. *"Location of the region with heavy precipitation in different stages of cyclonic vortex".* Proceedings of the Hydrometeorological Center of the USSR, Moscow, 232, 114-123, Russia, Hydrometeoisdat, Leningrad, ISSN: 0371 -7089, (1981).

Fedorova Natalia. *"Zones of heavy precipitation in the cloudiness field being identified by meteorological earth's satellites."* Meteorology and Hydrology, 9, Russia, Moscow, 32-38, (1980).

Fedorova Natalia. *"Detection Des Sones de Precipitations A L'Aide Des Images de La Nebulosite Obtenues Par Satellites."* Note technique de L'éstablissment d'etudes et de recherches météorologiques, Paris, França, 405, 1-11, (1978).

Fedorova Natalia. *"Precipitation zone in fields of cloud and wind".* In: Proceedings of the State Research Center for the Study of Natural Resources 1, 113-121, Russia, Hydrometeoisdat, Leningrad, (1976).

Gan M. A., Rao V. B. *"Surface cyclogenesis over South America."* Monthly Weather Review 119, 1293–1302, (1991).

Gan M. A., Rao V. B. "*The influence of the Andes Cordillera on Transient Disturbances."* Monthly Weather Review 122, 1141–1157, (1994).

GEMIACKI Lizandro. *"Influence of the frontal systems in the dry season in the Northeast of Brazil",* M. Sc. Teses, Universidade Federal de Alagoas, Maceió, MET-UFAL-MS-035, (2005).

REEDER MJ, SMITH RK. "*Mesoescale Meteorology."* In Meteorology of Southern Hemisphere. Karoly DJ, Vincent DG (ed.). American Meteorological Society: Boston, 201–241. (1998).

Kidder, Stanley Q.; Haar, Thomas H. Vonder., *Satellite meteorology: an introduction.*

San Diego, California: Academic Press, 466 p., 1995.

Satyamurty, P., Nobre C. A, Silva Dias, P.L., *"South America."* In: Meteorology of Southern Hemisphere. D. J. Karoly, D. G. Vincent Ed., Amer. Meteor. Soc., 119-139, (1998).

Sinhori Natalia Parisotto *"Estrutura vertical das frentes frias durante a passagem do sul até o nordeste brasileiro, entre 2000 e 2012".,* M.Sc. Thesis, Federal University of Alagoas, Maceio, Brazil: MET-UFAL-MS-XX, (2015)

Popova T.P., Fedorova, N.*"Using satellite images of clouds to predict precipitation."* Proceedings of the Hydrometeorological Center of the USSR, Moscow, 165, 69-76, Russia, Hydrometeoisdat, Leningrad, ISSN: 0371 -7089, (1975).

Vetlov I.P., Veltischev N.F (Editors), *"Guide to the use of satellite data in analysis and weather forecasting",* Hydrometeoisdat, Leningrad, 300pp, 551.507.362 (022), 1982

PART II

SHORT TERM FORECASTING IN BRAZILIAN NORTHEAST

CHAPTER II.1

TEMPERATURE AND MAXIMUM TEMPERATURE FORECASTING

NATALIA FEDOROVA, BRUCE FRANCISCO, PONTES DA SILVA AND VLADIMIR LEVIT

1. Temperature forecasting

Temperature forecasting is based on the thermodynamic energy equation (equation: II-1-1), (Petterssen, 1956 a,b; Djuric, 1994; Fedorova 1999, 2008b):

$$\frac{\partial T}{\partial t} = -\left(u\frac{\partial T}{\partial x} + v\frac{\partial T}{\partial y}\right) + \frac{RT(\gamma_d - \gamma)}{\rho g}\omega + \frac{H}{C_p\rho} \qquad \text{(II-1-1)}$$

Where:
ρ – air density
ω – p-component of wind (vertical motion)
γ – vertical temperature gradient
γ_d – vertical temperature gradient for dry adiabatic processes
H – diabatic term
R – universal gas constant
Cp – heat capacity of air at constant pressure
g – gravitational acceleration
T – temperature
t – time
u – zonal wind component
v – meridional wind component
The terms of this equation (II-1) will be described below in details

1.1 1 term – temperature advection

Basic methods of temperature advection calculation

The horizontal advection is (equation: II-1-2) (Petterssen, 1956 a,b; Djuric, 1994; Fedorova 1999, 2008b):

$$\frac{\partial a}{\partial t}_h = -\left(u\frac{\partial a}{\partial x} + v\frac{\partial a}{\partial y}\right) = -\vec{V}\cdot\nabla a \tag{II-1-2}$$

Where: ***a*** is a variable as a function of space and time,
u- and ***v***- are the ***x***- and ***y***- components of wind, respectively.
This equation is practical for a manual evaluation of advection.

Alternative equation (II-1-3) for advection calculation is:

$$-\vec{V}\cdot\nabla a = -V\frac{\partial a}{\partial l} = -V_n\frac{\partial a}{\partial n} = V\left|\nabla a\right|\cos\alpha \tag{II-1-3}$$

Where:
l – direction along the wind
V_n – normal wind component
α – advection angle between l and V_n

Using the geostrophic wind (II-1-4), the following equation (II-1-5) may be used for advection calculation of ***a*** variable:

$$u_g = -\frac{g}{f}\frac{\partial z}{\partial y} \quad v_g = \frac{g}{f}\frac{\partial z}{\partial x} \tag{II-1-4}$$

$$-\vec{V}_g\cdot\nabla a = \frac{g}{f}\left(\frac{\partial z}{\partial y}\frac{\partial a}{\partial x} + \frac{\partial z}{\partial x}\frac{\partial a}{\partial y}\right) = \frac{g}{f}J(a,z) = -\frac{g}{f}J(z,a) \tag{II-1-5}$$

Where:
z – geopotential (height in geopotential meters)

u_g – ***x*** component of geostrophic wind
v_g – ***y*** component of geostrophic wind
f – Coriolis parameter
J – Jacobian operator
g – gravitational acceleration

Simple method of temperature advection calculation

In operational practice, a simple method can be used (equation: II-1-6), (Petterssen, 1956 a,b; Fedorova 1999, 2008b). First, the trajectory length (***Tr)*** can be calculated as

$$\boldsymbol{Tr = V \times h} \qquad \text{(II-1-6)}$$

Where:
V is prevailing wind velocity in the direction from which the air parcel was moving
h is trajectory duration (for example, 12 or 24h)
The second step is to calculate the temperature advection (AdvT) as the temperature difference at the start and end points of the trajectory (equation II-1-7).

$$\boldsymbol{Tadv = Ts - Te} \qquad \text{(II-1-7)}$$

Where:
Ts – temperature at the start point, at the point from which the air parcel was moving
Te - temperature at the end point, for which the trajectory was calculated, forecast point

Identification of temperature advection by wind change with height

Using the change in wind direction with height from radiosonde data, it is possible to estimate the advection of cold or warm air (Djuric, 1994; Fedorova 2001). This method is based on the definition of *thermal wind* (V_T) as the difference of the wind vectors between two levels (1 and 2,), with level 1 higher than level 2 (equation II-1-8):

$$V_T \equiv V1 - V2 \qquad \text{(II-1-8)}$$

The conclusion about thermal advection will be based on the geostrophic approximation **V** ≅ **V**g. According to Djuric (1994), the thickness advection can be calculated in terms of wind at two levels. Then the thickness advection is given by (equation II-1-9):

$$-\boldsymbol{V}\cdot\nabla h \cong \frac{f}{g}\boldsymbol{k}\cdot(\boldsymbol{V}1\times\boldsymbol{V}2) \qquad \text{(II-1-9)}$$

Where:
V is the average wind in the layer
*V*1 and *V*2 are wind vectors in two levels, 1 is higher
f – Coriolis parameter
g – gravitational acceleration
h – height between two levels

This equation shows the *warm* advection when the wind turns to the *left/counterclockwise* with height. Conversely, there is *cold* advection when the wind turns to the *right/clockwise* with height (all in the Southern Hemisphere, Figure II-1-1).

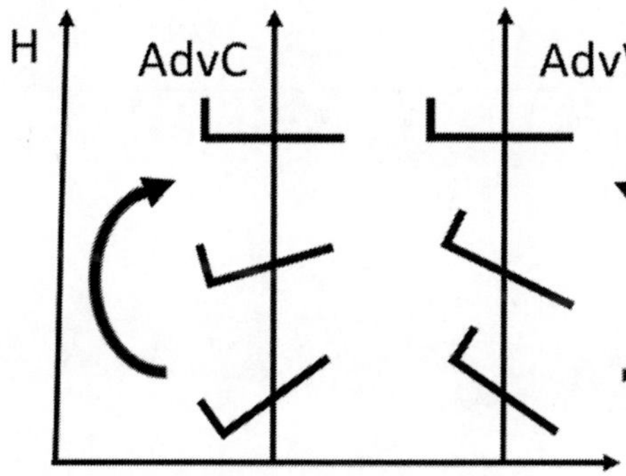

Figure II-1-1– Identification of temperature advection by wind direction change with height in the Southern Hemisphere
Adv C – cold air advection; change in wind direction with height *right/clockwise*.
Adv W – warm air advection; change in wind direction with height *left/counterclockwise*.
Round arrows indicate variation of wind direction with height clockwise and counterclockwise.

The examples in figures II-1-2 and II-1-3 show the identification of a thermal advection sign in Salvador and Maceio during the passage of the frontal zone over the BNE. Complete sets of radiosondes are presented in the Appendix to this chapter. The whole process will be described in detail in example 4 of this Chapter.

On 13 Jun 2022, cold advection in Salvador was determined by a clockwise wind direction change from SSE at 900-800 hPa to S at 750-650 hPa (Figures II-1-2a, b). The same advection in Maceio was between 1000 and 700 hPa, which was identified by a clockwise wind direction change from NNE to E (Figures II-1-3a, b).

On 15 Jun 2022, warm advection in Salvador was identified by a counterclockwise wind direction change from SSE at 1000-800 hPa to SE at 750-600 hPa (Figure II-1-2c, d). Warm advection in Maceio between 1000 and 700 hPa was defined by a counterclockwise wind direction change with height from S to ES (Figures II-1-3c, d).

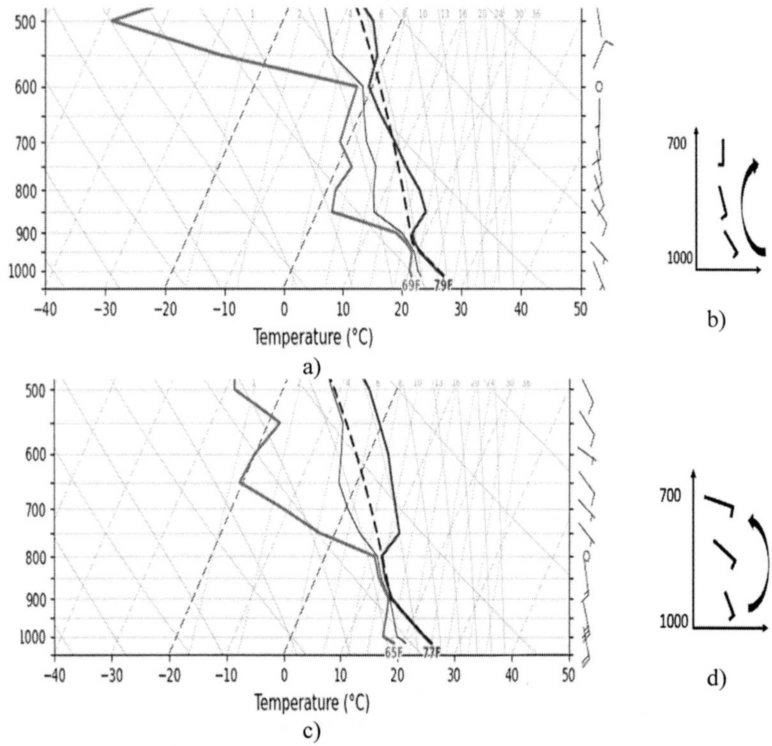

Figure II-1-2– Identification of temperature advection by wind direction change with height in Salvador by radiosonde data:
a, b) 13 Jun 2022, 12 UTC, clockwise wind direction change, cold advection,
c, d) 15 Jun 2022, 12 UTC, counterclockwise wind direction change, warm advection. Source: Tropical Tidbis: https://www.tropicaltidbits.com

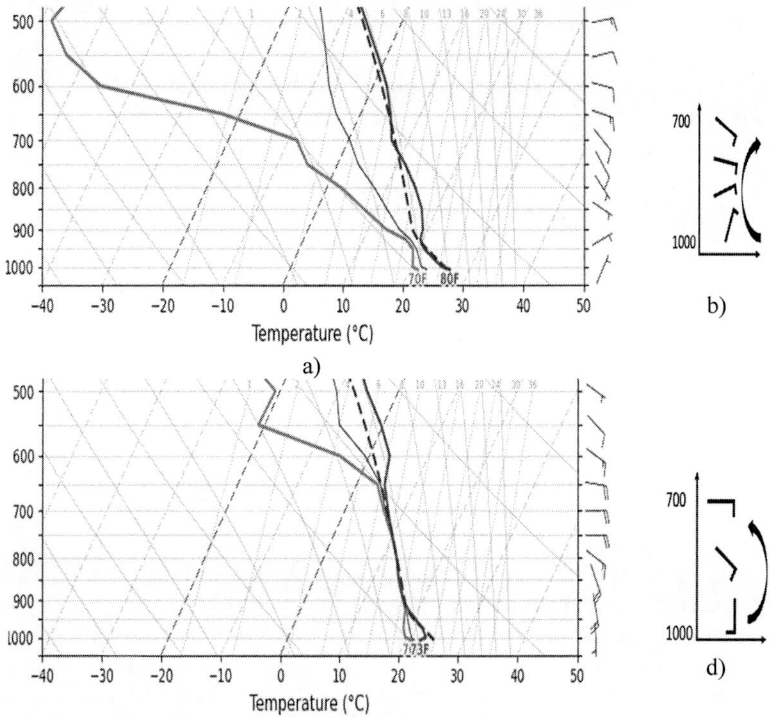

c)

Figure II-1-3 – Identification of temperature advection by wind direction change with height in Maceio by radiosonde data:
a, b) 13 Jun 2022, 12 UTC, clockwise wind direction change, cold advection,
c, d) 16 Jun 2022, 12 UTC, counterclockwise wind direction change, warm advection.
Source: Tropical Tidbis: https://www.tropicaltidbits.com

In addition to the methods described above, the identification of thermal advection from satellite data must be taken into account for temperature forecast. This information is presented in Chapter I-9.

Synoptic scale systems create temperature advection. This advection is the main factor in local temperature variations. Significant temperature advection is associated with cold and warm frontal zones. The warm frontal zone does not pass through the BNE. A cold front presents a weak temperature gradient in this region; however, there is a significant variation in the horizontal distribution of the equivalent potential temperature and its advection (Fedorova et al., 2016; Chapter I-7).

1.2 2 term – vertical motions

Vertical motions can be calculated, using equation below (II-1-10) (Petterssen, 1956 a,b; Djuric, 1994; Fedorova 1999, 2008b). This equation shows the temperature variation as a function of air instability.

$$\left(\frac{\partial T}{\partial t}\right)_{\omega} = \frac{RT(\gamma_d - \gamma)}{\rho g}\omega = a_1(\gamma_d - \gamma)\omega =$$
$$= -a_2(\gamma_d - \gamma)w \qquad \text{(II-1-10)}$$

Where:
ω – ***p*** component of wind (vertical motion in p- coordinates; (hPa s^{-1})
w – vertical component of wind (m s^{-1})
γ – lapse rate of temperature
γ_d – adiabatic lapse rate of temperature
a_1, a_2 – coefficients
g – gravitational acceleration

In stable air and conditionally unstable air, when $\gamma<\gamma_d$, the upward movement causes a temperature decrease and downward movement causes an increase in temperature (Fedorova et al., 2017). Figure II-1-4 shows the sinking and lifting processes. The initial vertical distribution of temperature and dew point is represented by the lines T_1 and T_{d1} respectively in both figures (II-1-4 a and b).

The example shows an air lifting of 200 hPa (1000 to 800 hPa) at low level, and 50 hPa at middle (700 to 650 hPa) and high (500 to 450 hPa) levels (Figure II-1-4 a). Due to the lifting, the temperature will be equal to the dew point temperature at the Lifting Condensation Level (LCL, 900 hPa) and, if lifting continues, the air will be humid (T=Td above 900 hPa) (Figure II-1-4 a). During this lifting, the temperature will decrease from T_1 to T_2. The temperature will decrease at 900 hPa from 12 to 8 °C, at 700 hPa from -2 to -4 °C and at 500 hPa -20 to -22 °C.

During the sinking process, the temperature will increase, from T_1 to T_2, and the humidity will decrease (the distance between vertical distribution of T_1 and T_{d1} (T_1-T_{d1}) will increase to T_2- T_{d2}). The example in Figure II-1-4b shows 100 hPa sinking from 600 to 700 hPa, and 50 hPa at higher (450 to 500 hPa) and lower (800 to 850 hPa and 950 to 1000 hPa) levels. The temperature will increase at 900 hPa from 12 to 14 °C, at 700 hPa from -2 to 2 °C, and at 500 hPa from -20 to -17 °C. At the initial moment (T_1-T_{d1}) was about 3 °C at all levels. After the air sinking (T_2-T_{d2}) will be 7 °C (13-6) at 900 hPa, 12 °C (2-(-10)) at 700 hPa, and 9 °C ((-17)-(-26)) at 500 hPa.

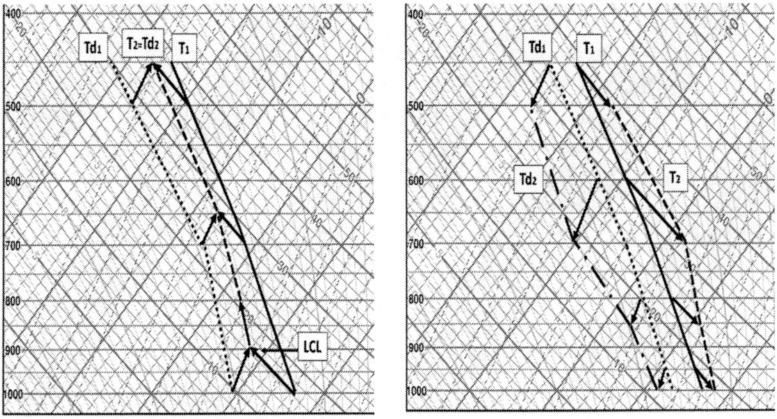

a) b)

Figure II-1-4 – Temperature variation due to upward movements (a) and downward movements (b).
The arrow shows vertical movements.
The solid line and *dotted* line show the air temperature (T_1, γ- lapse rate of temperature) and dew point (Td_1) at the initial moment, respectively.
The dashed line and *dash-dot* line show the air temperature (T_2) and dew point (Td_2) after vertical movements, respectively.
The dry adiabatic line (γ_d- adiabatic lapse rate of temperature) - *solid grey straight line, left sloping line.*
The wet adiabatic line - *solid grey curved line, left slanted line.*
The mixing ratio line - *dashed grey straight line, right slanted line.*
Source: Fedorova et al., 2017.

Vertical motions as well as temperature advection are generated by synoptic scale systems and processes, so synoptic analysis is the first step in temperature forecasting.

1.3 3 term – diabatic processes

The temperature change near the surface occurs due to the following diabatic processes (Fedorova, 1999, 2008b):

- turbulent heat exchange between the surface and the air mass (turbulent heat exchange);
- surface irradiation balance (radiation exchange);
- condensation and evaporation.

In general, air transformation works "like a brake" against air advection. Thus, the relationship between air advection and transformation will be shown below.

The change in surface temperature depends on wind speed; significant turbulent heat exchange occurs in slowly moving air masses. In fast moving air masses, there is not enough time for this exchange. The graph in figure II-1-5 shows the number (°C) of the transformation (Ttr) which depends on the temperature advection (Tadv, °C) and the wind speed V (km/h). For example, if the temperature advection is 12 °C and the wind speed is 20 km/h, then the temperature transformation will be ~4 °C. If in a similar situation the wind speed is 60 km/h, then the temperature transformation will be only 1 °C.

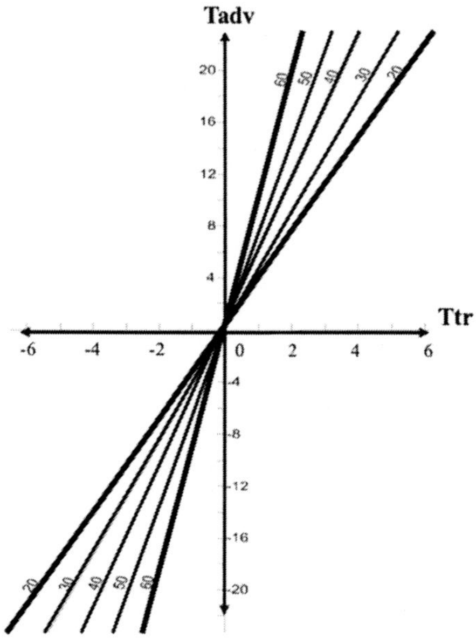

Figure II-1-5 – Surface temperature variation by transformation (Ttr; °C) for various wind speeds V (km/h) (lines with wind speed numbers from 20 to 60 km/h) and temperature advection (Tadv; °C).
Source: Fedorova 1999, 2008b.

The types and number of clouds on the trajectory of air masses also affect the temperature transformation. Low level clouds are more influential than middle and high-level clouds. This influence value can be seen from the coefficient number in equation (II-1-11).

$$N = N_b + 0.75N_m + 0.25N_a \quad \text{(II-1-11)}$$

Where:
N_b – low clouds amount
N_m – middle clouds amount
N_a – high clouds amount

The temperature variations due to the transformation of the air mass, according to the change in the amount of cloudiness (δN) along the trajectory, are shown in Figure II-1-6, where δN is the difference between the cloud amount at the end (N_{t+1}) and the beginning (N_t) of the trajectory (equation II-12). A more significant temperature transformation is observed in a situation with a strong change in the amount of cloudiness along the trajectory. This figure II-1-6 shows the transformation for different wind velocities similar to that described above. An increase in cloudiness corresponds to a decrease in transformation.

$$\delta N \approx 0.5\left(\delta N_{t+1} - \delta N_t\right) \tag{II-1-12}$$

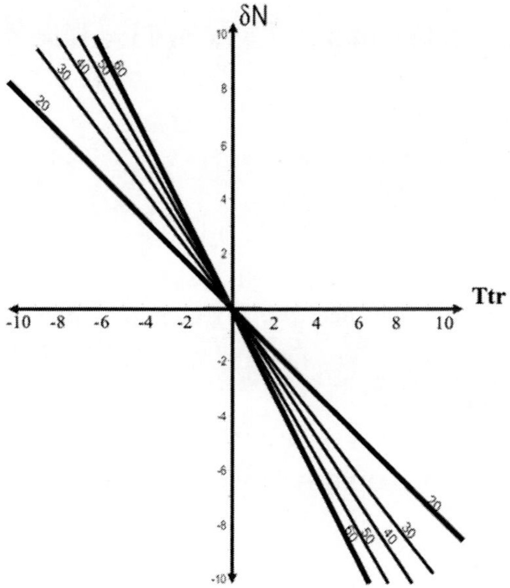

Figure II-1-6 – Surface temperature variations by transformation (Ttr; °C) for various wind speeds V (km/h) (lines with wind speed numbers from 20 to 60 km/h) and change in the amount of cloudiness (δN) along the trajectory.
Source: Fedorova, 1999, 2008b.

1.4 Forecasting T from the thermodynamic energy equation

Summarizing all the information described in item 1 above, an equation was created (equation II-1-13) for forecast temperature from the thermodynamic energy equation (Petterssen, 1956 a,b; Fedorova 1999, 2008b; Fedorova et al., 1999, 2005, 2017):

$$Tp = Ta + Tadv + Ttr \tag{II-1-13}$$

Where:
Tp – forecast temperature at the weather station
Ta – current temperature at the weather station
Tadv – temperature advection
Ttr – temperature transformation

Tmax and Tmin depend on the time of the initial data and the time of the forecast.

1.5 Example 1 – Warm advection:

The initial data is:

1) Ta = 20 °C,
2) Along the trajectory of the air parcel V = 20 km/h
3) T adv = +12 °C/24h (heating).

Calculation:

1) from the graph: T tr = 3.5 °C/24h
2) Transformation will decrease advection, i.e. the modification of Ta will be:

Ta = 12 - 3.5 = 8.5 (°C/24h)

3) Tp = 20 + 8.5 = 28.5 °C

1.6 Example 2 – Cold advection:

The initial data is:

1) Ta = 20 °C,
2) Along the trajectory of the air parcel V = 20 km/h
3) T adv = -12 °C/24h (cooling).

Calculation:

1) from the graph: T tr = 4 °C/24h

2) Transformation will decrease advection, that is, the modification of Ta will be: -12 - (-3.5) = - 8.5 (°C/24h)
3) Tp = 20 - 8.5 = 11.5 °C

2. Maximum temperature forecast

A dry adiabatic lapse rate layer near the surface is formed during the day when the maximum temperature is observed (Petterssen, 1956 a,b; Fedorova et al., 1999, 2005, 2017). An example can be seen in Figure II-1-7. The BNE sounding data in this figure shows a dry adiabatic gradient distribution between 1000 and 900 hPa; very humid (0<T-Td<3 °C) layer (900-750 hPa) and subsidence above the 750 hPa level.

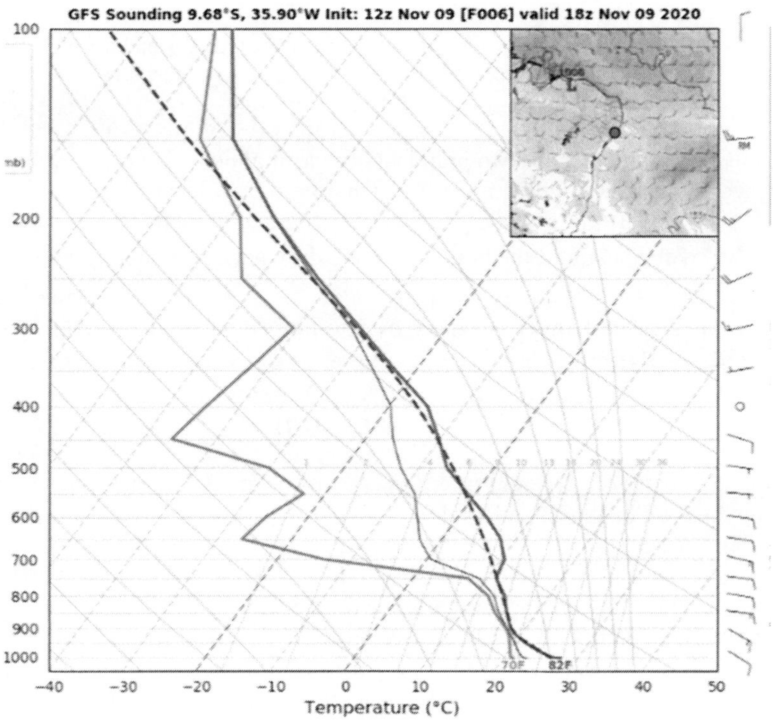

a)

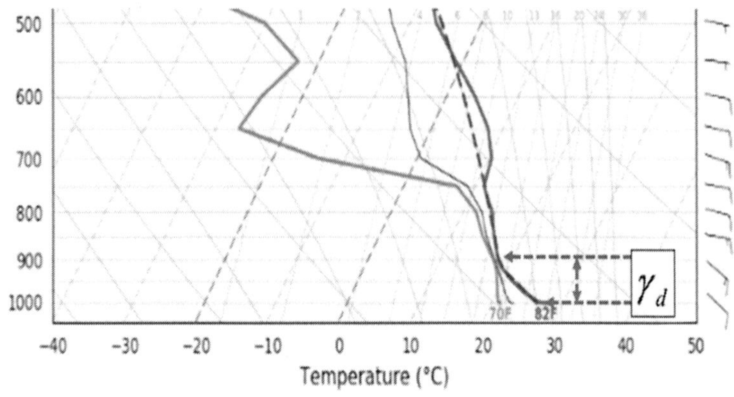

b)

Figure II-1-7 – Dry adiabatic gradient distribution (γ_d) on GFS sounding in the BNE, 9 November 2020: a) complete sounding and b) fragment of sounding.
Source: Tropical Tidbits: https://www.tropicaltidbits.com

Based on this result, two practical forecasting methods were developed in which the maximum temperature (Tmax) can be calculated:

1) by the height of the dry adiabatic gradient distribution; or
2) by the height of the boundary layer during surface heating.

2.1 Method I: Identify Tmax by the height of the dry adiabatic gradient distribution, using the level of convective condensation (Petterssen, 1956 a,b; Fedorova, 2008a; Fedorova et al., 1999, 2005, 2017).

The layer with the dry adiabatic lapse rate distribution is formed below the Convective Condensation Level (*CCL*). *CCL* is the height at which a portion of air heated from below will ascend adiabatically until it becomes saturated. During the day, the air is heated from the earth's surface, convection is formed and the CCL shows the base of cumulus clouds. The CCL can be found by ascending from Td at 1000 hPa along the mixing ratio line until it intersects with the vertical temperature distribution line (Figure II-1-8). More intensive convection develops during the hours of maximum temperature. Therefore, Tmax can be obtained by descending from the CCL along the dry adiabatic line.

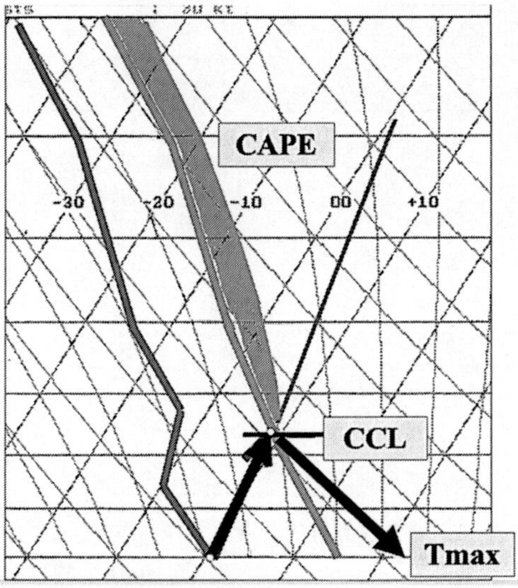

Figure II-1-8 – Tmax identification by CCL.
Step 1: go up from Td at 1000 hPa by mixing ratio line (up arrow) to crossing with temperature vertical distribution line
Step 2: go down from CCL along the dry adiabatic line
Source: Fedorova 2008a, Fedorova et al., 2017.

2.2 Method II: Identify layer height with surface heating (Petterssen, 1956 a,b; Fedorova et al., 1999, 2005, 2017).

Average layer heights with Tmax (Hmax, hPA) in all seasons in Brazil (Table II-1-1) are described in (Fedorova et al., 1999, 2005). Hmax seasonally varies slightly from 870 hPa in summer to 906-910 hPa in other seasons. The height of this layer in summer corresponds to the weather conditions in the BNE region.

Table II-1-1 – The layer heights with Tmax (Hmax, hPa) in all seasons. Source: (Fedorova et al., 1999, 2005, 2017).

Season	Winter	Spring	Summer	Autumn
Hmax	910	906	870	910

Cloudiness and cloudiness level change Hmax more significantly (Table II-1-2). This altitude is lower (920 hPa) on days with more than 60% low cloud cover and higher (880 hPa) on cloudless days or when there is high cloud cover.

Table II-1-2 – The Height (h_d, hPa) of the dry adiabatic lapse rate layer (Hmax) for each month relative to cloud cover and cloud level.
Source: Fedorova et al., 1999, 2005, 2017.

Month	Cloudiness		
	Without clouds or high clouds	Middle clouds or low clouds (≤ 6/10 of cover)	Low clouds (>6/10 of cover)
January	860	850	900
February	850	860	890
March	870	840	900
April	880	900	930
May	910	920	950
June	940	940	960
July	900	900	950
August	920	910	940
September	890	870	940
October	870	940	920
November	830	910	920
December	860	930	850
Mean	880	900	920

2.3 Example 3 - Tmax forecasting

The initial data can be seen in Figure II-1-9:

1) Ta = 26 °C.
2) January.
3) High (around 350 hPa) and low (around 925 hPa) level clouds, CAPE < 1000 J/kg, cloud cover less than 6/10 of the sky (Figure II-1-9a).

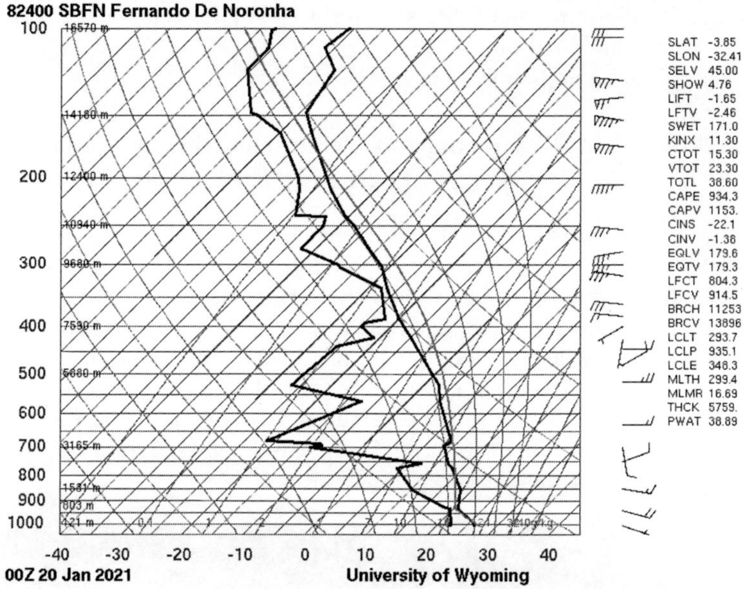

a)

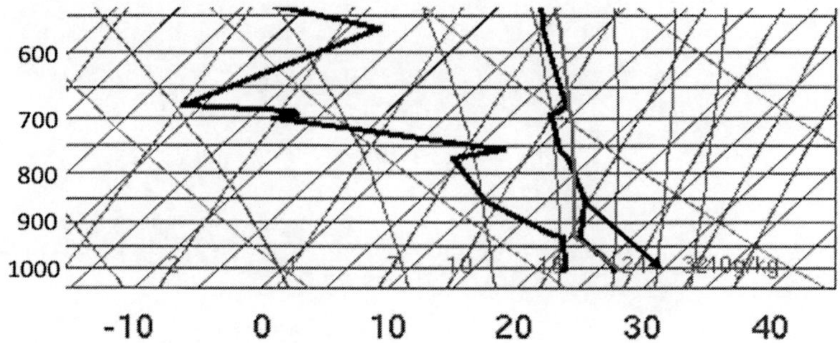

b)

Figure II-1-9 – Tmax forecast from radiosonde data in the BNE (Fernando de Noronha) on 20 January 2021, 00UTC:
a) full radio sounding and b) a fragment of radio sounding; the *arrow from 850 hPa to 1000 hPa along the dry adiabatic line shows Tmax 30oC at 1000 hPa.*
Source: University of Wyoming.

Calculation:

1) According to tabular data, the height of the dry adiabatic lapse rate layer (Hmax) is 850 hPa.

2) On the vertical profile (Figure II-1-9b) descend along the dry adiabatic line from 850 hPa.
3) Tmax found: 30 °C.

2.4 Example 4 –The passage of a cold front through the eastern BNE, 13 - 18 June 2022

A frontal zone passed through the eastern BNE from 13 to 16 June 2022. Satellite data in the visible channel show a baroclinic cyclone cloud band in the South Atlantic moving towards the BNE (Figure II-1-10). Rain and temperature changes were associated with this frontal extremity. The connection between the cloudy band of the cold front and the baroclinic cyclone was observed in the period from 13 to 15 June, 2022. The weakening of this connection occurred on 16 June, and a reminiscent of cloudiness at the frontal extremity in the BNE can be seen on 17 and 18 June. A similar process has been described in Chapter I-7.

a) 13 June 2022, 12 UTC

b) 14 June 2022, 12 UTC

VSCHN

c) 15 June 2022, 12 UTC

d) 16 June 2022, 12 UTC

e) 17 June 2022, 12 UTC

f) 18 June 2022, 12 UTC

Figure II-1-10 – The passage of the frontal zone through the eastern BNE from 13 to 18 June 2022 was identified from satellite images in the visible channel.
Source: METEOSAT, 11.

Weather conditions in the south of the BNE in Salvador city (13.01°S, 38.51°W) (Bahia State) indicate an increase in cloudiness and a change in wind direction/magnitude during the period from 13 to 14 June 2022. A migrating baroclinic anticyclone behind the cold front arrived in Salvador city the next day (15 June).

In the middle of the BNE, Maceio city (9.55°S, 5.77°W) (Alagoas State) was also affected by the same weather change with increased cloud cover and wind variation from 13 to 14 June. The cold front over Alagoas slowed down and became stationary from 15 to 17 June. According to surface data, a sharp change in temperature was observed both in Salvador and Maceio (Figure II-1-11b). A sudden drop in temperature was observed from 14 to 15 June in Salvador and from 14 to 17 in Maceio. A more intense change in temperature can be seen from the maximum temperature in Maceio, when it decreased from 31 to 24 °C.

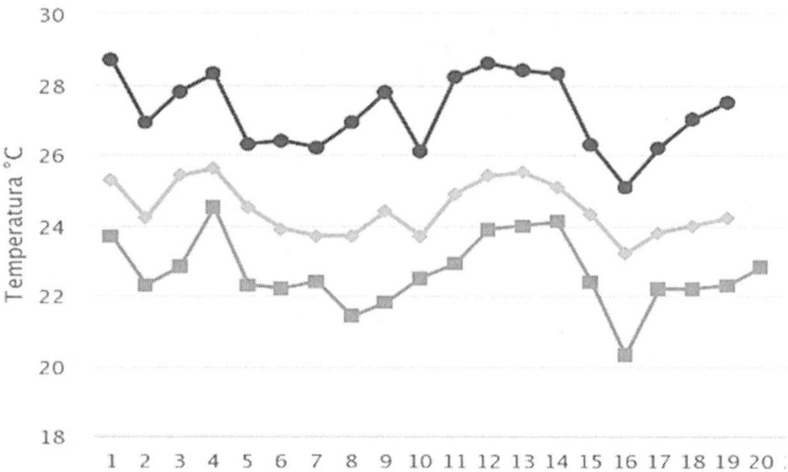

a)

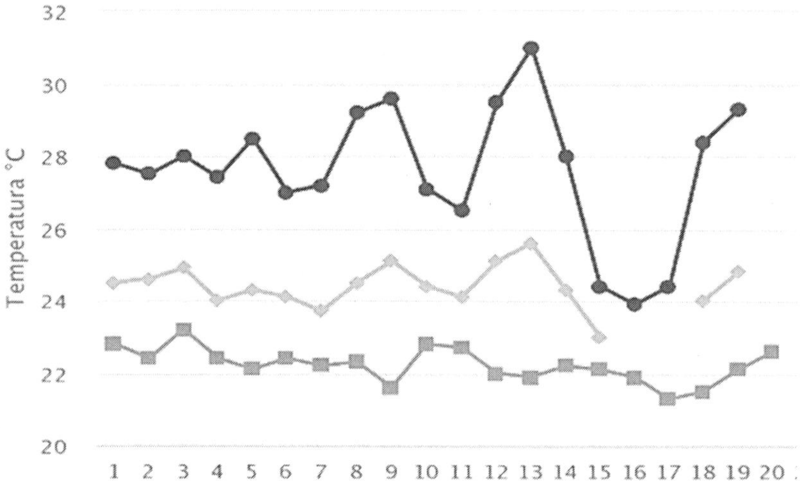

b)

Figure II-1-11 – Temperature change (maximum, instantaneous, minimum) during the passage of the frontal zone through the eastern BNE from 13 to 17 June 2022:
a) in the south of the BNE, in Salvador
b) in the middle of the BNE, in Maceio
Source: INMET, 2022.

Cold advection event in Maceio and Salvador on June 15, 2022.
Initial temperature data in Maceio/Salvador on June 14, 2022, 12 UTC:

1) Ta = 25.7/25.6 °C.
2) V = 35/35 km/h along the trajectory of an air parcel.
3) T adv = -1/0 °C/24h (cooling).

Calculation:

1) from the graph: T tr = -0.5/0 °C/24h.
2) Transformation will decrease advection, that is, the modification of Ta will be: -1-(-0.5)/0-(0) = -0.5/0 (°C/24h).
3) Tp = 25.7-0.5/25.6-0 = 25.2/25.6 °C.

Warm advection event in Maceio on June 19, 2022.
Initial temperature data in Maceio on June 18, 2022, 12 UTC:

1) Ta = 23.9 °C.
2) V = 17 km/h along the trajectory of an air parcel.
3) T adv = +1 °C/24h (warming).

Calculation:

1) from the graph: T tr = 0 °C/24h.
2) Transformation will decrease advection, that is, the modification of Ta will be: 1-0= 1 (°C/24h).
3) Tp = 23.9+1 = 24.9.

Tmax forecasting in Maceio and Salvador on June 14 and 16, 2022
Initial temperature data in Maceio/Salvador on June 13, 2022, 12 UTC:

1) Ta = 27.3/25.4 °C.
2) June.
3) No clouds or high-level clouds/ Middle clouds or low clouds (≤ 6/10 of cover).

Calculation:

1) From the tabular data, the height of the dry adiabatic gradient layer: 940/940 hPa.
2) In the vertical profile go down through the dry adiabatic of 940/940 hPa.
3) Tmax found: 28/28.3 °C.

Initial temperature data in Maceio/Salvador on June 15, 2022, 12 UTC:

1) Ta = 27.3/25.4 °C.
2) June.

3) No clouds or high-level clouds/ Middle clouds or low clouds (≤ 6/10 of cover).

Calculation:

1) From the tabular data, the height of the dry adiabatic gradient layer: 960/960 hPa .
2) In the vertical profile go down through the dry adiabatic of 960/960 hPa.
3) Tmax found: 23.9/25.1 °C.

3. Summary and practical recommendations

Summarizing the results of the temperature forecast in the BNE, the following remarks could be made.

Temperature advection on days with a frontal zone is significant. Therefore, it is necessary to use temperature advection and transformation. These terms of the thermodynamic energy equation play an important role in temperature forecast. These calculations are very important (play a fundamental role) for the extratropical region, in the southern part of Brazil (south of the latitude of the Sao Paulo state) and less frequently on days with frontal activity in the south of Bahia State.

Temperature advection is negligible (usually less than 2 °C/24h) in the tropical region, including the BNE. Vertical motion and cloudiness play a more important role in this region.

Practical recommendations are as follows:

1. Using the value of advT for day and night, a forecast of Tmax and Tmin can be developed.
2. Knowing Tmax and the daily temperature amplitude, Tmin can be forecasted.
3. The daily temperature amplitude depends on:
   - geographic position
   - seasons
   - cloud type

Daily temperature amplitude tables can be completed for each city using climatological data.

4. Appendix

South of the BNE: Salvador city (13.01°S, 38.51°W) (Bahia State)
Middle of the BNE, Maceio city (9.55°S, 5.77°W) (Alagoas State)

Radiosondes in Salvador and Maceio during the passage of the frontal zone over the BNE, 13 - 18 June 2022

Salvador

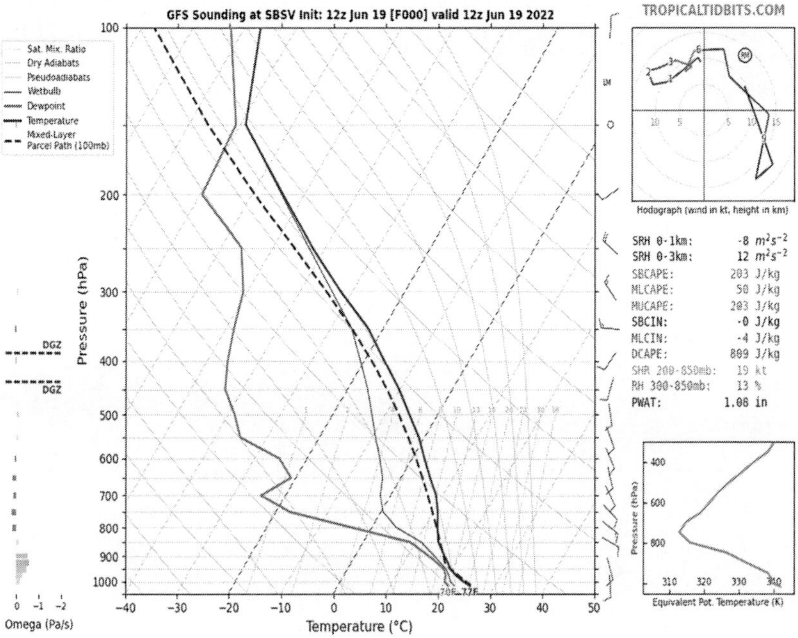

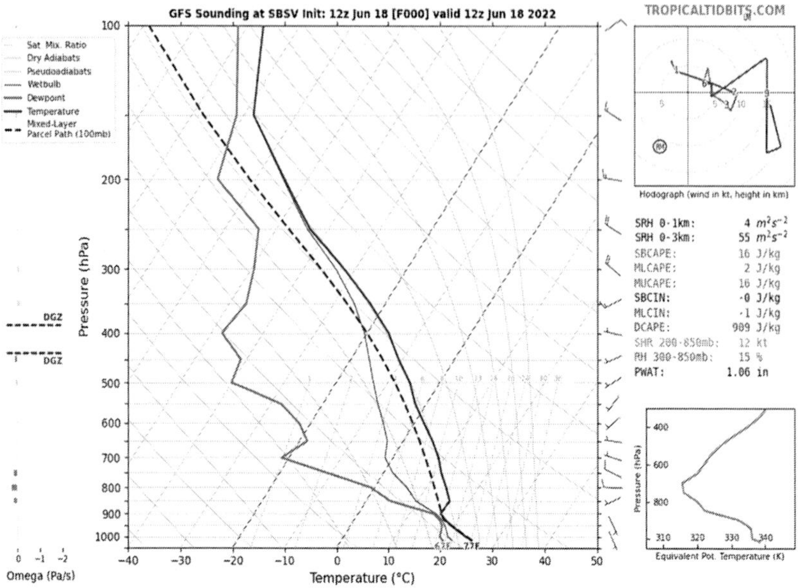
GFS Sounding at SBSV Init: 12z Jun 18 [F000] valid 12z Jun 18 2022
TROPICALTIDBITS.COM
Sat. Mix. Ratio
Dry Adiabats
Pseudoadiabats
Wetbulb
Dewpoint
Temperature
Mixed-Layer Parcel Path (100mb)
DGZ
DGZ
Pressure (hPa)
Temperature (°C)
Omega (Pa/s)
Hodograph (wind in kt, height in km)
SRH 0-1km: 4 m²s⁻²
SRH 0-3km: 55 m²s⁻²
SBCAPE: 16 J/kg
MLCAPE: 2 J/kg
MUCAPE: 16 J/kg
SBCIN: -0 J/kg
MLCIN: -1 J/kg
DCAPE: 909 J/kg
SHR 200-850mb: 12 kt
RH 300-850mb: 15 %
PWAT: 1.06 in
Pressure (hPa)
Equivalent Pot. Temperature (K)

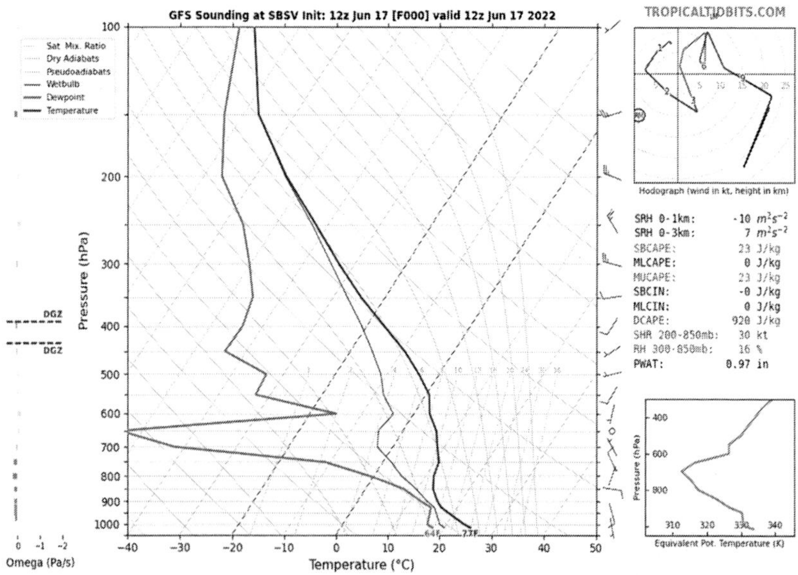
GFS Sounding at SBSV Init: 12z Jun 17 [F000] valid 12z Jun 17 2022
TROPICALTIDBITS.COM
Sat. Mix. Ratio
Dry Adiabats
Pseudoadiabats
Wetbulb
Dewpoint
Temperature
DGZ
DGZ
Pressure (hPa)
Temperature (°C)
Omega (Pa/s)
Hodograph (wind in kt, height in km)
SRH 0-1km: -10 m²s⁻²
SRH 0-3km: 7 m²s⁻²
SBCAPE: 23 J/kg
MLCAPE: 0 J/kg
MUCAPE: 23 J/kg
SBCIN: -0 J/kg
MLCIN: 0 J/kg
DCAPE: 920 J/kg
SHR 200-850mb: 30 kt
RH 300-850mb: 16 %
PWAT: 0.97 in
Pressure (hPa)
Equivalent Pot. Temperature (K)

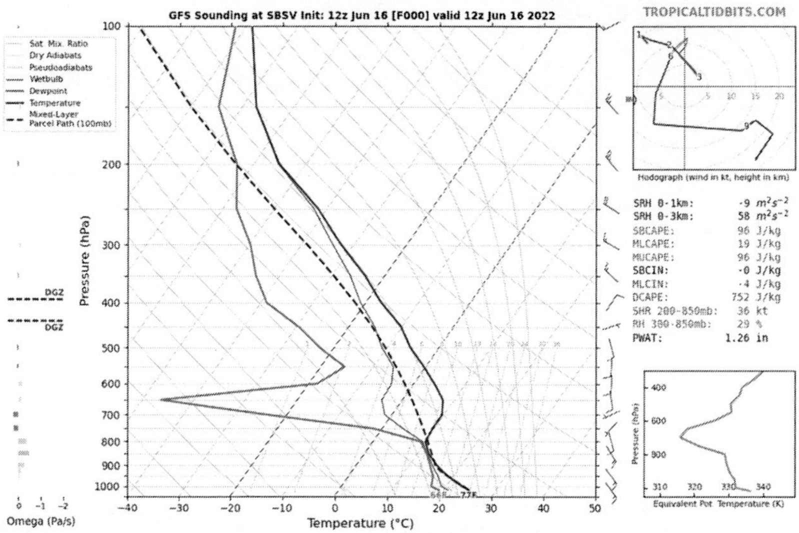
GFS Sounding at SBSV Init: 12z Jun 16 [F000] valid 12z Jun 16 2022
TROPICALTIDBITS.COM
Sat. Mix. Ratio
Dry Adiabats
Pseudoadiabats
Wetbulb
Dewpoint
Temperature
Mixed-Layer Parcel Path (100mb)
DGZ
DGZ
Pressure (hPa)
Temperature (°C)
Omega (Pa/s)
Hodograph (wind in kt, height in km)
SRH 0-1km: -9 m²s⁻²
SRH 0-3km: 58 m²s⁻²
SBCAPE: 96 J/kg
MLCAPE: 19 J/kg
MUCAPE: 96 J/kg
SBCIN: -0 J/kg
MLCIN: -4 J/kg
DCAPE: 752 J/kg
SHR 200-850mb: 36 kt
RH 300-850mb: 29 %
PWAT: 1.26 in
Equivalent Pot. Temperature (K)

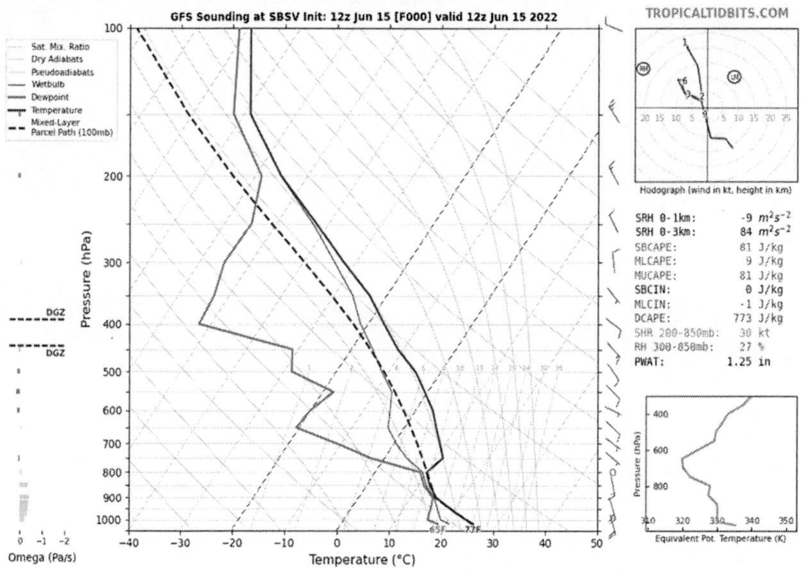
GFS Sounding at SBSV Init: 12z Jun 15 [F000] valid 12z Jun 15 2022
TROPICALTIDBITS.COM
Sat. Mix. Ratio
Dry Adiabats
Pseudoadiabats
Wetbulb
Dewpoint
Temperature
Mixed-Layer Parcel Path (100mb)
DGZ
DGZ
Pressure (hPa)
Temperature (°C)
Omega (Pa/s)
Hodograph (wind in kt, height in km)
SRH 0-1km: -9 m²s⁻²
SRH 0-3km: 84 m²s⁻²
SBCAPE: 81 J/kg
MLCAPE: 9 J/kg
MUCAPE: 81 J/kg
SBCIN: 0 J/kg
MLCIN: -1 J/kg
DCAPE: 773 J/kg
SHR 200-850mb: 30 kt
RH 300-850mb: 27 %
PWAT: 1.25 in
Equivalent Pot. Temperature (K)

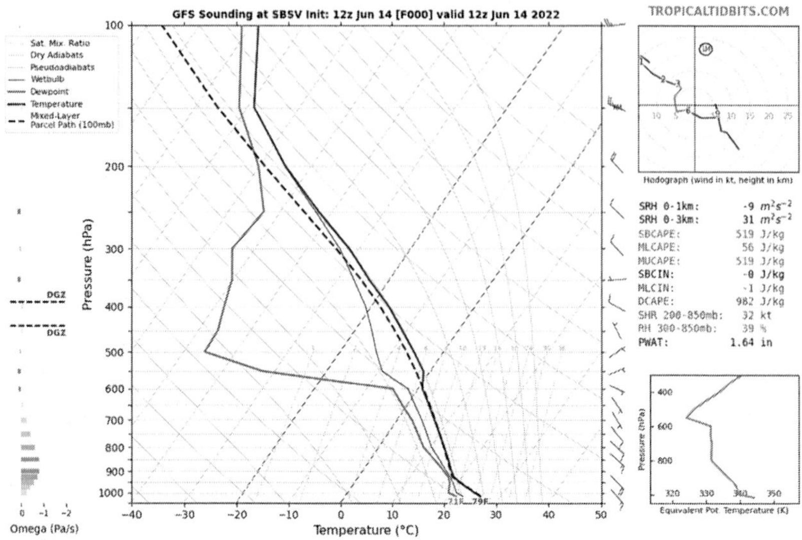
GFS Sounding at SBSV Init: 12z Jun 14 [F000] valid 12z Jun 14 2022
TROPICALTIDBITS.COM
Sat. Mix. Ratio
Dry Adiabats
Pseudoadiabats
Wetbulb
Dewpoint
Temperature
Mixed-Layer Parcel Path (100mb)
DGZ
DGZ
Omega (Pa/s)
Pressure (hPa)
Temperature (°C)
71F
79F
Hodograph (wind in kt, height in km)
SRH 0-1km: -9 m^2s^{-2}
SRH 0-3km: 31 m^2s^{-2}
SBCAPE: 519 J/kg
MLCAPE: 56 J/kg
MUCAPE: 519 J/kg
SBCIN: -0 J/kg
MLCIN: -1 J/kg
DCAPE: 982 J/kg
SHR 200-850mb: 32 kt
RH 300-850mb: 39 %
PWAT: 1.64 in
Pressure (hPa)
Equivalent Pot. Temperature (K)

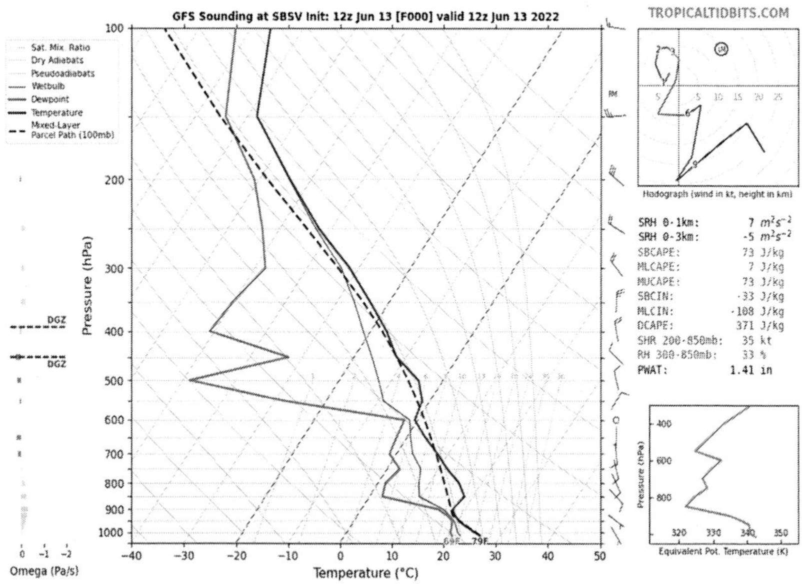
GFS Sounding at SBSV Init: 12z Jun 13 [F000] valid 12z Jun 13 2022
TROPICALTIDBITS.COM
Sat. Mix. Ratio
Dry Adiabats
Pseudoadiabats
Wetbulb
Dewpoint
Temperature
Mixed-Layer Parcel Path (100mb)
DGZ
DGZ
Omega (Pa/s)
Pressure (hPa)
Temperature (°C)
69F
79F
Hodograph (wind in kt, height in km)
SRH 0-1km: 7 m^2s^{-2}
SRH 0-3km: -5 m^2s^{-2}
SBCAPE: 73 J/kg
MLCAPE: 7 J/kg
MUCAPE: 73 J/kg
SBCIN: -33 J/kg
MLCIN: -108 J/kg
DCAPE: 371 J/kg
SHR 200-850mb: 35 kt
RH 300-850mb: 33 %
PWAT: 1.41 in
Pressure (hPa)
Equivalent Pot. Temperature (K)

Maceio

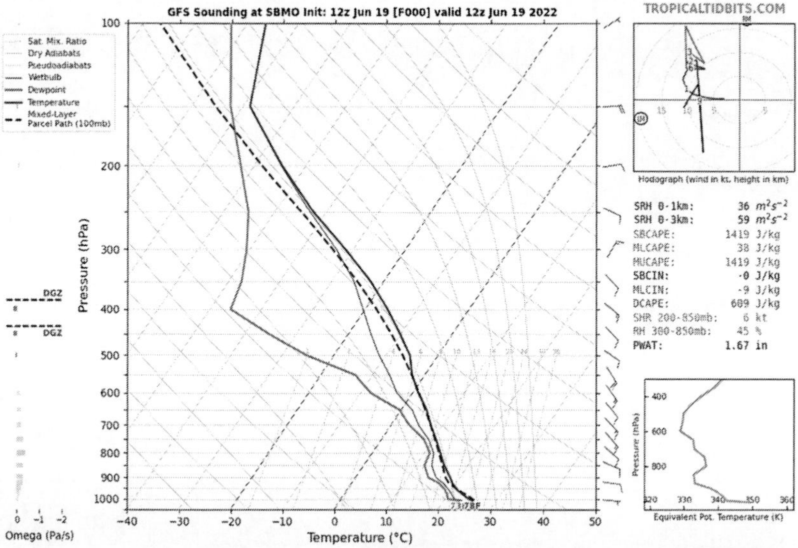
GFS Sounding at SBMO Init: 12z Jun 19 [F000] valid 12z Jun 19 2022
TROPICALTIDBITS.COM
Sat. Mix. Ratio
Dry Adiabats
Pseudoadiabats
Wetbulb
Dewpoint
Temperature
Mixed-Layer Parcel Path (100mb)
DGZ
Pressure (hPa)
Temperature (°C)
Omega (Pa/s)
Hodograph (wind in kt, height in km)
SRH 0-1km: 36 m^2s^{-2}
SRH 0-3km: 59 m^2s^{-2}
SBCAPE: 1419 J/kg
MLCAPE: 38 J/kg
MUCAPE: 1419 J/kg
SBCIN: -0 J/kg
MLCIN: -9 J/kg
DCAPE: 609 J/kg
SHR 200-850mb: 6 kt
RH 300-850mb: 45 %
PWAT: 1.67 in
Equivalent Pot. Temperature (K)

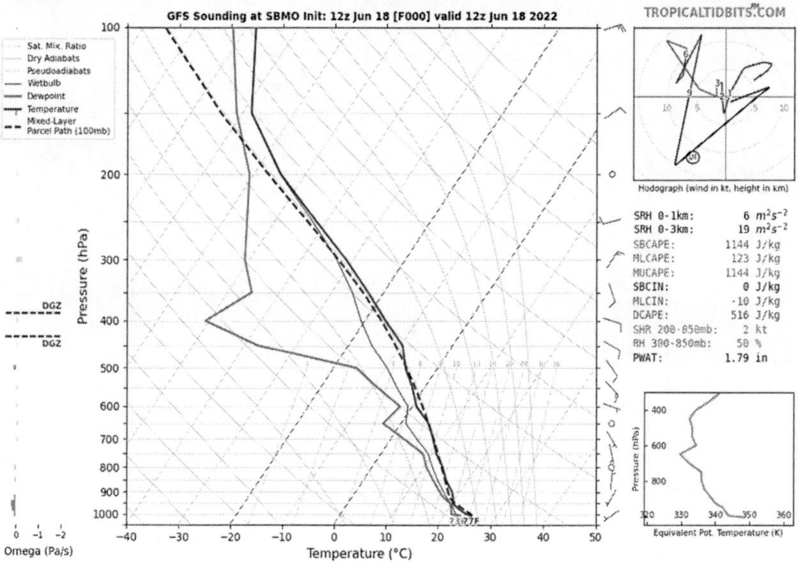
GFS Sounding at SBMO Init: 12z Jun 18 [F000] valid 12z Jun 18 2022
TROPICALTIDBITS.COM
Sat. Mix. Ratio
Dry Adiabats
Pseudoadiabats
Wetbulb
Dewpoint
Temperature
Mixed-Layer Parcel Path (100mb)
DGZ
Pressure (hPa)
Temperature (°C)
Omega (Pa/s)
Hodograph (wind in kt, height in km)
SRH 0-1km: 6 m^2s^{-2}
SRH 0-3km: 19 m^2s^{-2}
SBCAPE: 1144 J/kg
MLCAPE: 123 J/kg
MUCAPE: 1144 J/kg
SBCIN: 0 J/kg
MLCIN: -10 J/kg
DCAPE: 516 J/kg
SHR 200-850mb: 2 kt
RH 300-850mb: 50 %
PWAT: 1.79 in
Equivalent Pot. Temperature (K)

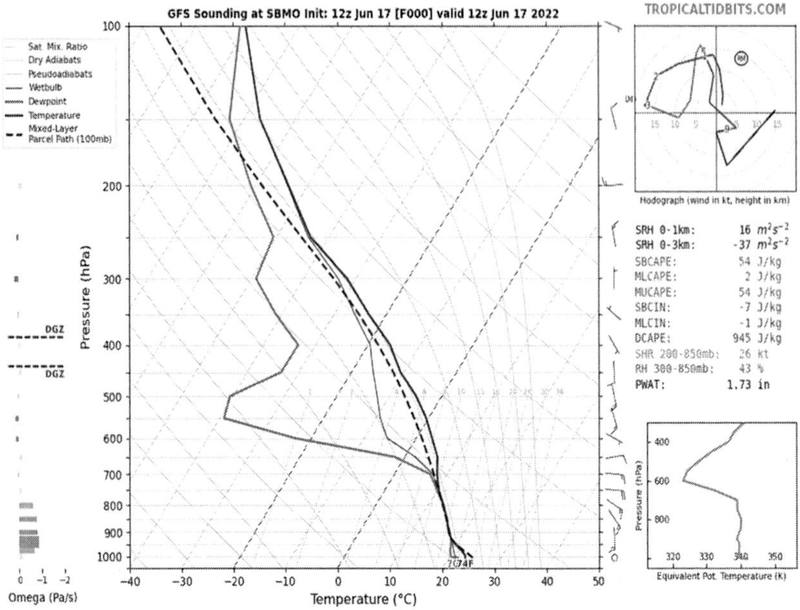
GFS Sounding at SBMO Init: 12z Jun 17 [F000] valid 12z Jun 17 2022
TROPICALTIDBITS.COM
Sat. Mix. Ratio
Dry Adiabats
Pseudoadiabats
Wetbulb
Dewpoint
Temperature
Mixed-Layer Parcel Path (100mb)
Pressure (hPa)
Temperature (°C)
Omega (Pa/s)
DGZ
DGZ
Hodograph (wind in kt, height in km)
SRH 0-1km: 16 m²s⁻²
SRH 0-3km: -37 m²s⁻²
SBCAPE: 54 J/kg
MLCAPE: 2 J/kg
MUCAPE: 54 J/kg
SBCIN: -7 J/kg
MLCIN: -1 J/kg
DCAPE: 945 J/kg
SHR 200-850mb: 26 kt
RH 300-850mb: 43 %
PWAT: 1.73 in
Equivalent Pot. Temperature (K)

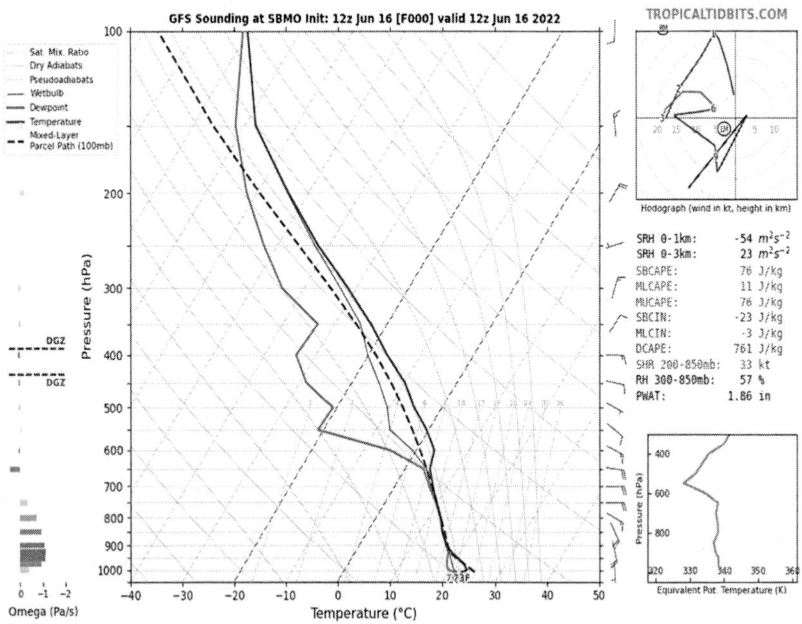
GFS Sounding at SBMO Init: 12z Jun 16 [F000] valid 12z Jun 16 2022
TROPICALTIDBITS.COM
Sat. Mix. Ratio
Dry Adiabats
Pseudoadiabats
Wetbulb
Dewpoint
Temperature
Mixed-Layer Parcel Path (100mb)
Pressure (hPa)
Temperature (°C)
Omega (Pa/s)
DGZ
DGZ
Hodograph (wind in kt, height in km)
SRH 0-1km: -54 m²s⁻²
SRH 0-3km: 23 m²s⁻²
SBCAPE: 76 J/kg
MLCAPE: 11 J/kg
MUCAPE: 76 J/kg
SBCIN: -23 J/kg
MLCIN: -3 J/kg
DCAPE: 761 J/kg
SHR 200-850mb: 33 kt
RH 300-850mb: 57 %
PWAT: 1.86 in
Equivalent Pot. Temperature (K)

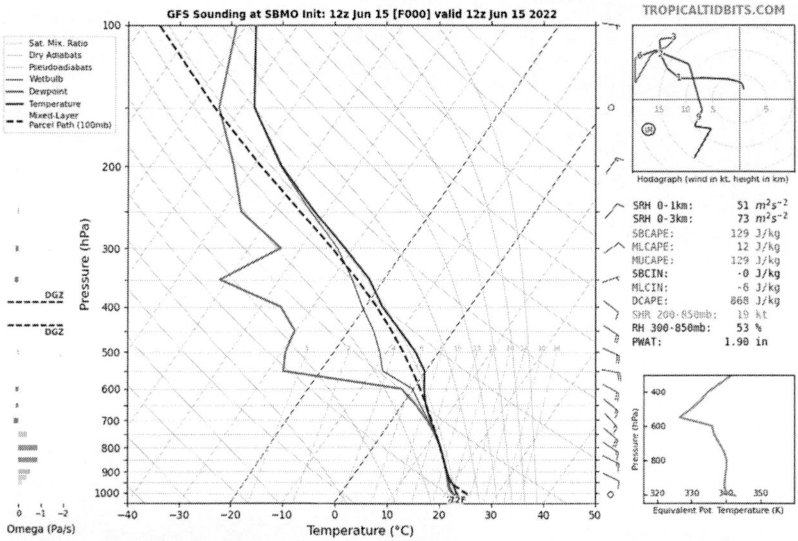

GFS Sounding at SBMO Init: 12z Jun 15 [F000] valid 12z Jun 15 2022
TROPICALTIDBITS.COM
Sat. Mix. Ratio
Dry Adiabats
Pseudoadiabats
Wetbulb
Dewpoint
Temperature
Mixed-Layer Parcel Path (100mb)
Pressure (hPa)
Temperature (°C)
Omega (Pa/s)
DGZ
Hodograph (wind in kt, height in km)
SRH 0-1km: 51 m^2s^{-2}
SRH 0-3km: 73 m^2s^{-2}
SBCAPE: 129 J/kg
MLCAPE: 12 J/kg
MUCAPE: 129 J/kg
SBCIN: -0 J/kg
MLCIN: -6 J/kg
DCAPE: 868 J/kg
SHR 200-850mb: 19 kt
RH 300-850mb: 53 %
PWAT: 1.90 in
Equivalent Pot. Temperature (K)

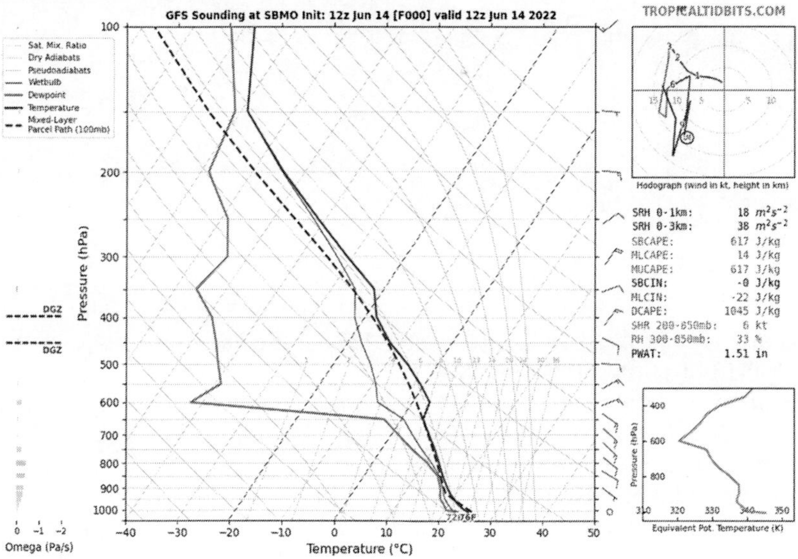

GFS Sounding at SBMO Init: 12z Jun 14 [F000] valid 12z Jun 14 2022
TROPICALTIDBITS.COM
Sat. Mix. Ratio
Dry Adiabats
Pseudoadiabats
Wetbulb
Dewpoint
Temperature
Mixed-Layer Parcel Path (100mb)
Pressure (hPa)
Temperature (°C)
Omega (Pa/s)
DGZ
Hodograph (wind in kt, height in km)
SRH 0-1km: 18 m^2s^{-2}
SRH 0-3km: 38 m^2s^{-2}
SBCAPE: 617 J/kg
MLCAPE: 14 J/kg
MUCAPE: 617 J/kg
SBCIN: -0 J/kg
MLCIN: -22 J/kg
DCAPE: 1045 J/kg
SHR 200-650mb: 6 kt
RH 300-850mb: 33 %
PWAT: 1.51 in
Equivalent Pot. Temperature (K)

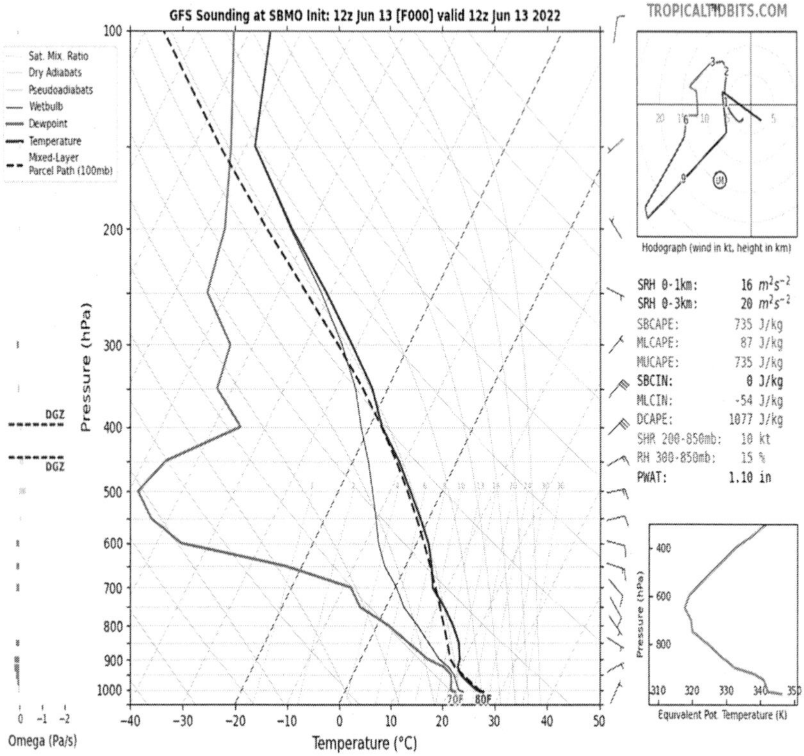

5. References

Djuric Dusan, *Weather Analysis*. New Jersey: Prentice Hall, 304p, ISBN: 0-13-501149-3, 1994.

Fedorova Natalia. *Meteorologia Sinótica. Volume 1.* Pelotas: Ed. Universitária/UFPel, 260p, ISBN: 85-7192-108-3, 1999.

Fedorova Natalia. *Meteorologia Sinótica Volume 2.* Pelotas: Ed. Universitária/UFPel, 242p, ISBN: 85-7192-145-7, 2001.

Fedorova, N., Carvalho, M.H., Gonçalves, A,M., Alves E.P., Signorini, E, Pinheiro, G.C., Marques ,J., Oliveira, V.M., Almeida, A.J. "Verificação de um método de previsão de temperatura máxima para Pelotas" *Revista Brasileira de Meteorologia* 14, no. 2., (1999): 37-45.

Fedorova, N., Carvalho, M.H., Levit, V., Marcelino, B.C., Gonçalves, A,M., Alves E.P., Signorini, E, Pinheiro, G.C., Marques ,J., Oliveira, V.M., Almeida, A.J., Botelho, C.B., "*A maximum temperature forecasting method for the city of Pelotas, Rio Grande do Sul.*" Meteorologica 30, n 1-2: 91-99, (2005).

Fedorova Natalia. *Sinótica I: Fenômenos atmosféricos, dados sinóticos e massas de ar.* Material didático: sinopses, figuras, equações Maceió: Ed. Universitária/ UFAL, 175 p, ISBN: 978-85-7177-419-3, 2008a.

Fedorova Natalia. *Sinótica II: Conceitos básicos de cinemática e de dinâmica da atmosfera e análise dos campos meteorológicos.* Material didático: sinopses, figuras, equações Maceió: Ed. Universitária/ UFAL, 180 p, ISBN:978-85-7177-420-9, 2008b.

Fedorova N., Levit, V., Cruz C.D. "*On Frontal Zone Analysis in the Tropical Region of the Northeast Brazil*", Pure and Applied Geophysics 173, 1403-1421, DOI 10.1007/s00024-015-1166-y. (2016).

Fedorova Natalia, Pontes Da Silva Bruce Francisco, Levit Vladimir. *Análise e previsão prática das nuvens pelo diagrama termodinâmico SKEW-T/LOG-P,* 227p. ISBN: 978-85-923360-1-1, 2017.

Petterssen, Sverre. *Weather analysis and Forecasting.* v. 1, New York, Toronto, London: McGraw-Hill, 428p., 1956a.

Petterssen, Sverre. *Weather analysis and Forecasting.* v. 2, New York, Toronto, London: McGraw-Hill, 266p., 1956b.

CHAPTER II.2

ELABORATION OF A FORECASTED PROFILE

NATALIA FEDOROVA AND VLADIMIR LEVIT

1. Simulated profiles – Sp

The simulated profiles (***Sp***) are vertical profiles of T and Td constructed using data from numerical weather forecasting models. ***Sp*** are used due to lack of radiosonde data in the region. Various numerical models can be used to construct these profiles (Oliveria et al., 2020). The data used to construct ***Sp*** are the u and v components of wind, humidity (relative humidity or Td), and air temperature. The profiles were plotted using data from 10 standard levels (1000, 950, 900,850, 800, 700, 600, 500, 400, 300hPa).

Figures II-2-1, II-2-3 and II-2-4 show ***Sp*** elaborated by data from the reanalysis I and II of the National Centers for Environmental Prediction (NCEP) with a resolution of 2.5° in latitude and 2.5° in longitude. Simulated profiles can be also created with the data from the European Center for Medium-Term Weather Forecasting (ECMWF), which are available at synoptic times on the internet portal: http://apps.ecmwf.int/datasets/data/interim-full- daily/levtype=pl (Oliveria et al., 2020). This data has better resolution, up to 0.25° X 0.25°.

2. Elaboration of a forecasted profile – *Fp*

2.1 Method I – using trajectories of the air particle

The forecast of rainfall and other adverse phenomena is based on forecasted profiles (F***p***). To elaborate on ***Fp***, data on the trajectories of an air particle (Djuric, 1994; Fedorova, 1999, 2008) at standard levels can be used. The Hybrid Single Particle Lagrangian Integrated Trajectory (HYSPLIT) model is a complete system for the calculation of air particle trajectories. The HYSPLIT template (online template) is available on the website: http://www.ready.noaa.gov/HYSPLIT_traj.php or www.arl.noaa.gov/hysplit/hysplit/. The HYSPLIT model was used to calculate the trajectories of air particle for the forecasting of 12, 24, 36 and 48h. The direction of the trajectory is BACKWARD, that is, from the starting point of the trajectory to the forecast point.

The construction of ***Fp*** was based on the use of air temperature (T) and dew point temperature (Td) at the starting points of the air particle trajectories at all standard levels (Levit et al., 2010, 2015; Fedorova et al., 2013, 2019).

The steps for the forecasted profile elaboration are following:

Step 1. Construction of the air particle trajectory at standard levels by the HYSPLIT model. The output of the model is plotted horizontally (latitude, longitude) and vertically (vertical movements), and makes the data available every hour in tabular form. An example of a trajectory approximately at the level of 7000 m can be seen in Figure II-2-1a in both the horizontal and vertical planes. An example of the presentation of tabular results can be seen in Figure II-2-2, where the data is available every hour in a table in ASCII format.

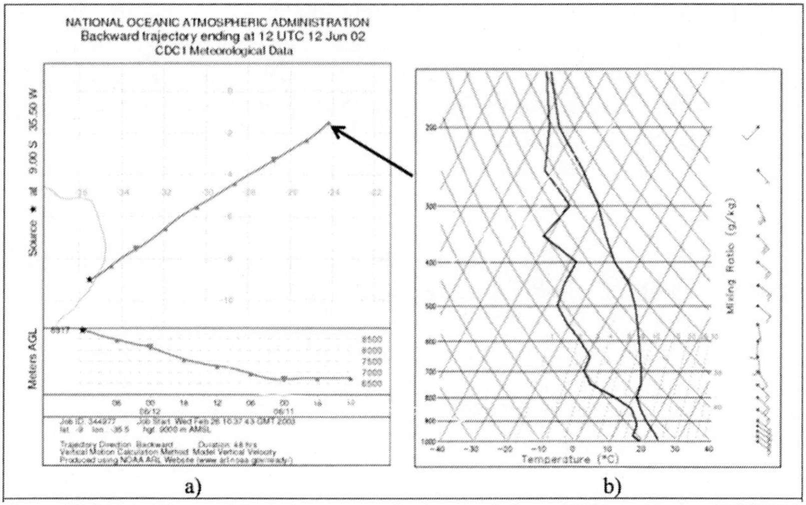

Figure II-2-1 – Air particle trajectory at the level of about 7000 m by the HYSPLIT model (a) and simulated vertical profiles at the starting point of this trajectory (b). The black arrow indicates the starting point of the air particle trajectory.
Source: HYSPLIT; adapted from Fedorova et al., 2019

```
 5      1
GDAS     13     10     15      0      0
GDAS     13     10     22      0      0
GDAS     13     10     29      0      0
GDAS     13     11      1      0      0
GDAS     13     11      8      0      0
 1 BACKWARD OMEGA
13     11      3     18    -9.600   -36.000   9380.0
 1 PRESSURE                                                Horário  Latitude  Longitude  Altitude   Pressão
 1      1     13     11      3     18      0      0     0.0    -9.600   -36.000   9380.0    300.1
 1      1     13     11      3     17      0      1    -1.0    -9.817   -36.686   9183.2    308.5
 1      1     13     11      3     16      0      2    -2.0   -10.165   -37.348   9103.7    319.1
 1      1     13     11      3     15      0      3    -3.0   -10.675   -37.957   9138.3    308.2
 1      1     13     11      3     14      0      2    -4.0   -11.347   -38.535   9204.1    306.2
 1      1     13     11      3     13      0      1    -5.0   -12.139   -39.089   9334.6    302.0
 1      1     13     11      3     12      0      0    -6.0   -13.121   -39.553   9367.2    293.0
 1      1     13     11      3     11      0      1    -7.0   -14.275   -39.915   9264.3    298.7
 1      1     13     11      3     10      0      2    -8.0   -15.535   -40.142   8893.8    311.8
 1      1     13     11      3      9      0      3    -9.0   -16.816   -40.212   8495.7    331.7
 1      1     13     11      3      8      0      2   -10.0   -18.080   -40.116   8360.2    351.1
 1      1     13     11      3      7      0      1   -11.0   -19.326   -39.832   8303.2    354.7
 1      1     13     11      3      6      0      0   -12.0   -20.531   -39.406   8211.5    362.9
```

Figure II-2-2 – Trajectory of the air particle at the level of 300hPa by the HYSPLIT model in table form. An example of a 12h plot location is shown in the black box. Source: HYSPLIT; adapted from Fedorova et al., 2019

Step 2. Construction of the vertical profiles of T and Td at the starting point of the air particle trajectory at all standard levels (Figure II-2-1b).

It is important to emphasize that the starting point is the point from which the air particle arrives, and the end point of the trajectory coincides with the point for which the forecast will be made. That is, the air properties of the start point will be transferred to the end trajectory point, to the forecast point. Due to the lack of radiosonde stations, the vertical profile at the starting point of the trajectory can be constructed from the numerical model data, i.e., just like ***Sp***.

Step 3. Obtaining information about T and Td through the vertical profile at the starting point of the trajectory (Step 2) at the level of this trajectory.

It should be emphasized that the air particle does not move only at one level, it can rise or fall. For example, it can be seen how an air particle rises along a trajectory in the vertical trajectory plot in figure II-2-3a. At the beginning of the trajectory, the air particle was at 6800 m level and, at the end of the trajectory, at 8917 m level. The properties of the air particle at the starting point (T and Td at the initial level of 6800 m) will be transferred to the forecast point at 8917 m level. For comparison, T and Td are shown at the nearest standard level at 400 hPa (II-2-3b). If T and Td vary greatly with height, their differences at standard and actual levels can be significant. For vertical profiles without significant variations in T and Td with height, standard levels can be used.

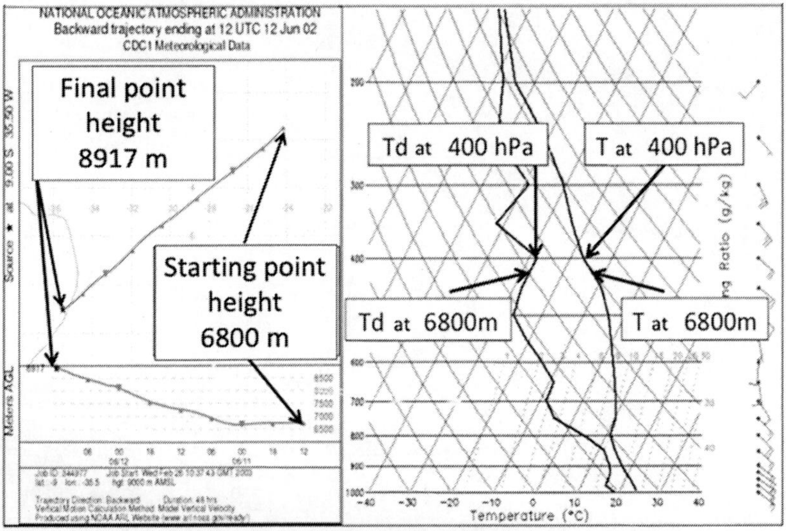

Figure II-2-3 – The heights of the start and final points of the air particle trajectory at the level of about 7000m (a) and vertical profile at the start point of the trajectory showing where the T and Td data came from to the final point (b). For comparison, T and Td values at a standard level of 400 hPa are shown.
Source: HYSPLIT; adapted from Fedorova et al., 2019

Step 4. Transfer the T and Td data from the starting point and level (step 3) to the forecasted vertical profile, by including T and Td in the forecasted profile at the final level.

Step 5. Steps 1 to 4 must be repeated for all standard levels.

The use of more levels gives better ***Fp*** quality, usually 10 levels are used (950, 900, 850, 800, 700, 600, 500, 450, 400 and 300 hPa).

The complete process of building ***Fp*** is shown in Figures II-2-4 and II-2-5. The trajectory data were used in the form of a table in Figure II-2-4 and in the form of a vertical profile in Figure II-2-5.

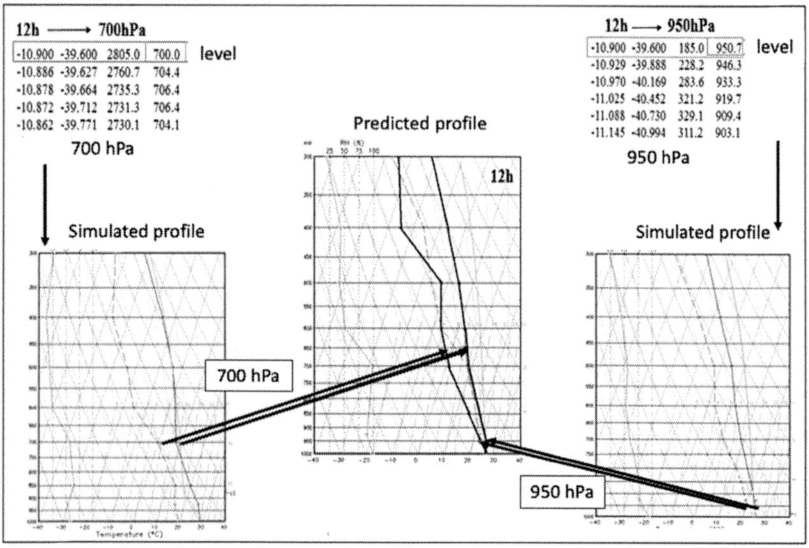

Figure II-2-4 – Building a forecasted profile based on trajectory data in the form of a table:

- tabular results of trajectories (coordinates of the starting point) at levels of 700 hPa and 950 hPa (tables above the figure);
- simulated profiles at starting points for the same levels;
- transfer of T and Td data from the starting point and level of the simulated profile to the forecasted vertical profile, shown by arrows.

Source: Adapted from Fedorova et al., 2019

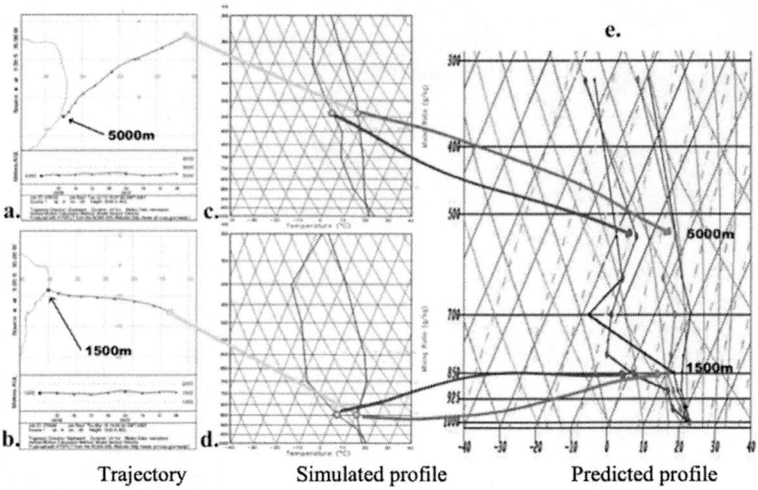

Figure II-2-5 – Forecasting of vertical profile:
a) air parcel trajectory of the HYSPLIT model at the 5000m level;
b) air parcel trajectory of the HYSPLIT model at the 1500m level;
c) vertical profile by NCEP reanalysis data at the beginning point of the trajectory at the 5000m level;
d) vertical profile by NCEP reanalysis data at the beginning point of the trajectory at the 1500m level;
e) predicted vertical profile which were produced using temperature and humidity data at the corresponding levels from vertical profiles in figures c and d.
Only 2 levels are presented, but indeed 9 levels were used.
Source: Adapted from Fedorova et al., 2013

Summarizing all these steps, a diagram of the predicted profile was constructed (Figure II-2-6).

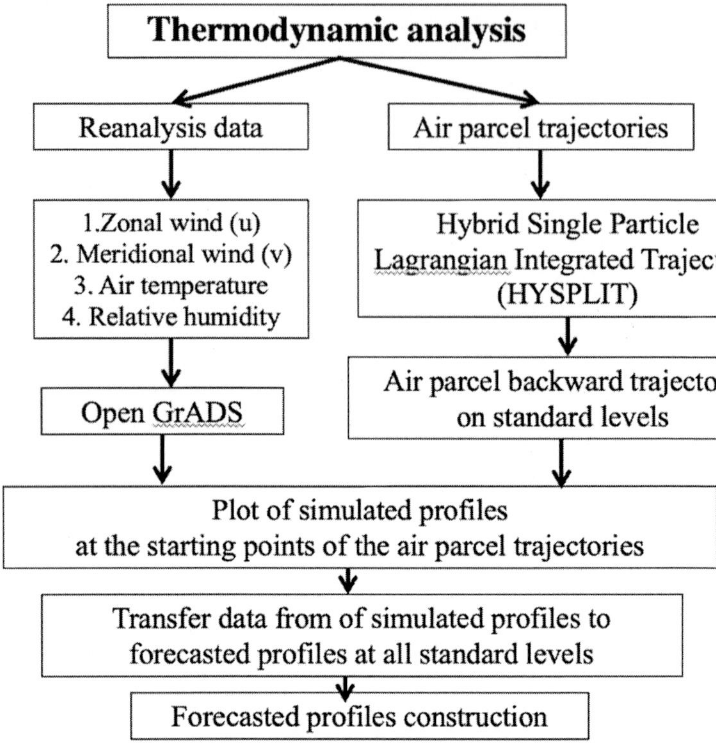

Figure II-2-6 – Steps for the forecasted profile elaboration

2.2 Method II – using vertical motions and advection T and Td

The forecasted profile can be built using information about modification in the real profile (vertical profile from radiosonde data or a simulated profile) by vertical motions and advection (Fedorova et al., 2017). Temperature variation due to upward and downward movements and temperature advection has been described in detail previously (Chapter II-1).

The steps for the forecasted profile elaboration are following:

Step 1. Obtaining vertical profile T and Td from radiosonde data or numerical model (Simulated profile (***Sp***)) at the place where the forecast will be elaborated (at the forecasted point).

An example of the forecasted profile elaboration can be seen in Figure II-2-7. The simulated profile at the starting time (time 1) is represented by the line T_1 (solid line) and T_{d1} (dashed line).

Step 2. Construction of a vertical profile of T and Td with information on temperature advection (Fedorova et al., 2017).

This example (Figure II-2-7) shows warm advection at low levels of 1000-700 hPa. The values of warm advection are following: at 700 hPa is 2°C, at 800 hPa is 3°C, at 900 hPa is 4°C, and at 1000 hPa is 6°C. Above 700 hPa there was no temperature advection and hence no thermal modification. Therefore, the forecasted profile (vertical temperature distribution at T_2) passes along the *dashed* line (T_2): between 1000-700 hPa the line passes at the ends of the arrows and over the continuation of this dashed line until it intersects with T_1 at 650 hPa. The full forecasted profile includes a dashed line (T_2) between 1000-650 hPa and a solid line above 650 hPa.

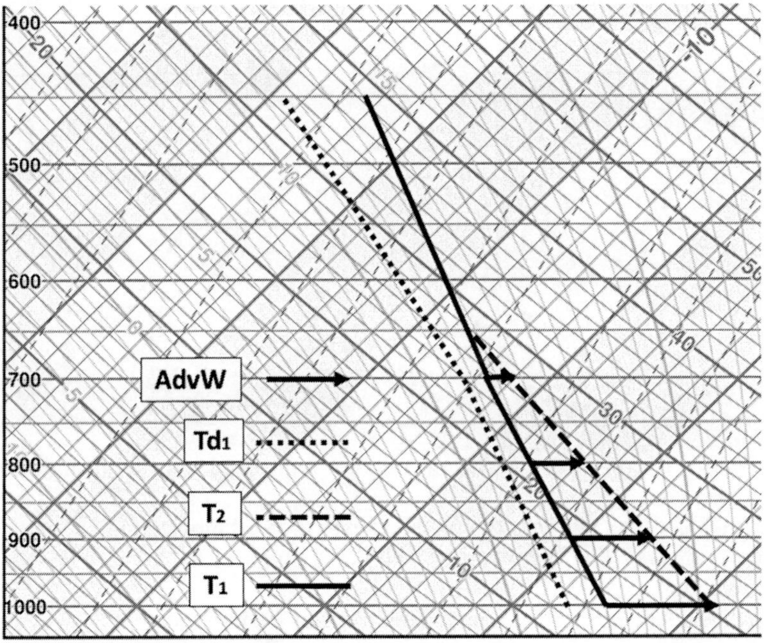

Figure II-2-7 – Simulated profile at time 1 (T_1 and T_{d1}) and construction of the forecasted profile using warm air temperature advection (AdvW) at 1000, 900, 800 700 hPa.

T_1 – *solid* line, temperature at starting time 1;

T_2 – *dashed* line, temperature at forecasted time 2;

Td_1 – *dotted* line, dew point at starting moment.
Arrow shows warm advection
Source: Adapted from Fedorova et al., 2017

Step 3. Modification of the vertical profile T and Td with information about vertical motions Fedorova et al., 2017

The example shows the modification in the vertical temperature profile obtained in step 2 (*dashed* line in Figure II-2-8) due to upward movements. The upward movements were as follows: by 60 hPa from 900 hPa to 840 hPa and by 20 hPa from 750 hPa to 730 hPa. The point T will rise along the dry adiabatic line, and the point Td - along the mixing ratio line.

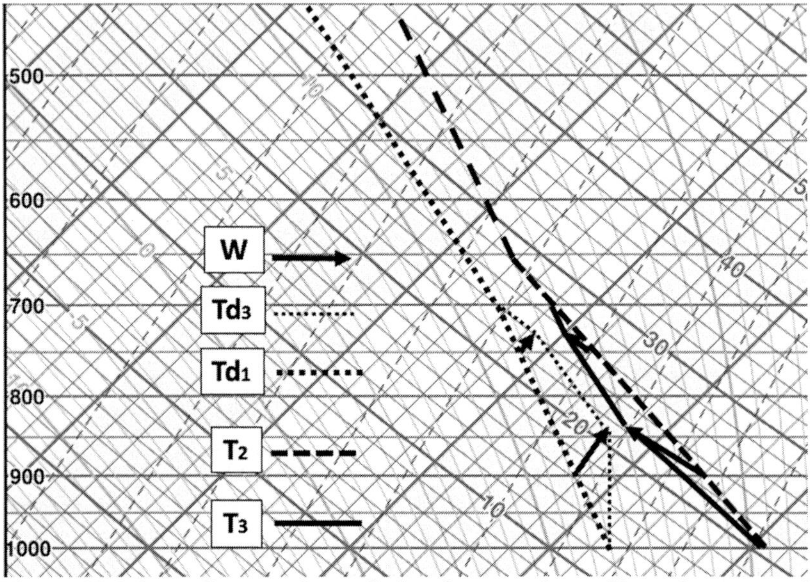

Figure II-2-8 – Construction of the forecasted profile using upward movements.
T_3 – *solid* line, temperature profile due to upward movements.
T_2 – *dashed* line, temperature profile in forecasted time 2 due to warm advection.
Td_1 – *dotted* line of big points, dew point in starting moment.
Td_3 – *dotted* line of small points, dew point after upward movements.
The arrows show upward movements, W.

The arrows show these upward movements. The forecasted T_3 profile will pass through the end of the arrows from T_2 (*solid* line) and Td_3 profile will pass through the end of the arrows from Td_1 (*small*, dotted line).

If the upward movement is ***more intense,*** by 100 hPa from 900 hPa and by 60 hPa from 750 hPa (Figure II-2-9), lifting is carried out as follows: from the initial level, the point T rises along the dry adiabatic line, and the rise of Td along the mixing ratio line up to the intersection and above the point rise along the wet adiabatic line. The example in figure II-2-9 shows that points T and Td rise from 900 hPa up to their intersection at 830 hPa, condensation occurs at this level (T=Td) and above this point up to 800 hPa the rise occurs along the wet adiabatic line. The condensation of water vapor rising from 750 hPa occurs at the level of 720 hPa, and above this point the rise by 40 hPa occurs along the wet adiabatic line.

As in the previous example, the forecasted T_3 profile will pass through the end of the arrows from T_2 (*solid* line), and the Td_3 profile will pass through the end of the arrows from Td_1 (*small* dotted line). Finally, the forecasted T profile will pass through *dashed* line and the Td profile will pass through the *dotted* line (small points in the layer of 1000-650 hPa and big points above 650 hPa).

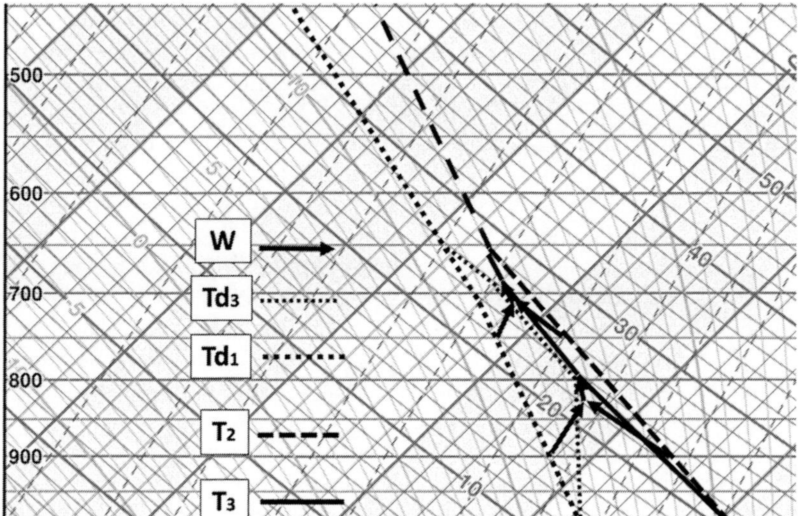

Figure II-2-9 – Construction of the forecasted profile using intense upward movements

T_3 – *solid* line, temperature profile due to upward movements.

T_2 – *dashed* line, temperature profile in forecasted time 2 due to warm advection.

Td_1 – *dotted* line of big points, dew point in the initial moment.
Td_3 – *dotted* line of small points, dew point after upward movements.
The arrows show upward movements, W.

It is necessary be concluded that predicted profiles are the basis of all types of thermodynamic analysis and, consequently, the forecast of all adverse meteorological phenomena.

3. References

Djuric Dusan, *Weather Analysis*. New Jersey: Prentice Hall, 304p, ISBN: 0-13-501149-3, 1994.

Fedorova Natalia, Pontes Da Silva Bruce Francisco, Levit Vladimir. *Complexos Convectivos de Mesoescala sobre o Nordeste do Brasil.* Maceió: Ed. Clube dos Autores, 248 p.; ISBN: 978-65-900634-0-3, 2019.

Fedorova Natalia, Pontes Da Silva Bruce Francisco, Levit Vladimir. *Análise e previsão prática das nuvens pelo diagrama termodinâmico SKEW-T/LOG-P,* Maceió: Ed. Clube dos Autores, 227p. ISBN: 978-85-923360-1-1, 2017.

Fedorova Natalia. *Sinótica II: Conceitos básicos de cinemática e de dinâmica da atmosfera e análise dos campos meteorológicos.* Material didático: sinopses, figuras, equações. Maceio: Ed. Universitaria/ UFAL, 180 p, ISBN: 978-85-7177-420-9, 2008.

Fedorova Natalia. *Meteorologia Sinótica I.* Pelotas: Ed. Universitária/UFPel, 260p, ISBN: 85-7192-108-3, 1999.

Fedorova Natalia, Levit Vladimir, Oliveira da Silva Aliton, Bonfim dos Santos Deydila Michele *"Low visibility formation and forecasting on the northern coast of Brazil." Pure and Applied Geophysics*, 170: 689-709, (2013) DOI: 10.1007/s00024-012-0565-6

Levit Vladimir, Bonfim Dos Santos Deydila Michele, Fedorova Natalia *"Haze formation on the northern coast of Brazil."* In: *5th International conference on fog, fog collection and dew*, Munster, Germany, 11-14, (2010).

Levit Vladimir, Fedorova Natalia, Cordeiro Edwans Silva, Cavalcante Lucas Carvalho Vieira. *Capitulo 8. Previsão de trovoadas no estado de Alagoas.* In: Atmosfera e Sociedade: Meio Ambiente e Elementos Meteorológicos, Volume IV, Edufal, Maceio –Al, 127-138. ISBN 978-85-7177-919-8, CDU 551.5.504. 2015.

Oliveria Camila, Melo Davidson, Silva Glenda, Lins Mayara, Barros Thiago, Edited by Natalia Fedorova, Vladimir Levit. *Primeiros Passos em Análise Meteorológica: OpenGrADS é Fácil,* 85p. ISBN: 978-65-00-02977-2, 2020.

HYSPLIT. http://www.ready.noaa.gov/HYSPLIT_traj.php; www.arl.noaa.gov/hysplit/hysplit/.

CHAPTER II.3

PRECIPITATION FORECASTING

NATALIA FEDOROVA AND VLADIMIR LEVIT

1. General Considerations for Precipitation Forecasting

Precipitation forecasting can be developed using various methods: synoptic analysis, numerical models, satellite data and thermodynamic methods. Synoptic analysis is based on the joint analysis of the results of numerical models, surface and satellite data. The thermodynamic analysis uses radiosonde data and vertical profiles developed from the results of numerical models.

Synoptic large-scale processes in tropical and extratropical regions affect the weather formation in **Northeast Brazil (NEB)**. There are differences in the methods of analyzing the processes that are formed in these regions, so they will be considered separately.

2. Extratropical region

The main extratropical large-scale synoptic systems affecting the NEB are Frontal Zones (Fz) and the Brazilian Northeast Jet Stream (BNEJS). These systems are detailed in Chapters 6 and 7 of Part I.

Below are various synoptic maps developed using numerical models, to identify *(step 1)* and forecast *(step 2)* these systems.

Step 1. Identification of synoptic systems

Synoptic systems are identified by the following characters on maps and satellite images:

1) Fz (Fedorova et al., 2016; Muanza, 2018; Muanza et al., 2021):
- periphery of a weak trough of a baroclinic cyclone from pressure maps at surface level;
- zone of streamlines confluence at low levels (925 and 850 hPa);

- band of negative values of relative vorticity at the level of 1000 hPa;
- is localized between the advection (Adv θ_e) regions of cold and warm equivalent potential temperature (θ_e);
- cloudiness band on the infrared (IR) satellite image.

2) BNEJS (Fedorova et al., 2018a, 2018b):
- strong air current on streamline maps at high level (200 hPa);
- wind speed in the core of the air current ≥ 20m s^{-1}.

Step 2. Forecast of synoptic systems

Using the forecasted maps and the same characteristics mentioned for the systems identification in step 1, evaluate the location and intensity of the same systems.

Special attention should be paid to

1. Frontogenesis and cyclogenesis process (Chapter I.9.3).
   Pay particular attention to the formation of a frontal wave and a cyclone and their evolution at a latitude of 20ºS near the coast of South America.
2. Speed of cyclone and frontal zone (Chapter I.9.5).
3. Identification of parts of the cyclone and frontal zone that will affect the forecast area. Usually, the periphery of cold fronts reaches the NEB (Chapters I.9.4, I.7).
4. Clouds evolution on satellite images (Chapter I.9.6).

This information will make it possible to forecast the type of cloudiness of extratropical synoptic scale systems and their evolution.

Additional information for Steps 1 and 2: Estimation of heavy precipitation according to its typical location by satellite data (Chapter I.9.7).

3. Tropical region

The list of synoptic systems of tropical origin, typical for the BNE region, is presented below. Also mentioned are various synoptic maps developed using numerical models to identify *(step 1)* and forecast *(step 2)* these systems.

Step 1. Identification of synoptic systems

Synoptic systems are identified using the following maps and satellite images:

1) ITCZ (Chapter I-1):
- difluence of streamlines at high (200 hPa) levels.
- confluence of streamlines at low (925, 850 hPa) levels.
- cloudiness band along the equator on IR satellite images; this band is heterogeneous; and it consists of conglomerates of convective clouds.

2) UTCV (Chapter I-4):
- cyclonic circulation at high levels (200, 300 hPa), which can be identified by streamlines.
- cloudiness on IR satellite images; in some events, the cloudiness has the shape of a vortex, but in many cases it does not.

3) MTCV (Chapter I-5):
- cyclonic circulation at middle levels (700 - 500 hPa), which can be identified by streamlines.
- cloudiness on IR satellite images; the cloudiness does not show the shape of a vortex.

4) Wave Disturbances in Trade Winds (WDTW) (Chapter I-3):
- trough on pressure maps at surface level.
- trough at low levels (1000, 925, 850 hPa) by streamlines; this though usually appears in trade winds.
- cloudiness on IR satellite images; in some events, the cloudiness presents the shape of an MCC.

5) MCC: cloudiness on IR satellite images (Chapter II-6).

Step 2. Forecast of synoptic systems

First, the evolution and displacement of the synoptic systems that were identified in the study region in step 1 must be estimated using the same variables. For this, forecasted maps of the same variables are used. The same characteristics specified in step 1 are used for this goal. Simulated satellite clouds can be derived from Tropical Tidbits data (Simulated IR Brightness temperature). Humidity forecast maps will also be useful. For this, maps of different levels for different synoptic systems are used. For example, high level humidity maps would be used for the UTCV forecast, and low levels humidity maps would be used for the WDTW forecast.

A list of forecasted maps with additional comments is presented below:

1) ITCZ:
- forecasted maps of streamlines at high (200 hPa) and low (925, 850 hPa) levels.
- simulation of satellite clouds can be obtained from Tropical Tidbits data.

2) UTCV:

- forecasted maps of streamlines at high levels (200, 300 hPa).
- the location of the UTCV center locat will allow to find the UTCV part (western, eastern, … part), that affects the study region.
- simulated satellite imagery and humidity maps at high levels will help determine the presence of a vortex shape.

3) MTCV:

- forecasted maps of streamlines at middle levels (700 - 500 hPa).
- simulated satellite imagery and mid-level humidity maps help to identify cloudiness.

4) Wave Disturbances in Trade Winds (WDTW):

- forecasted pressure maps at surface level.
- forecasted maps of streamlines at low levels (1000, 925, 850 hPa).
- simulated satellite images and low-level humidity maps will be helpful to identify cloudiness.

5) MCC:

information about MCC forecast is presented in (Chapter II-6):

The result of this step is the definition of a synoptic system that influences the formation of typical vertical motions and humidity and, consequently, precipitation.

4. Precipitation forecasting by thermodynamic methods

As a result of the analysis of synoptic systems (all Chapters of Part I), the forecaster will receive information about the typical of these systems vertical motion, humidity, normal precipitation, and adverse phenomena in the forecast region. A more detailed forecast can be elaborated using thermodynamic analyses of the forecasted profiles (Chapter II-3). Information about the typical features of the structure of synoptic systems will be used to verify and modify the forecasted profile. These forecasted vertical profiles will be analyzed by thermodynamic methods. Thermodynamic methods are similar in tropical and extratropical regions; therefore *steps 3 and 4* for them will be described together for these regions.

Step 3. Elaboration of predicted profile (Chapter II-2)

Step 4. Identification of clouds, precipitation, and phenomena by forecasted profile (Fedorova et al., 2017).

The thermodynamic diagrams (Figures II-3-1 and II-3-2) show the levels at which precipitation will form. The upward motions will cause the air parcel to rise to the condensation level. This rising will occur for T and Td along dry adiabatic line and mixing ratio line, relatively, up to its intersection. This intersection point T, equal to Td, is the condensation point (Tc). Above this point, the rise will occur along the wet adiabat. Precipitation will form between Tc point and the end point of the rising ("End" at figures II-3-1 and II-3-2).

Precipitation can be calculated as the difference between the mixing ratio at the Tc point due to lifting and the mixing ratio at the end point (*End*) of the rising. This process will be illustrated with examples for dry and wet air and, also, for lifting and sinking.

Examples of precipitation calculation:

1) Dry air (T-Td) = 10°C and upward motions (200 hPa) (Figure II-3-1)

Beginning point is located at the level of 700 hPa, T = 10°C, Td = 0°C, W = 200 hPa (from 700 hPa up to 500 hPa).

Firstly, the *condensation point Tc* will be found:

- from 10°C at 700 hPa will rise along the dry adiabatic;
- from 0° at the same level will rise along the mixing ratio line; and
- Tc will be determined at their interaction.

The mixing ratio at the *Tc* point is 5,5 g/kg and at the *End* point is 3.3 g/kg. Precipitation calculation result: 5,5-3,3 = **2,2 g/kg**

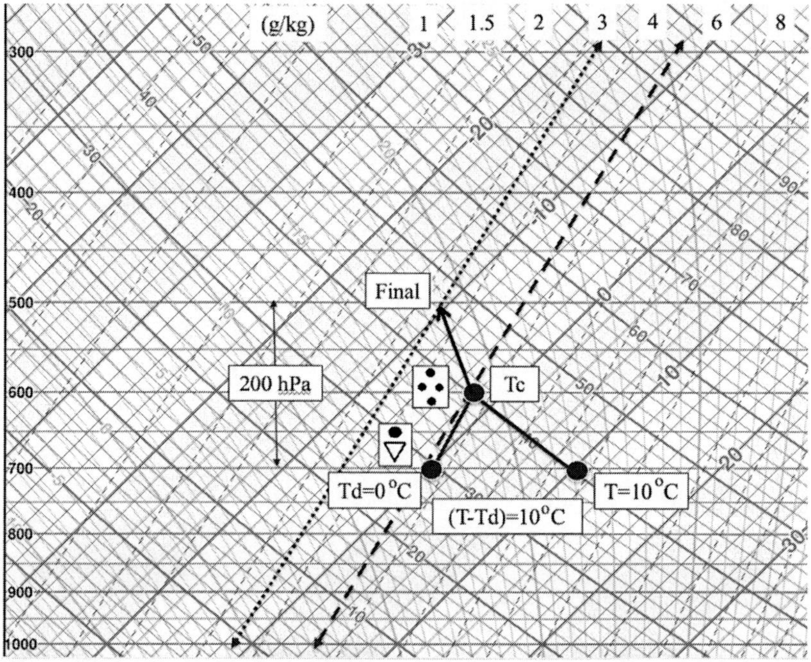

Figure II-3-1 – Example of precipitation calculation with upward movement in dry air.
Dry adiabatic line (γ_d - adiabatic lapse rate of temperature) - *solid grey straight line, left sloping line.*
Wet adiabatic line - *solid grey curved line, left slanted line.*
Mixing ratio line - *dashed grey straight line, right slanted line.*
Tc- *condensation point.*

2) Humid air (T-Td) = 4°C and upward movement (200 hPa) (Figure II-3-2)

Beginning point T = 10°C, Td = 6°C, W = 200 hPa (from 700 hPa up to 500 hPa). Precipitation calculation result: 9 – 4.8 = **4.2 g/kg**

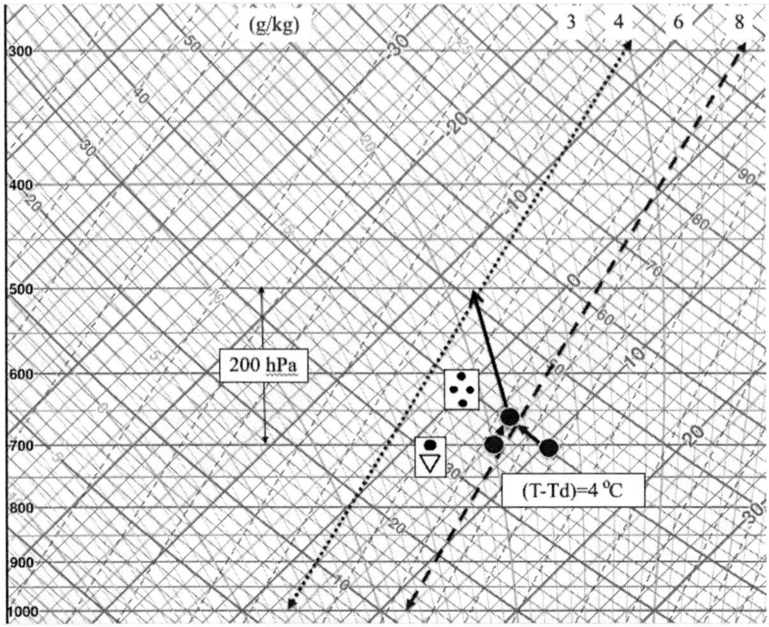

Figure II-3-2 – Example of precipitation calculation with upward movement in humid air.

3) Dry air (T-Td) = 10°C and weak upward movement (50 hPa) (Figure II-3-3)

Beginning point at 700 hPa level, T = 10°C, Td = 0°C, W = 50 hPa (from 700 hPa up to 750 hPa). Final point (T-Td) = 5°C. Precipitation calculation result: without precipitations.

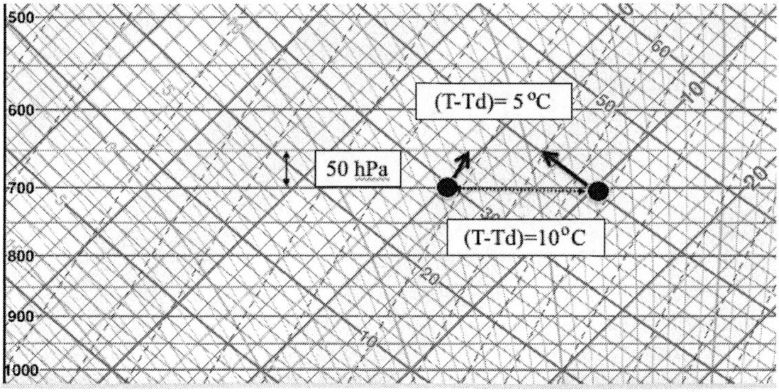

Figure II-3-3 – Example of precipitation calculation with weak upward movement in dry air.

4) Humid air (T-Td) = 4°C and sinking (100 hPa) (Figure II-3-4).

Beginning point T = 10°C, Td = 6°C, W = 100 hPa (from 700 hPa up to 800 hPa). Final dry air with (T-Td) = 12°C. Precipitation calculation result: without precipitations.

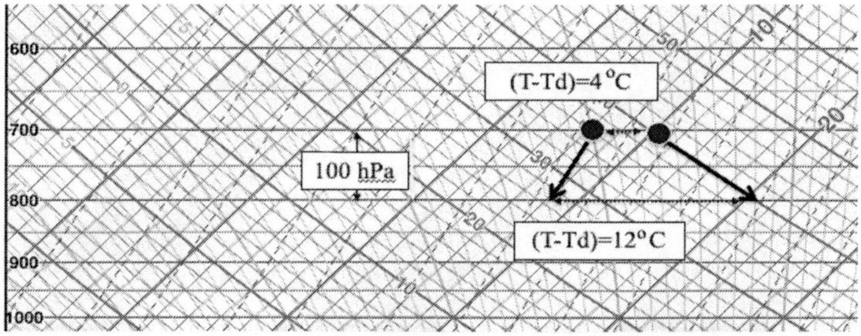

Figure II-3-4 – Example of precipitation calculation of humid air sinking.

5. References

Fedorova Natalia, Pontes Da Silva Bruce Francisco, Levit Vladimir. *Análise e previsão prática das nuvens pelo diagrama termodinâmico SKEW-T/LOG-P,* 227p. ISBN: 978-85-923360-1-1, 2017.

Fedorova Natalia, Levit Vladimir., Campos Antonio Marcos Vianna. *"Brazilian Northeast Jet Stream: frequency, wind speed and direction."* Meteorological Applications, 25: 254-260, (2018a), DOI: 10.1002/met.1688

Fedorova Natalia, Levit Vladimir., Campos Antonio Marcos Vianna.

"*Brazilian Northeast Jet Stream: Association with synoptic scale systems.*" Meteorological Applications, 25: 261-268, (2018b) DOI: 10.1002/met.1693

Fedorova Natalia, Levit, Vladimir., Cruz Cezar Duarte. "*On Frontal Zone Analysis in the Tropical Region of the Northeast Brazil*," Pure and Applied Geophysics 173: 1403-1421. (2016) DOI 10.1007/s00024-015-1166-y.

Muanza Gomes António Muanza. *Padrões de circulação atmosférico associados á ligação da extremidade frontal com a ZCIT.* MSc. Thesis, Federal University of Alagoas, Maceio, Brazil. (2018).

Muanza Gomes António Muanza; Natalia Fedorova; Vladimir Levit "*Connection of the frontal extremities whith intertropical convergence zone and analysis of associated phenomena.*"

Northeast Geociences Magazine *7, No 1, 2021*, https://doi.org/10.21680/2447-3359.2021v7n01ID

CHAPTER II.4

THUNDERSTORMS FORECASTING

NATALIA FEDOROVA AND VLADIMIR LEVIT

The thunderstorm forecast is based on climatological data together with synoptic and thermodynamic analysis. Below, these analysis steps will be described separately and, finally, an algorithm (decision tree) for thunderstorm forecasting will be presented.

1. Thunderstorm climatology

The climatology of thunderstorms is presented in the middle region of the BNE (~10ºS). Thunderstorm frequency was determined using data from the METeorological Aerodrome Report -METAR and Airspace Control Department (ACD or Departamento de Controle do Espaço Aéreo - DECEA) data obtained from the Meteorology Network (Rede de Meteorologia- REDEMET) https://www.redemet.aer.mil.br/?i=facilidades&p=api-redemet.

Data from the TRMM (Tropical Rainfall Measuring Mission) satellite were also used during the period of satellite operation (1998-2015) from the Lightning Imaging Sensor (LIS). LIS records the time of lighting strike, measures the radiation energy and estimates the location with high detection efficiency during day and night. Sensor LIS observes lightning activity between 35ºN and 35ºS latitudes. This data is taken from the website: http://lightning.nsstc.nasa.gov.

Thunderstorms are most frequent during the warmest months (January, February, and March (50-58 events for 24 years and 2-2,5 events per month) (Table II-4-1; Brito et al., 2008, 2011; Cordeiro 2014; Levit et al., 2016; Cordeiro et al. 2013, 2018; Oliveira Silva, 2022). This frequency falling a little in April (40 events for 24 years and 1.7 events per month). Thunderstorms were not observed or were observed very rarely from June to October. In general, an average of 11 thunderstorms were recorded per year.

Table II-4-1 – Thunderstorm frequencies for 24 years (1998-2021), using data from TRMM (1998-2015) and from METAR/DECEA (1998-2021) in the International Airport Maceio - Zumbi dos Palmares (MCZ) (9° 31' S, 35° 47' W).
Me-medium
Source: METAR/DECEA; TRMM; Brito et al., 2008, 2011; Levit et al., 2016; Cordeiro et al., 2018; Oliveira Silva, 2022.

Month --- year	J	F	M	A	M	J	J	A	S	O	N	D	Σ	*Me*
1998	3	0	2	1	0	0	0	0	0	0	0	0	**6**	**0.5**
1999	1	1	4	0	2	0	0	0	0	0	1	4	**13**	**1.1**
2000	5	3	0	0	2	0	0	0	0	0	1	0	**11**	**0.9**
2001	0	0	2	0	0	0	1	0	0	2	0	7	**12**	**1.0**
2002	4	4	2	2	0	0	0	0	0	0	3	1	**16**	**1.3**
2003	4	5	2	2	0	0	0	2	0	0	0	1	**16**	**1.3**
2004	5	2	2	0	1	0	0	0	0	0	1	1	**12**	**1.0**
2005	4	2	5	1	1	1	0	0	0	0	0	3	**17**	**1.4**
2006	0	4	5	6	2	0	0	0	0	0	2	0	**19**	**1.6**
2007	1	8	3	6	1	0	0	0	0	0	0	1	**20**	**1.7**
2008	2	5	8	1	3	0	0	0	0	0	0	2	**21**	**1.8**
2009	0	1	1	2	3	0	0	0	0	0	0	2	**9**	**0.8**
2010	7	1	4	3	0	0	0	0	0	3	0	2	**20**	**1.7**
2011	5	3	3	4	1	0	0	0	0	0	0	0	**16**	**1.3**
2012	3	1	0	0	0	0	0	0	0	0	0	0	**4**	**0.3**
2013	1	0	1	1	0	0	0	0	0	0	2	0	**5**	**0.4**
2014	2	1	0	5	0	0	0	0	0	0	1	1	**10**	**0.8**
2015	1	2	0	0	1	0	0	0	0	0	0	1	**5**	**0.4**
2016	1	3	1	0	0	0	0	0	0	0	0	0	**5**	**0.4**
2017	0	1	1	0	1	0	0	0	0	0	0	0	**3**	**0.3**
2018	1	0	0	0	0	0	0	0	0	0	0	0	**1**	**0.1**
2019	1	1	6	0	0	0	0	0	0	0	0	0	**8**	**0.7**
2020	1	2	5	5	1	0	0	0	0	0	0	0	**14**	**1.2**
2021	1	0	1	1	0	0	0	0	0	0	1	1	**5**	**0.4**
Σ	**53**	**50**	**58**	**40**	**19**	**1**	**1**	**2**	**0**	**5**	**12**	**27**	**268**	
Me	**2.2**	**2.1**	**2.4**	**1.7**	**0.8**	**0**	**0**	**0.1**	**0**	**0.2**	**0.5**	**1.1**	**11.2**	

2. Synoptic systems associated with thunderstorms

Analysis of synoptic systems associated with thunderstorms was developed using data from the National Centers for Environmental Prediction do

National Center for Atmospheric Research (NCEP/NCAR), reanalysis2 model, resolution 2,5° X 2,5°. The data were taken from the website: https://psl.noaa.gov/data/gridded/data.ncep.reanalysis2.pressure.html.

In addition, ECMWF (European Centre for Medium-Range Weather Forecasts) Reanalysis v5 (ERA5) data were used. ERA5 model of the Copernicus Climate Change Service (C3S) has a resolution of 0,25° X 0,25°. This information was obtained from the website:
https://cds.climate.copernicus.eu/cdsapp#!/search?type=dataset.

Summarizing the previous results of Brito et al. (2008, 2011), Rodrigues et al. (2008), Cordeiro (2014), Levit et al. (2016), Cordeiro et al. (2013, 2018) and Oliveira Silva (2022), a list of synoptic systems associated with thunderstorms in the BNE was created. The predominant systems are represented at high (300-200 hPa), middle (700-400 hPa) and low (1000-850 hPa) levels.

The list of predominant synoptic systems associated with thunderstorms in the BNE is as follows:

- *High levels*: Trough, UTCV and BNEJS (Chapters I-4 and I-6). All these systems are interconnected; UTCV are formed from a Trough and BNEJS located on both sides of the Trough. Anticyclonic circulation was rarely observed. As a rule, thunderstorms occurred in the NW sector of the UTCV and less often around, in the N, NE, E, SW areas (Figure II-4-1a). Very rarely, thunderstorms were observed in the S, SE and E segments of UTCV. The most frequent thunderstorms are observed on the rear side of the Upper Tropospheric Trough (UTT), less often it is forms on the vanguard side of the UTT (Figure II-4-1b).
- *Middle levels*: cyclonic (Low or trough) and anticyclonic (High or ridge) circulation were recorded with similar frequency; at the same time, MTCV (Chapter I-5) was rarely observed. Therefore, these levels do not provide information for a thunderstorm forecast.
- *Low levels*: Typical synoptic systems for thunderstorm formation are the Trade Winds and Wave Disturbance on the Trade Winds (WDTW) (Chapter I-3), the periphery of the frontal zone and/or the connection of the frontal periphery with the ITCZ (Chapter I-7).

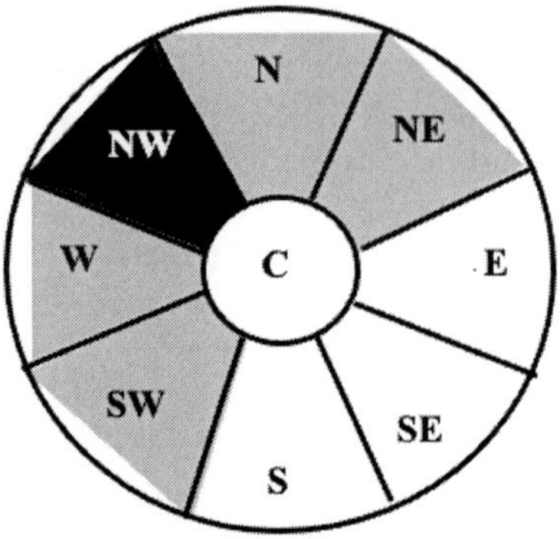

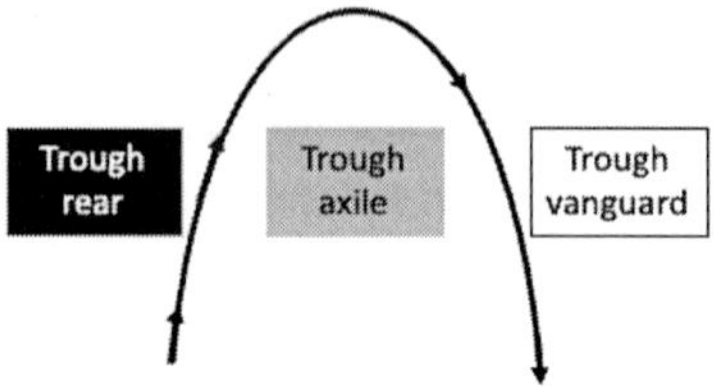

Figure II-4-1 – Example of precipitation calculation of humid air sinking. Frequency of thunderstorm occurrence in UTCV (a) and in the high-level Trough (b): black areas show very high frequency and rare occurrence in white areas. Source: Cordeiro, 2014; Levit et al, 2016; Cordeiro et al. 2018

3. Thermodynamic characteristics of the troposphere during thunderstorm days

Thermodynamic analysis developed based on the guidelines put forward in Alcântara (1969) and Fedorova et al. (2017).

Moderate instability is very typical for thunderstorm days in the BNE. The results of the Reanalysis 2 show more intense instability than the ERA5 model in the same synoptic systems (Tables II-4-2 and II-4-3).

On thunderstorms days, more intense instability was associated with UTCV, UTT, BNEJS, ITCZ and ITCZ with a frontal zone: CAPE > 1300 J/kg and LI <4.5 (Cordeiro et al., 2014, 2018). Moderate instability was observed on the frontal zones (CAPE ~ 1000 J/kg and LI ~ -4). At the same time, high instability was recorded on the days with the connection of the frontal periphery with the ITCZ (CAPE ~ 1500 J/kg and LI ~ -5). In all these situations, the ERA5 model shows less instability (CAPE: 600 - 1100 J/kg and LI<-2.5).

WDTW create intense convection (CAPE: 1300 - 1500 J/kg and LI: -4; -7) and this instability has been confirmed by both models.

Low instability was on thunderstorm days in MTCV (CAPE ~ 400 J/kg and LI ~2); usually the obtained CAPE values are not associated with the intensive development of convection and show weak convection, but the corresponding LI values show the convection development.

Thunderstorms formation with low CAPE+ (> 400 J/Kg) was described by Cordeiro et al. (2018) and confirmed (CAPE+ >500 J/Kg) by Oliveira

Silva (2022). At the same time, all studies show LI <-2 (in rare cases <-1) on thunderstorms days. This means that LI index indicated the possibility of intensive convection development. The LI<-4 index was recorded more often, and according to this indicator, very heavy rain was forecasted with severe thunderstorms.

Table II-4-2 – Thermodynamic characteristics of the troposphere on thunderstorms days: average values and CAPE+ limits (J/Kg).
In parentheses - years analyzed.
N- number of events; Ave-average, L-limits
ITCZ - InterTropical Convergence Zone
UTCV - Upper Tropospheric Cyclonic Vortex
UTT - Upper Tropospheric Trough,
BNEJS - Brazilian Northeast Jet Stream
FZ - Frontal Zone
MTCV - Middle Tropospheric Cyclonic Vortex
TW - Trade Winds
WDTW - Wave Disturbance on Trade Winds
Sources: Cordeiro 2014, Cavalcante, 2016, Oliveira Silva 2022.

	Cordeiro 2014 (1998-2012) 15 years			Cavalcante, 2016 (2013-2015) 3 years			Oliveira Silva 2022 (2018-2021) 4 years		
Synoptic system	N	Ave	L	N	Ave	L	N	Ave	L
ITCZ	15	1480	1000-2200	1	1370	1370	1	1100	1100
UTCV	94	1320	700-2300	6	1770	1060-1900	6	700	400-1000
UTT	40	1370	1000-2400				4	800	500-1000
UTCV+ BNEJS				1	1660	1660	3	750	500-1000
BNEJS				2	1310	1000-1600	4	600	300-900
FZ + ITCZ				3	1400	1000-1900	5	850	500-1500
FZ	10	1030	400-1800						
FZ + WDTW				2	1500	1100-1800			
WDTW				2	1400	1200-1600	1	1300	1300

TW							4	700	500- 1000
MTCV	2	400	200- 600						

Table II-4-3 – Thermodynamic characteristics of the troposphere on thunderstorms days: average values and LI limits.
Abbreviations are the same as in Table II-4-2
Sources: Cordeiro 2014, Cavalcante, 2016, Oliveira Silva 2022.

	Cordeiro 2014 (1998-2012) 15 years			Cavalcante, 2016 (2013-2015) 3 years			Oliveira Silva 2022 (2018-2021) 4 years		
Synoptic systems	N	Ave	L	N	Ave	L	N	Ave	L
ITCZ	15	-5	-4 -7	1	-4	-4	1	-4	-4
UTCV	94	-5	-2 -9	6	-6	-4 -9	6	-3	-1 -5
UTT	40	-5	-2 -8				4	-2.5	-2 -3
UTCV +BNEJS				1	-6	-6	1	-3	-2 -4
BNEJS				2	-5	-4 -6	4	-2.5	-1 -5
FZ + ITCZ				3	-5	-3 -7	5	-3	-2 -5
FZ	10	-4	-1 -6						
FZ + WDTW				2	-6	-6 -7			
WDTW				2	-5	-4 -6	1	-4	-4
TW							4	-2	-1 -3
MTCV	2	-2	-1 -3						

It is important to note that in order to calculate CAPE+ and LI in mountainous areas, it is necessary to take into account the process with the rise of air masses. For this, potential and latent instabilities are used (Alcantra, 1968; Cavalcante, 2016). Due to the influence of the relief, the instability characteristics increased significantly, approximately twice (Levit et al., 2016, Cavalcante, 2016).

4. Forecast Decision Tree for Thunderstorm Forecasting

Forecast Decision Tree for thunderstorm forecasting was described by Brito et al. (2011), Cordeiro (2014), Levit et al. (2016), Cavalcante (2016), Cordeiro et al. (2018), Oliveira Silva (2022). Summarizing all previous information, a new Decision Tree for thunderstorm forecasting has been elaborated (Figure II-4-2).

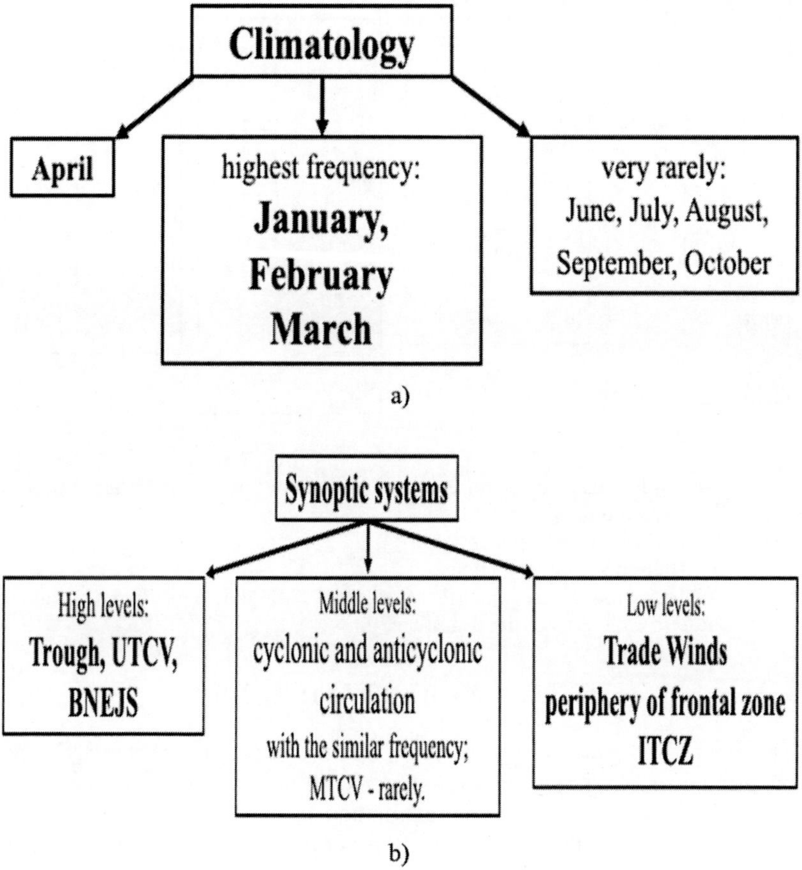

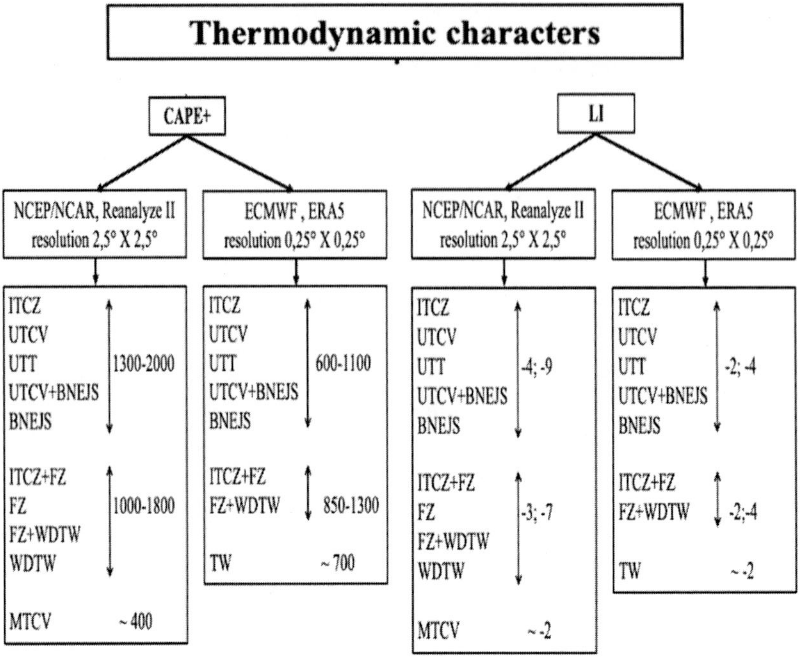

c)

Figure II-4-2 – Forecast decision tree for thunderstorm forecasting by three steps:
a) Step 1- Climatologic analysis,
b) Step 2- Analise synoptic systems,
c) Step 3 - Thermodynamic characters.
Source: summarizing Brito et al., 2011; Cordeiro 2014, Levit et al, 2016, Cavalcante, 2016, Cordeiro et al. 2018, Oliveira Silva 2022

5. Example on 13 March 2019

The example shows a typical thunderstorm situation in the BNE that occurred on 13 March 2019. First, this month has the highest frequency of thunderstorms. A typical synoptic situation with UTCV was observed on this day. UTCV at 200 hPa was located over the middle of the BNE (Figure II-4-3c). Below this vortex, at a level of 500 hPa, the High center was located (Figure II-4-3b). Very typical trade winds with weak cyclonic circulation and intense confluence were observed over the same region at 1000 hPa level (Figure II-4-3a). Very intense convection was identified by IR satellite images in the same region (Figure II-4-3d).

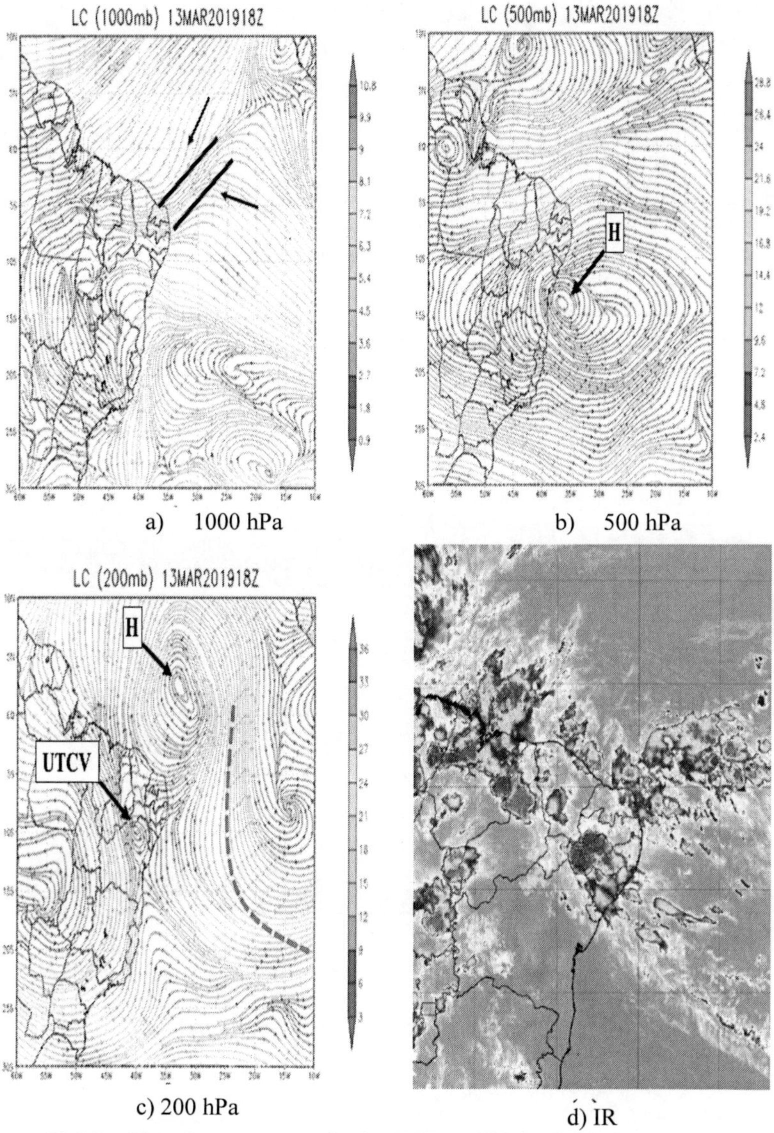

Figure II-4-3 – Thunderstorm event in the BNE on 13 March 2019:
a), b) and c) streamlines at 1000, 500 and 200 hPa, respectively
d) IR satellite image
a) double line – ITCZ

c) *UTCV- Upper Tropospheric Cyclonic Vortex; dashed line- jet stream from the Northern Hemisphere to the Southern Hemisphere.*
Source: Oliveira Silva, 2022

The vertical structure of the troposphere shows a humid layer at low levels (1000-800 hPa), a dry layer above (800-400 hPa) and a CAPE of about 1000 J/Kg, LI= -4 (Figure II-4-4). High humidity above 350 hPa is associated with *Cirrus* clouds over *Cumulus* e *Cumulonimbus* clouds. All these data indicate the instability and convection development, which is already recorded by satellite images.

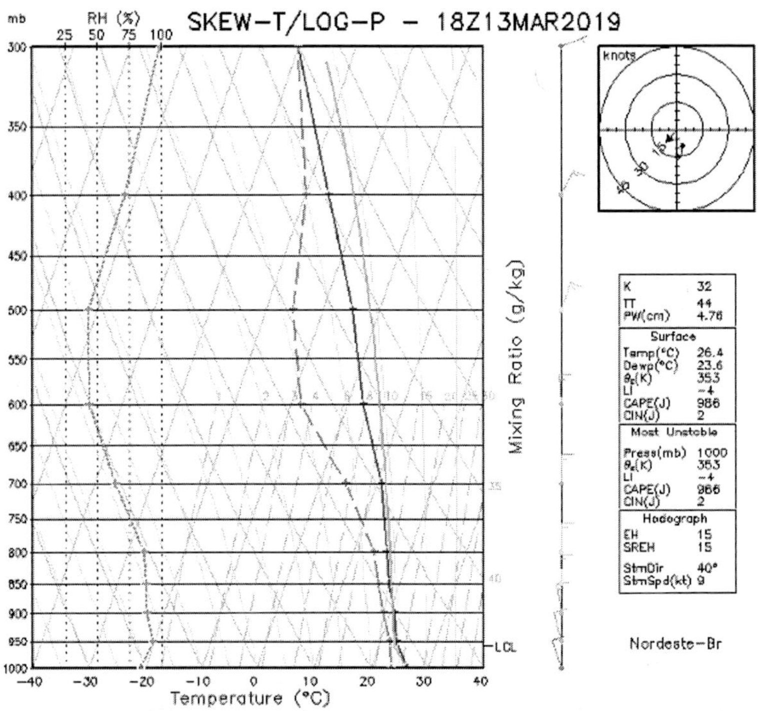

Figure II-4-4 – Thunderstorm event in the BNE on 13 March 2019: vertical profile of T and Td simulated by the ERA5 model and thermodynamic characteristics.
Vertical profile of the temperature (right black solid line) and dew point (dashed line) in the middle of the BNE coastal region (9°S, 36°W).
Source: Oliveira Silva, 2022

The map in Figure II-4-5 shows multiple electrical discharges in the Alagoas State between 17 and 23 UTC. The number of electrical discharges increased away from the coast over the highlands.

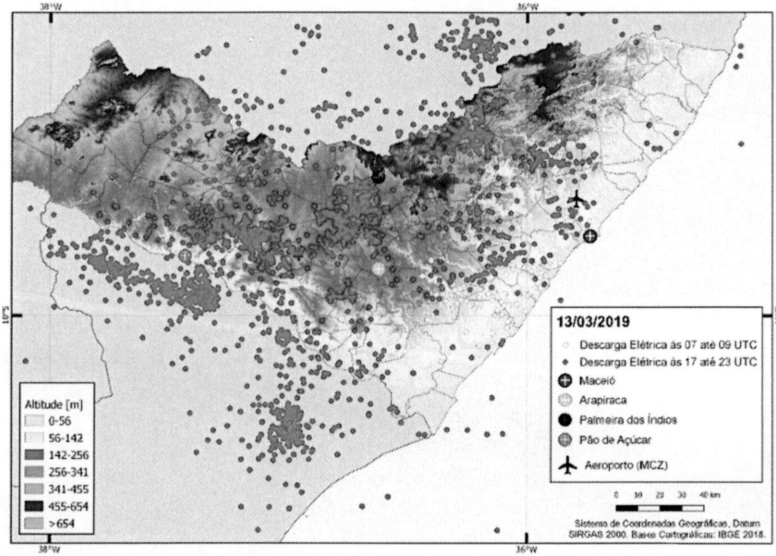

Figure II-4-5 – Thunderstorm event in the BNE on 13 March 2019: map of relief (m) and electrical discharges (gray points) in the Alagoas State.
Source: Oliveira Silva, 2022; ELAT-INPE, 2019

6. References

Alcântara, Fabio. *Manual de Análise do Diagrama "Skew-T, log P"*. Ministério da Aeronáutica, Diretoria de Rotas Aéreas. 122 p, 1969.

Brito, B.M.; Levit, V.; Fedorova, N.; Molion, L.C.B.; Tenório, R.S.; Rodrigues, R.N.; Silva, B.F.P. "*Análise do comportamento das trovoadas no estado de Alagoas, previsão à curto prazo.*" Revista Brasileira de Meteorologia 26, n.2: 243-256, (2011).

Brito Bruno Miranda. "*Utilização do Modelo HYSPLIT para a Previsão de Trovoada no Aeroporto Zumbi dos Palmares, em Maceió, Alagoas.*" Dissertation. – Universidade Federal de Alagoas, Maceió, 2008.

Cavalcante, Lucas Carvalho Vieira. "*Análise dos eventos com trovoadas no estado de Alagoas entre 2013 e 2015, levando em conta a influência do relevo.*" Dissertation. – Universidade Federal de Alagoas, Maceió, 2016.

Cordeiro, Edwans Silva. "*Análise sinótica dos eventos com trovoadas para o estado de Alagoas dentre o período de 15 anos (1998-2012)*" Dissertation. – Universidade Federal de Alagoas, Maceió, 2014.

Cordeiro, E.S.; Fedorova, N.; Levit, V. *"Análise sinótica e termodinâmica dos eventos com trovoadas para o estado de Alagoas no período de 15 Anos (1998-2012)."* Revista Brasileira de Meteorologia 33, n. 4: 685-694, (2018). DOI: http://dx.doi.org/10.1590/0102-77863340010.

Cordeiro, E.S., Silva, N.M., Fedorova, N.; Levit V. "*Análise termodinâmica e frequência de trovoadas entre os anos de 2008 a 2012 em Maceió, Alagoas."* Revista Brasileira de Geografia Física 6, n. 5: 1208-6, (2013).

Fedorova Natalia, Pontes Da Silva Bruce Francisco, Levit Vladimir. *Análise e previsão prática das nuvens pelo diagrama termodinâmico SKEW-T/LOG-P,* 227p. ISBN: 978-85-923360-1-1, 2017.

Rodrigues, L.R.L., Fedorova, N., Levit, V. "*Trovoadas na costa leste do Nordeste do Brasil associadas a interação entre ondas nos ventos de leste e VCAN."* In: 15 Congresso Brasileiro de Meteorologia (CBMET), São Paulo: SBMET. (2008).

Levit Vladimir, Cordeiro, Edwans Silva, Cavalcante, Lucas Carvalho Vieira. "Capítulo 2. Método de previsão de trovoadas no Nordeste Brasileiro." In: *Meteorologia em tópicos*, v 3, 79-131, 2016.

Glenda Cristine de Oliveira Silva, Natalia Fedorova; Vladimir Levit *"Estudo de Casos de Trovoadas com Descargas Eletricas em Alagoas"*. Anuário do Instituto de Geociências (IGEO) 45: 1-12, (2022). DOI: 10.11137/1982-3908_45_38477

Chapter II.5

Short Term Forecasting of Weak Rain, Low Visibility, Fog and Mist in the BNE

Natalia Fedorova, Vladimir Levit, Joao Pedro Goncalves Nobre and Lucas Carvalho Vieira Cavalcante

1. General information

Fog is solid/liquid suspended water droplets that reduce horizontal visibility below 1000 meters. Horizontal visibility in fog affects various sectors of society, including aviation, maritime, road, waterway and other types of transport. Thus, fog forecasting and monitoring prevent accidents and socio-economic losses (including fatalities) (Fedorova & Levit, 2016).

In recent years, the development of Numerical Weather Prediction (NWP) and the emergence of instruments capable of measuring aspects related to the fog microstructure have contributed to a better understanding of this phenomenon. In parallel, parametric methods were also developed relating the physical aspects of fog to horizontal visibility (Bott & Trautmann, 2002; Gultepe et al., 2006).

Low visibility events are a common occurrence throughout all regions of the Brazilian Northeast (BNE), with the central region experiencing them more frequently and the coastal region experiencing them more rarely (Fedorova & Levit, 2016). Despite the relatively small number of fog events registered by ground stations at BNE airports, they are still enough to cause significant social and economic losses. For example, at Maceio Airport (SBMO), fog was responsible for a plane crash that resulted in the tragic death of the pilot and the destruction of numerous banking documents.

Fog occurred on average twice a year in the center of the coastal region, more often during the rainy season of the BNE from March to August (Fedorova and Levit, 2016). The highest frequency reached six events per year, mainly in June and July. Most of the fogs were weak (with minimum visibility between 600 and 900 m) or moderate (with minimum visibility from 200 to 500 m), and only one was intense (with minimum visibility of 190 m). The onset of fog was more common between 4 and 7 am, and the duration of the fog was short, approximately two hours. Fog was more common in the central BNE region. Between 2008 and 2016, Campina Grande Airport experienced 105 days of fog, while only 20 events occurred at Maceio Airport (Melo et al., 2019).

Light fog with a visibility of 1-2 kilometers occurred in June (around 33 events per month, 1100 hours per year). Low visibility of 2-4km was detected in the center of the coastal region of the BNE during the rainy season with the highest frequency of 236 hours per season (Fedorova et al., 2013). Light fog was usually recorded along with rain or drizzle.

Haz frequency was lower than light fog for 80 hours per year in the center of the BNE coastal region (Fedorova et al., 2008; Levit, et al., 2010).

Stratus clouds appearance in the center of the coastal region of the BNE was observed from April to August with the highest frequency of 8 days per month (Fedorova et al., 2008; Gomes et al., 2011; Fedorova & Levit, 2016). Throughout the day, stratus clouds combined with altostratus and cumulus clouds.

2. Synoptic systems

Low visibility events in the central region of the BNE are associated with the ridge at low levels (Melo et al., 2019, 2023; Melo, 2022). These events are characteristic features of the radiation fog.

The main circulation of air currents in conditions of fog, drizzle, mist events and stratus clouds in the coastal region, according to Fedorova et al. (2013, 2015) and Fedorova and Levit (2016), has the following features:

1. Low levels: weak cyclonic circulation;
2. Middle levels: anticyclonic circulation (High or Ridge);
3. High levels: no permanent circulation was found; cyclonic and anticyclonic circulations have the same frequency.

These air circulations are created by the following synoptic systems (Rodrigues et al., 2010; Fedorova et al., 2008, 2013, 2015; Fedorova and Levit, 2016):

1. Wave Disturbance in the Trade Winds (WDTW) (Chapter I.3);
2. Jet stream in the BNE - Brazilian Northeast Jet Stream (BNEJS) (Chapter I.6);
3. Middle Tropospheric Cyclonic Vortex (MTCV) (Chapter I.5);
4. Intertropical Convergence Zone (ITCZ) (Chapter I.1);

5. Circulation between the synoptic systems of both hemispheres (Chapter I.8).

These synoptic systems created air currents that contribute humidity accumulation at low levels through two main mechanisms: *humidity convergence in a weak trough* and *humidity accumulation under the subsidence layer.*

Humidity convergence at low levels occurred in two types of troughs associated with WDTW and ITCZ. The axis of the ITCZ trough often represents the meridional motion, and the axis of the WDTW trough is from northeast to southwest. A similar weak convergence at low levels takes place during the MTCV formation and the vortex development in the northern sector.

Humidity accumulation under the subsidence layer occurred with the BNEJS, MTCV, and circulation between both hemispheres. These synoptic systems generate sinking motions over the BNE and humidity accumulation. The mechanisms of sinking formation are different. In the BNEJS ageostrophic flow, the air subsidence form. Air subsidence below the MTCV is very typical for this system. Sinking motions are often observed at lower levels from the surface up to 550 hPa during the last stage of the WDTW. The circulation between both hemispheres occurred between a tropical depression, storm, or cyclone in the Northern Hemisphere (NH) and a frontal zone in the Southern Hemisphere (SH). Convergence at high levels between these two air currents creates a sinking motion, air subsidence and consequent humidity accumulation at low levels.

*WDTW*s were associated with *fog* (Silveira, 2003; Fedorova et al., 2008, 2013, 2015), *haze* events (Levit et al., 2010; Fedorova et al., 2013) and *stratus* clouds (Gomes et al., 2011). According to (Rodriques et al., 2010), 87% of throughs near the surface are associated with WDTW.

The *Trough on the southern periphery of the ITCZ* is associated with weak rain in the BNE (Rodriques et al., 2010). In some cases, the trough on the Trade Winds-WDTW is a trough on the southern periphery of the ITCZ.

MTCVs create light precipitation due to sinking in the southern part of the vortex. At the last stage of WDTW (before dissipation), lifting from the surface up to 550 hPa was recorded, and a sinking was detected above this level (Fedorova et al., 2017; Silva, 2015).

BNEJS was often associated with light rain and mist (Fedorova et al., 2018 a and b). These adverse phenomena occur (usually) on the hot entrance side and the cold exit side of the jet core in the BNEJS from SW to NE. These phenomena are common in the coastal region of the BNE and were more frequent in Recife, in the center of the BNE coast.

Interhemispheric air currents were observed together with BNEJS in all events, and on transverse southwestern air currents, BNEJS prevailed. These currents occurred during the warm season from October to December (more days with these currents), and February. Adverse phenomena (*weak rain* and *light fog*) were also observed on all days with BNEJS. These phenomena occurred on all days (4 days on average) with these BNEJS.

Circulation between both hemispheres (between a tropical depression, storm or cyclone in the NH and a frontal zone in the SH) create conditions for the formation of various phenomena (Fedorova & Levit, 2016; Fedorova et al., 2020). Fogs, mists, and weak rains were more often observed in the coastal and southern regions of the BNE. Fog and drizzle occurred in all events with circulation between two hemispheres, and fog was not recorded in 3 from 10 events. A tropical cyclone ranging between 35 and 50W and up to 20N in the NH affected the weather in the BNE. At this time, *the tropical storm stage* prevailed.

The predominance of different synoptic systems at low levels was identified in the coastal and central areas of the BNE, where weak cyclonic and anticyclonic circulations, respectively, were revealed (Melo et al., 2019, 2023; Melo 2022).

Some synoptic maps are necessary to identify which synoptic systems have caused the fog phenomenon, its depth, displacement, and duration. Streamline maps with wind magnitude (m s^{-1}) are used at low (1000, 950, 900, 850, 800 hPa), middle (700, 600, 500, 400 hPa) and high (300 and 200 hPa) levels of the troposphere at synoptic time (00 UTC, 06 UTC, 12 UTC, 18 UTC). Low level moisture convergence maps are used to estimate water vapor concentration in a particular region. The longitudinal section of the vertical wind speed is used to determine upward and downward movements, helping us to see the circulation. Two examples of fog originating from different systems were selected.

Example 1. Fog formation in the WDTW

The first example shows fog resulting from WDTW on June 11, 2008 in Alagoas. Maceio Airport Ground Stations (SBMO) recorded fog at 08:20 UTC and mist at 09 UTC. When evaluating synoptic maps, a trough near the BNE coast with a northeast-southwest axis was observed at a level of 1000 hPa, which is typical for the WDTW (Figure II-5-a). This WDTW was formed on a Trade Wind from anticyclone (High) in the mid-Atlantic. This indicates weak lifting at low levels, which is typical for fog in the BNE. Moisture convergence occurred along the trough axis (Figure II-5-1d). A trough is also observed at 500 hPa with a northwest-southeast axis (Figure

II-5-1b) in the air current from the Northern Hemisphere (5°N, close to the coast of Africa). The anticyclonic circulation over the BNE centered in Alagoas was recorded at a high level (300 hPa) (Figure II-5-1c). These synoptic systems at middle and high levels indicate a sinking air. Therefore, the vertical section along 36°W shows the air lifting at low levels and sinking higher in Maceio (9.6°S 36°W) (Figure II-5-1e). This air lifting caused a moisture convergence at the low levels (Figure II-5-1d), which contributed to the fog formation.

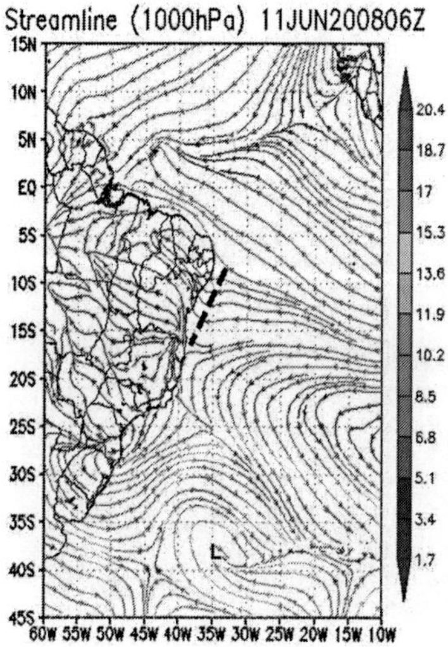

a) 1000 hPa

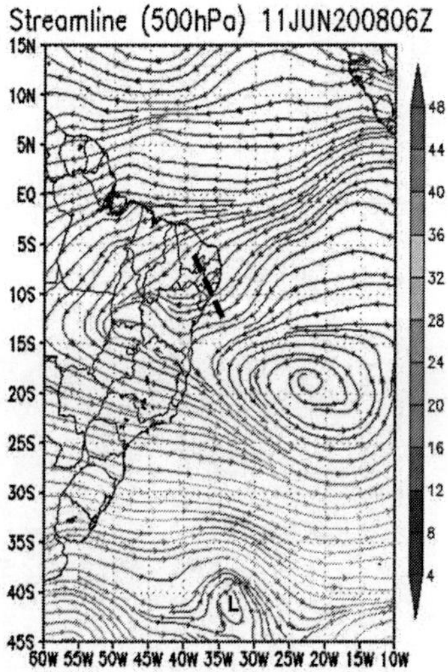

b) 500 hPa

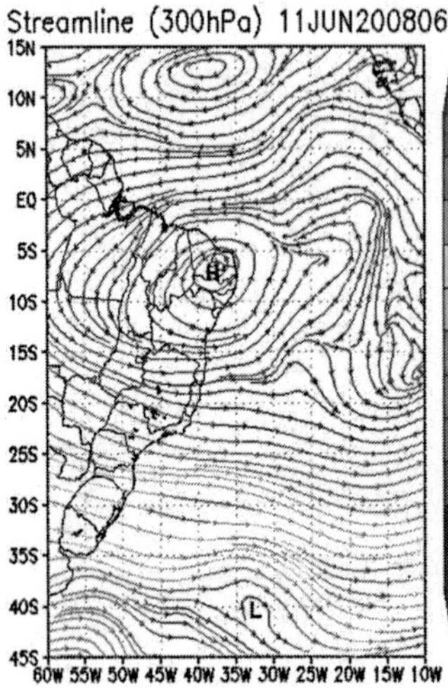

c) 300 hPa

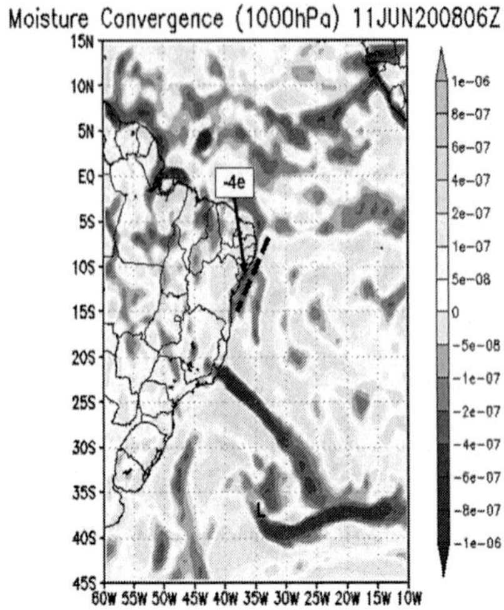

d)

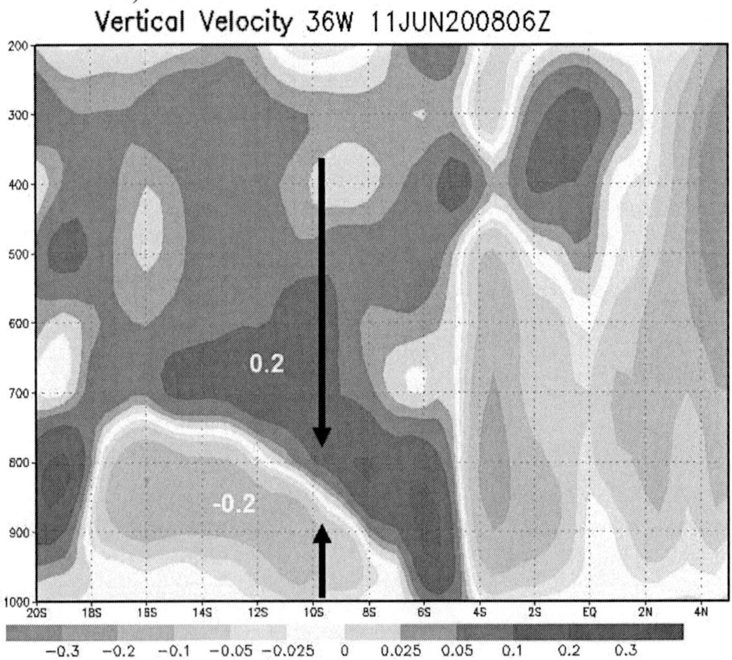

e) along 36°W

Figure II-5-1 – Fog formation in the WDTW on 11 July 2008, 06 UTC.
Streamline (m s^{-1}) maps at 1000 hPa (a), 500 hPa (b) and 300 hPa (c).

Moisture Convergence at 1000 hPa (d, mm s^{-1}*).*
Vertical section of vertical speed (e, Pa s^{-1}*) along 36°W.*
L - Low/cyclone center.
H - High/Anticyclone center.
Source: European Centre for Medium-Range Weather Forecasts (ECMWF) Re-Analysis-Interim/ECMWF (2022).

Example 2. Fog formation in the MTCV.

The second example shows a fog event in Alagoas in the MTCV area on March 7, 2012. This MTCV was first identified on 5 March, two days before the fog formed. The level of the main location of the MTCV was 600 hPa, where it was observed from 5 March, 12 UTC to 6 March, 18 UTC (Figure II-5-2a, b, c). This cyclone can also be seen at 700 hPa on 6 March, 06 UTC (Figure II-5-2d). Above this vortex, troughs were observed with a northwest-southeast axis at a level of 500 hPa and with a southwest-northeast axis at 300 hPa (Figure II-5-3).

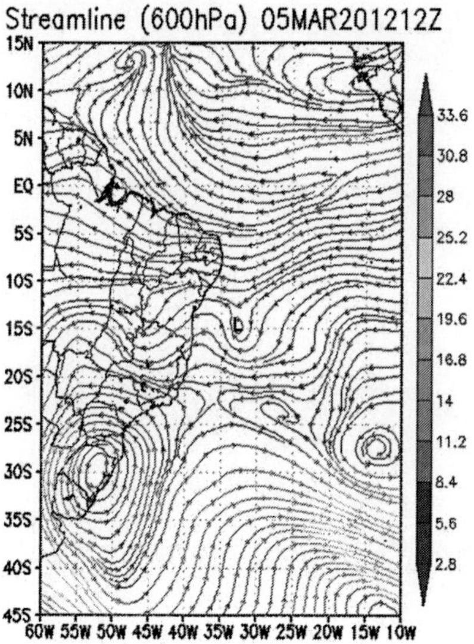

a)

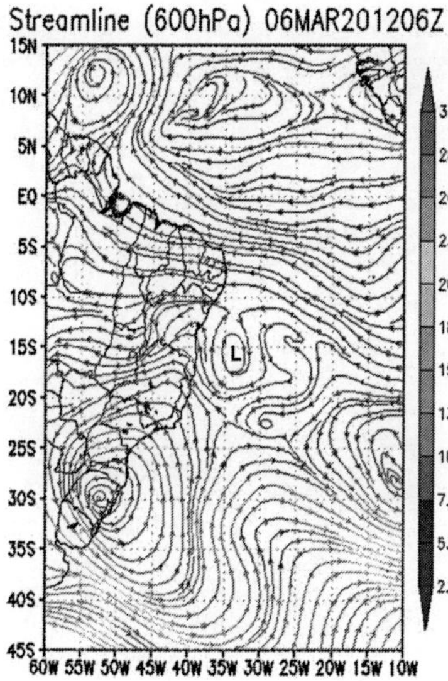

b)

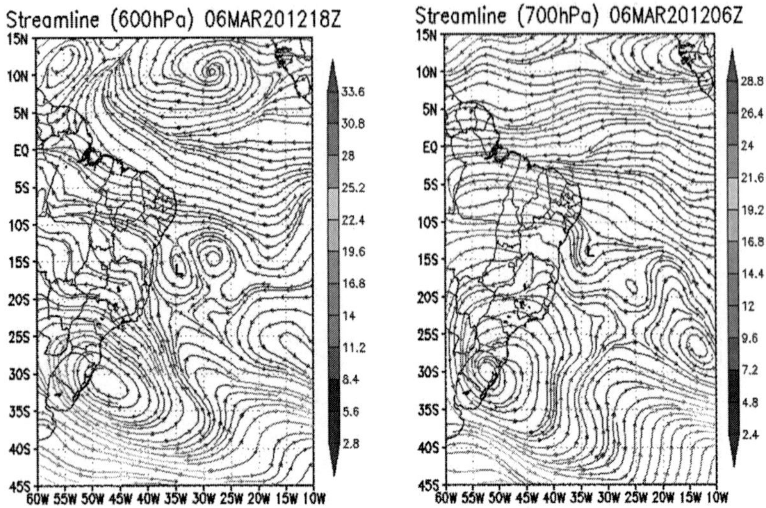

c) d)

Figure II-5-2 – Fog formation in the MTCV on 5-6 March 2012:
Streamline (m s^{-1}) maps at 600 hPa on 5 March 2012, 12 UTC (a); 6 March 2012, 06 UTC (b); 6 March 2012, 18 UTC (c).
Streamline (m s^{-1}) maps at 700 hPa on 6 March 2012, 06 UTC(d.)
L - Low/cyclone center.
Source: European Centre for Medium-Range Weather Forecasts (ECMWF) Re-Analysis-Interim/ECMWF (2022).

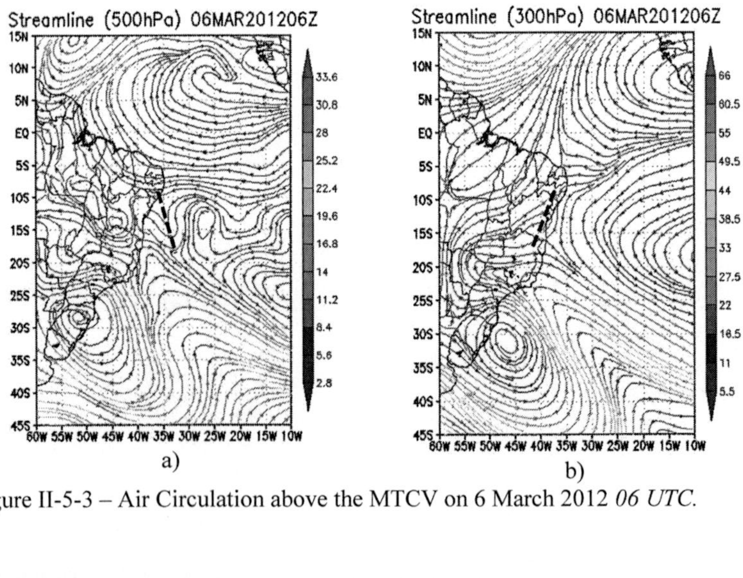

a) b)

Figure II-5-3 – Air Circulation above the MTCV on 6 March 2012 *06 UTC.*

Streamline (m s^{-1}) maps at 500 hPa (a) and 300 hPa (b).
Dotted line – axis of the trough.
Source: European Centre for Medium-Range Weather Forecasts (ECMWF) Re-Analysis-Interim/ECMWF (2022).

Vertical velocity cross section intersecting MTCV at 36°W shows the lifting at its center. While at middle levels (600 hPa) on the MTCV periphery, air sinking is observed (Figure II-5-4). This process contributes to the humidity accumulation near the surface and, as a result, the fog formation. Therefore, it is on the MTTSV periphery that fog can form.

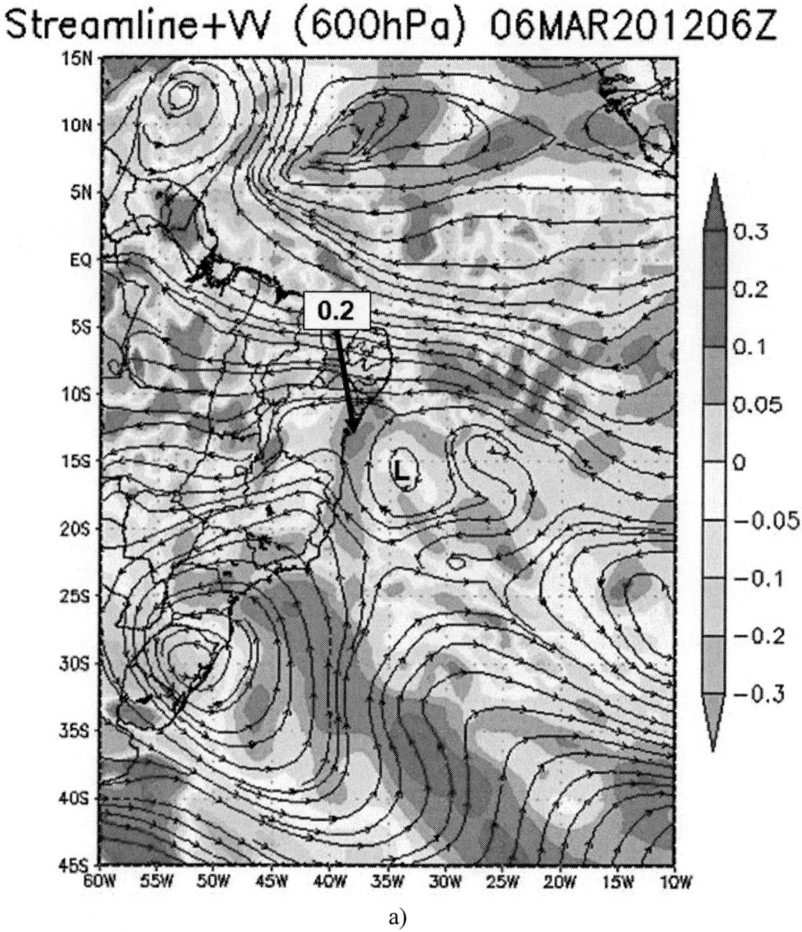

a)

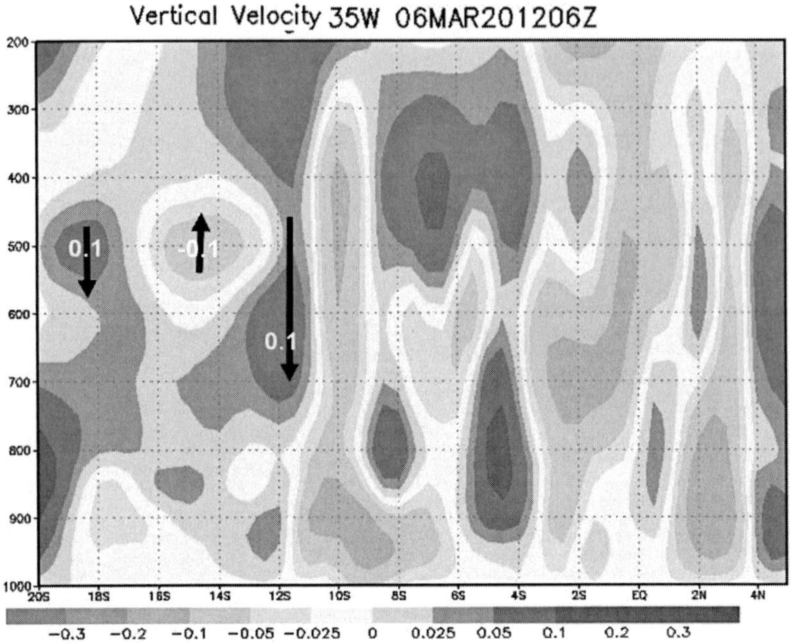

b)

Figure II-5-4 – Vertical velocity in fog event in the MTCV on 6 March 2012, 06 UTC
a) Streamline (m s^{-1}) maps at 1000 hPa and vertical velocity (mm s^{-1}).
b) Vertical section of vertical velocity (Pa s^{-1}) along 36°W.
L - Low/cyclone center. The numbers show the extreme values of the vertical velocity.
Source: European Centre for Medium-Range Weather Forecasts (ECMWF) Re-Analysis-Interim/ECMWF (2022).

3.Thermodynamic forecasting methods

During fog, drizzle, mist and stratus clouds in the coastal region and in the BNE center different characteristics of the atmospheric vertical structure appear.

The following characters of vertical distribution of T and Td are typical for coastal region (Fedorova and Levit, 2016), as shown in Figure II-5-5:

1. Humid layer up to 900 hPa
2. Conditionally unstable layer 1000-900 hPa and stable above.

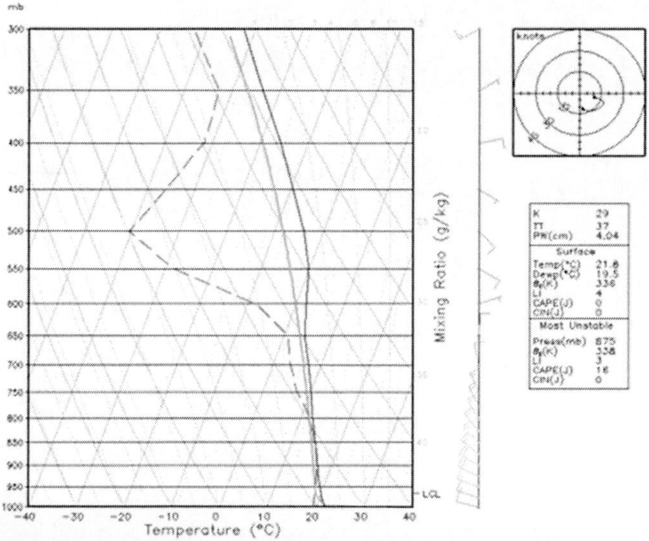

Figure II-5-5 – Typical vertical profile of temperature and dew point on foggy days in the coastal area of the BNE.
Source: Melo et al. (2019)

The depth of the humid layer was less (up to around 900 hPa) in the BNE center (Melo et al., 2019), Figure II-5-6.

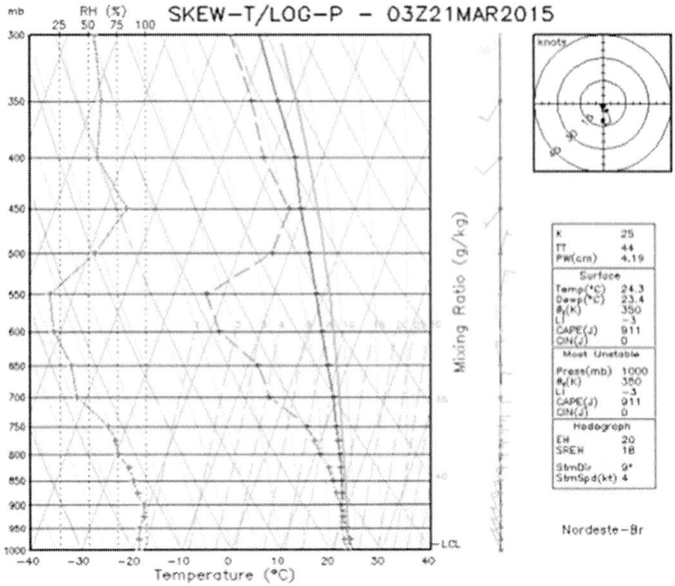

Figure II-5-6 – Typical vertical profile of temperature and dew point on foggy days in the central area of the BNE on 21 March 2015, 03 UTC.
Source: Melo et al. (2019)

The European Centre for Medium-Range Weather Forecasts (ECMWF) reanalyses (ECMWF, 2022) shows a higher moisture content in the lower layer than the National Centers for Environmental Prediction reanalysis II (NCAR, 2022) and therefore represents more realistic conditions for fog formation (Melo et al., 2019). Convective clouds were observed during fog events at Campina Grande Airport and in some fog cases at Maceio Airport (Melo et al., 2019, 2023).

4. PAFOG model and FogVIS tool

The Parameterized Fog (PAFOG) model (Bott and Trautmann, 2002) has been the subject of several fog studies in the BNE (Fedorova et al., 2013, 2015; Fedorova & Levit, 2016; Afonso, 2016; Nobre et al., 2019; Melo et al., 2019; Melo, 2022). This model was obtained from Prof. A. Bott in accordance with an international agreement between the Federal University of Alagoas-UFAL and the University of Bonn, Germany. This model proved to be an effective tool for fog phenomena detecting in the BNE region.

Another alternative, more recently developed by the Synoptic and Physical Meteorology Laboratory of the Federal University of Alagoas-UFAL, was the FogVisv1.0 tool (Nobre et al., 2019). FogVisv1.0 has proven to be an effective tool for detecting fog phenomena at Maceio Airport. Its use shows satisfactory results in comparison with the PAFOG model and real visibility measurements.

4.1 PAFOG

PAFOG is a one-dimensional parametric model. It contains four modules with information on various meteorological variables, and it predicts fog events in extratropical continental regions. Its modules are the next:

1. *dynamic module*: with geographic data (latitude, longitude, and altitude of the meteorological station); soil (type of soil); vegetation (coverage and height); meteorological station data (air temperature and surface dew point, relative humidity at 2 m above ground, horizontal visibility, and surface pressure).
2. *cloud microphysics module*: cloud cover at low, middle and high levels.
3. *module for radiation calculations*: radiosonde data (pressure at different levels between 1000 and 100 hPa), air and surface dew point temperatures, and geostrophic wind speed (calculated using the Weather Research and Forecast (WRF) model (Skamarock et al., 2008).
4. *low vegetation module*: having soil temperature and relative humidity at different depths (0.1, 0.2, and 0.3 m).

The PAFOG model was used for the tropical region of Brazil by Silva (2012), Fedorova et al. (2013, 2015), Fedorova & Levit (2016), Afonso (2016), Nobre et al. (2019), and Melo (2022). In general, these works have shown that this model is effective for fog forecasting.

A change in the model of vegetation parameters near Maceio Airport did not affect the horizontal visibility results predicted by the model (Afonso, 2016). The weak influence of vegetation parameters on the fog forecast results at Maceio Airport is because the fog formation mechanism on the northeast coast of Brazil differs from the typical radiation fog formation mechanism, and to a greater extent depends on dynamic processes in the atmosphere.

Thus, for the execution of the PAFOG model for forecasting the horizontal visibility is necessary:

1. analysis or reanalysis data (from Global Forecast System GFS, NOAA Physical Sciences Laboratory NCEP/DOE Reanalysis II or ECMWF Reanalysis v5/ERA5), ensuring the best atmospheric conditions to the model forecast initialization;
2. adaptation of modules 1, 2, and 3 of the model to the data obtained at the previous step;
3. setting the model forecast time, that is, the start and end dates of the rounds;
4. PAFOG model execution results (Figure 7).

A scheme of the fog forecast steps using the PAFOG model is shown in Figure II-5-7

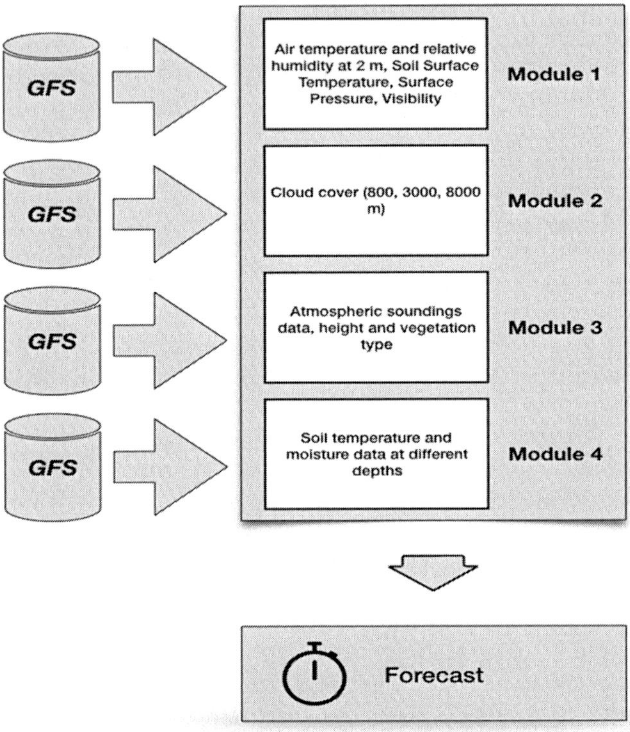

Figure II-5-7 – Schematic of fog forecasting steps using the PAFOG model. Source: Bott and Trautmann (2002).

An example of a visibility forecast using the PAFOG model initialized by the Climate Forecast System Reanalysis - CFSR (NCEP, 2022) model

24h, 18h, 12h and 6h before the event that occurred at Maceio Airport (Zumbi dos Palmares International Airport, Maceio, Alagoas) on June 11, 2010 is shown in Figure II-5-8. The fog was forecasted with 12 h antecedence.

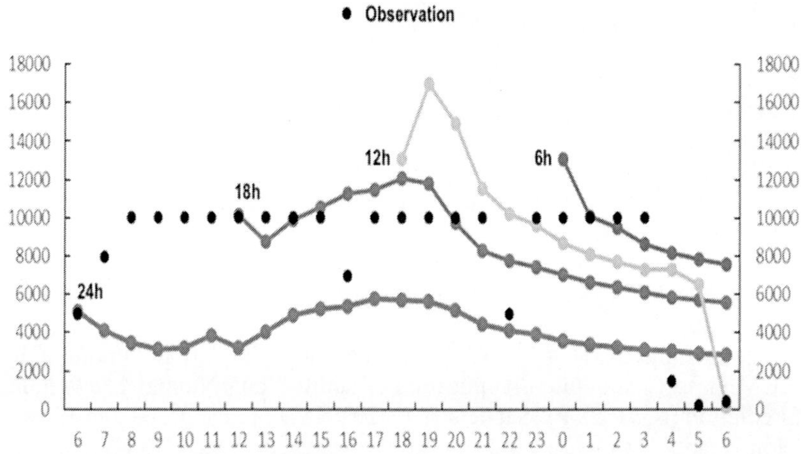

Figure II-5-8 – Visibility forecast (m), using PAFOG model and initialized with CFSR model, with 24h, 18h, 12h and 6h antecedence on June 11, 2010, for Zumbi dos Palmares International Airport, Maceio, Alagoas.
Source: Fedorova and Levit (2016).

4.2 FogVisv1.0

The FogVIS v1.0 tool is an application developed by the Synoptic and Physical Meteorology Laboratory of the Federal University of Alagoas on the base of Gultepe et al. (2006) calculations. It predicts the horizontal visibility at Zumbi dos Palmares airport. This tool consists of three modules:

1. Introduction with description of functions and parameterization.
2. Georeferencing, where the geographic coordinates of the airport are set.
3. Execution module, where the input data of the analysis or reanalysis, such as pressure (hPa), wind speed (ms^{-1}), specific humidity ($g.kg^{-1}$), air temperature and dew point (°C) near the surface is necessary for estimating horizontal visibility (m).

FogVisv1.0 has shown satisfactory results compared to the PAFOG model in estimating fog occurrence in the tropical region (Nobre et al.,

2019) and it is calculated from liquid water content (LWC) and droplet number concentration (N_d) as suggested by Gultepe et al. (2006). Equation (1) is intended to determine horizontal visibility during fog as a function of LWC and N_d, which, in turn, must be calculated from Equations (2) and (3).

$$Vis = \frac{1.002}{(LWC \cdot N_d)^{0.6473}} \quad (1)$$

This equation indicates that visibility is inversely related to both LWC and *Nd* and that the presence of variability in droplet sizes and its role in influencing visibility in fog (Gultepe et al., 2006).

N_d was calculated using the parametrization proposed by Gultepe and Isaac (2004), where T_c refers to air temperature near the surface in degrees Celsius.

$$N_d = -0.071T_C^2 + 2.213T_C + 141.56 \quad (2)$$

According to the RTTOV-12 Science and Validation Report of the EUMETSAT Satellite Application Facility on Numerical Weather Prediction (NWP SAF) (Sanders et al., 2017) the optical properties of water clouds are parameterized from liquid water content (LWC) in $g.m^{-3}$. However, NWP models provide cloud information in units of mass mixing ratio (or specific cloud liquid water content) in $kg.kg^{-1}$, i.e., ratio between the mass of liquid water and the mass of moist air. If we consider that the air follows the perfect gas law, then the conversion for liquid water cloud is:

$$LWC = MMR_l \frac{P \cdot 10^2}{R_{ma} \cdot T} \quad (3)$$

where MMR_l is the mass mixing ratio for liquid water cloud, P is the atmospheric pressure in hPa, T is the atmospheric temperature in K and R_{ma} is the moist air gas constant (in $Jg^{-1}K^{-1}$) given by:

$$R_{ma} = R_{da}\left(1 + \frac{1-\varepsilon}{\varepsilon} q\right) \quad (4)$$

where R_{da} is the gas constant for dry air and q is the specific humidity (given as the ratio between the mass of water vapor and the mass of moist air). The equation is demonstrated in Jacobson (2005, 2012) and coefficient ε is given by:

$$\varepsilon = \frac{M_{wv}}{M_{da}} \quad (5)$$

where M_{wv} and M_{da} are the molar mass for water vapor and for dry air in g.mol^{-1}, respectively.

The gas constant for dry air is given by Jacobson (2005, 2012):

$$R_{da} = \frac{R}{M_{da}} \quad (6)$$

where R is the universal gas constant (in J mol^{-1} K^{-1}).

The method described by Knox et al. (2017) was adopted to calculate the wet-bulb temperature (T_w, °C), which estimates the wet-bulb temperature through the weighted average of dry-bulb temperature (T, °C) and dew point (T_d, °C).

$$T_{W_{\frac{1}{3}}} = T - \frac{1}{3}(T - T_d) = \frac{2}{3}T + \frac{1}{3}T_d \quad (8)$$

The saturation vapor pressure (e_s, kPa) at wet-bulb temperature, was obtained from wet-bulb temperature data at the surface level (T_w, °C), if $T_w \geq 0$ °C, is given by (MOREIRA, 1999, 2019):

$$e_S = 0.6108 \cdot 10^{\left(\frac{7.5T_W}{237.3+T_W}\right)} \quad (9)$$

The actual vapor pressure (e_a kPa) was obtained from T_w and T, in °C, the atmospheric pressure (P_{atm} kPa), and the psychrometric constant of the instrument the *A*, adopted in FogVIS v1.0 as 0,00067 °C^{-1} (MOREIRA, 1999, 2019), as next:

$$e_a = e_s - A \cdot P_{atm} \cdot (T - T_w) \quad (10)$$

A diagram of the steps in executing FogVisv1.0 is shown in Figure II-5-9.

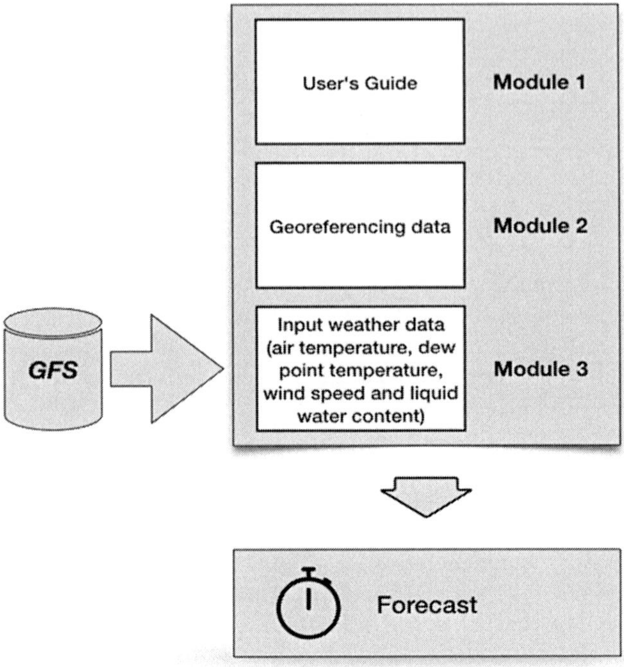

Figure II-5-9 – Outline of steps for fog forecasting using FogVisv1.0. Source: Nobre et al. (2019).

5. Complex methods of forecasting

Recommendations for fog, drizzle, haze and stratus clouds forecasting (Figure II-5-10):

First step: check for synoptic scale systems that can create mechanisms for subsidence layer formation and humidity convergence (Figure II-5-1e). If these conditions exist, go to the second step.

Second step: determine moisture and stability distributions from radiosonde and/or meteorological model data; compare with the pattern, shown in (Figure II-5-5 and II-5-6). If the vertical profiles T and Td correspond to the pattern, proceed to the third step.

Third step: evaluation of fog / horizontal low visibility formation using FogVis v1.0.

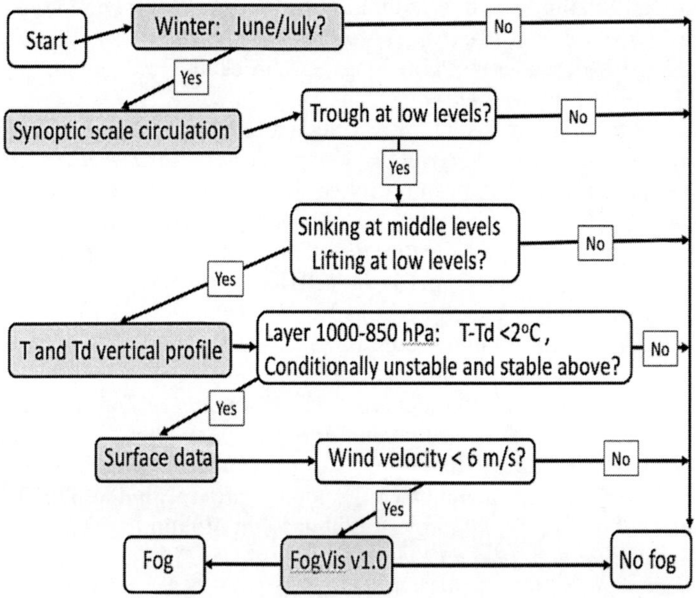

Figure II-5-10 – Low visibility forecasting tree.

6. References

Afonso, Joao Maria Sousa. *Análise de ocorrência de nevoeiros e dos resultados do modelo de previsão 'PAFOG' em diferentes regiões climáticas do Brasil.* MSc. Thesis, Federal University of Alagoas, Maceio, Brazil. 237 f, 2016.

Bott, Andreas and Thomas Trautmann. "*PAFOG—a New Efficient Forecast Model of Radiation Fog and Low-Level Stratiform Clouds.*" Atmospheric Research 64: 191–203, (2002).

Fedorova, N., Levit, V., Fedorov, D. *"Fog and stratus formation on the coast of Brazil", Atmospheric Research*, 87: 268-278, (2008).

Fedorova, N.; Levit, V.; Silva A.O.; Santos D.M.B. "*Low Visibility Formation and Forecasting on the Northern Coast of Brazil."* Pure and Applied Geophysics 170, n.4: 689-709, (2013).

Fedorova, N.; Levit, V.; Souza, J. L.; Silva, A. O.; Afonso, J. M. S.; Teodoro, I. *"Fog Events at Maceio Airport on the Northern Coast of Brazil During 2002-2005 and 2007."* Pure and Applied Geophysics, 172: 2727–2749, (2015). DOI 10.1007/s00024-014-1027-0.

Fedorova, Natalia, Levit Vladimir. *Fog in the tropical region. Fog formation in the tropical region of the northeast of Brazil.* LAP LAMBERT Academic Publishing: Saarbrücken, Germany. 82 p, 2016. ISBN: 978-3-659-87098-9

Fedorova, Natalia; Santos, D. M. B ; Segundo, M. M. L. ; Levit, Vladimir. *"Middle Tropospheric Cyclonic Vortex in Northeastern Brazil and the Tropical Atlantic."* Pure and Applied Geophysics, 174: 397-411, (2017).

Fedorova N., Levit V., Campos A.M.V. *"Brazilian Northeast Jet Stream: frequency, wind speed and direction."* Meteorological Applications, 25: 254-260, DOI: 10.1002/met.1688, (2018a).

Fedorova N., Levit V., Campos A.M.V. *"Brazilian Northeast Jet Stream: Association with synoptic scale systems."* Meteorological Applications, 25: 261-268, DOI: 10.1002/met.1693, (2018b).

Fedorova, N., Levit, V., Cavalcante, L. C. V. *Impact of Tropical Cyclones in the Northern Atlantic on adverse phenomena formation in northeastern Brazil.* In: Lupo, A., R. *Current Topics in Tropical Cyclone Research.* IntechOpen: London, United Kingdom, 2020.

Gomes Helber Barros; Fedorova Natalia; Levit Vladimir. "*Rare events of stratus clouds on the northeast coast of Brazil.*" Revista Brasileira de Meteorologia 26, n. 1: 9-18, (2011). FapUNIFESP (SciELO).

Gultepe, I. & Isaac, G. A. "*Aircraft observations of cloud droplet number concentration: Implications for climate studies.*" Nd. Quarterly Journal of the Royal Meteorological Society 130, Part A: 2377–2390. (2004). doi: 10.1256/qj.03.120

Gultepe, I., Müller, M. D. And Boybeyi, Z. "*A New Visibility Parameterization for Warm-Fog Applications in Numerical Weather Prediction Models.*" Journal of Applied Meteorology and Climatology 45: 1469-1480, (2006). DOI: https://doi.org/10.1175/JAM2423.1

Jacobson, M. Z. *Fundamentals of atmospheric modeling,* Cambridge Eds. Cambridge University Press, 2005. 2nd edition, Stanford University, California, 2012. ISBN: 9781139165389. DOI: https://doi.org/10.1017/CBO9781139165389

Knox, J.A.; Nevus, D.S.; Knox, P.N. "*Two simple and accurate approximations for wet-bulb temperature in moist conditions, with forecasting applications.*" Bulletin of the American Meteorology Society 98, Issue 9: 1897 – 1906, (2017).

Levit, V.; Santos, D. M. B.; Fedorova, N. *"Haze Formation on the Northern Coast of Brazil."* In: 5th International Conference on Fog, Fog Collection and Dew. Munster, Germany. Conference Book, 11-14, 2010. University of Munster: http://kli.uni-muenster.de

Melo, D. L.; Oliveira, C. G. ; Muanza, G. A. ; Fedorova, N. ; Levit, V. *"Fog and mist forecast by the PAFOG model at the Northeast Brazilian airports*". In: 8th International Conference on Fog, Fog Collection and Dew, Taipei, Taiwan. 8th International Conference on Fog, Fog Collection and Dew, Taipei, Taiwan: IFDA, 6-6. 2019.

Melo, D. L.; Oliveira, C. G. ; Fedorova, N. ; Levit, V. *"Fog Analysis and Forecast by PAFOG Model in Brazilian Northeast*" Revista Brasileira de Meteorologia 38, n. 1, 1-13, DOI: http://dx.doi.org/10.1590/0102-77863810215, (2023)

Melo, D. L. "*Análise dos processos físicos de formação da visibilidade baixa no nordeste brasileiro e previsão pelos modelos PAFOG e FOGVIS",* Teses, Universidade Federal de Alagoas, Maceió, 2022.

Moreira, J. R. S. *Fundamentos e Aplicacoes da Psicometria*. Sao Paulo: RPA Editorial LTDA, 1999, 194p. 2 edition: Moreira, J. R. S. and Neto, A.H. Blucher, 280 p, 2019. ISBN-10: 8521218395. ISBN-13: 978-8521218395.

Nobre, J. P. G.; Fedorova, N.; Levit, V.; Santos, A. S.; Lyra, M. J. A. "*New Methodology for Evaluation of the Fog Events for the Zumbi dos Palmares Airport, in Maceió (Alagoas)".* Anuário do Instituto de Geociências (UFRJ. IMPRESSO) 42: 527-535, (2019).

Rodrigues, L. R. L.; Fedorova, N.; Levit, V. *"Adverse meteorological phenomena associated with low-level baric troughs in the Alagoas State, Brazil, in 2003."* Atmospheric Science Letter 11: 204-209, (2010).

Skamarock, W. C., Klemp, J. B., Dudhia, J., Gill, D. O., Barker, D. M., Wang, W., & Powers, J. G. (2008). "*A description of the advanced research WRF",* version 2. NCAR Technical Note, NCAR/TN-468+ STR, 88.

Saunders, R.; Hocking, J.; Rundle, D.; Rayer, P.; Havemann, S.; Matricardi, M.; Geer, A.; Lupu, C.; Brunel, P., Vidot, J. "*RTTOV-12 science and validation report. NWPSAF-MO-TV-41."*. EUMETSAT, 2017.

Silveira, V. P. *Análise dos casos de nevoeiros e nuvens stratus no aeroporto de Maceió (Alagoas).* 91p. M. Sc. Teses, Universidade Federal de Alagoas, Maceió, 2003.

Silva Thais Gomes *Vórtices ciclônicos de médios níveis: estrutura, processos de formação e tempo associado.* M. Sc. Teses, Universidade Federal de Alagoas, Maceió, 2015.

Silva Aliton Oliveira *Análise e previsão e visibilidade baixa no aeroporto de Maceió.* M. Sc. Teses, Universidade Federal de Alagoas, Maceió, 2012.

ECMWF. ERA-Interim: 1979-2010:

https://www.ecmwf.int/en/forecasts/datasets/reanalysis-datasets/era-interim. Accessed 10 set. 2022.
NCAR. NCEP/DOE Reanalysis II. Disponível em: https://psl.noaa.gov/data/gridded/data.ncep.reanalysis2.html. Acesso em: 24 set. 2022.
NCEP. CFS version 2 Download. Disponível em: https://cfs.ncep.noaa.gov/cfsv2/downloads.html. Acesso em: 12 ago. 2022.

CHAPTER II.6

MESOSCALE CONVECTIVE COMPLEXES IN THE BNE

MATHEUS JOSÉ ARRUDA LYRA,
NATALIA FEDOROVA, VLADIMIR LEVIT AND
MARCIO HENRIQUE DOS SANTOS SILVEIRA

1. Introduction

Mesoscale Convective Complexes (MCC) are characterized by a rounded conglomerate of cold and dense Cumulonimbus clouds with vertical explosive growth, lasting 6 or more hours and associated with intense rainfall accompanied by strong wind gusts. These systems were identified and defined by Maddox (1980) during studies of the central United States of America (USA) based on characteristics that can be observed on satellite images.

The first studies in South America were carried out by Velasco and Fritsch (1987), where they mapped the MCC regions for the entire South American continent. However, Velasco and Fritsch (1987) did not identify MCCs over the Brazilian Northeast (BNE) during their study (Figure II-6-1).

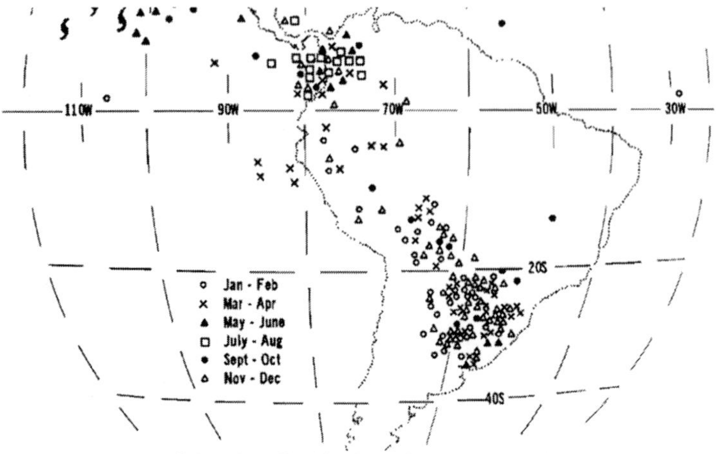

Figure II-6-1 – Map of monthly distribution of MCC in South America.
Source: Adapted from Velasco and Fritsch (1987)

Velasco and Fritsch (1987) used criteria like Maddox (1980) to identify MCCs over South America, but with different temperature limits: -40°C to -42°C instead of -32°C, and -62°C to -64°C instead of -52°C, while maintaining area values. Subtropical MCCs in South America develop somewhat later and longer and are also 60% larger than those in the United States, while tropical MCCs are similar in size and last 1 to 3 hours or less (Velasco and Fritsch, 1987; Silveira, 2010).

Some mesoscale meteorological systems can be observed in the BNE: MCCs, squalls lines and Cb clouds, associated with heavy precipitation. The study of the MCC in this region is carried out by the research group of the laboratory "Physical and Synoptic Meteorology" of the Institute of Atmospheric Sciences of the Federal University of Alagoas (ICAT/UFAL). Several studies have been carried out on these systems: Fedorova et al. (2008); Silva et al. (2008); Milhahn Junior (2016); Lyra et al. (2018); Fedorova et al. (2019); Lyra et al. (2020, 2022).

2. MCC identification and life cycle

Maddox (1980) suggested some physical characteristics for the MCC identification and monitoring.

The first characteristic (1) concerns the size and temperature of the cloud top of the system:

(a) Part outside the core of the system with temperatures ≤ -32 °C (in enhanced infrared channel (IR)) and an area ≥ 100,000 km² (e.g.

compared to the BNE states, half of the total area of the Paraná State).

(b) Core with temperatures ≤ -52 °C and an area ≥ 50,000 km² (equal to area of the Rio Grande do Norte State).

The second characteristic (2) is the durability of the system, which must persist for at least six hours at the temperatures and dimensions indicated in (a) and (b).

The maximum length of the MCC is the third characteristic (3) of the system evolution. This occurs when the characteristics specified in (a) reach their maximum size.

Another characteristic (4) is the shape of the system, which must have an eccentricity (smallest axis/longest axis) ≥ 0.7. That is, a convective system of a rounded shape ("cloud shield") can be observed on satellite images.

Finally, termination or dissipation of the system, last feature (4), occurs when requirements (a) and (b) of feature (1) can no longer be satisfied.

Figure II-6-2 shows four examples of MCCs identified by GOES-12 and GOES-13 satellite enhanced infrared images, highlighting the temperature of the convective cloud tops.

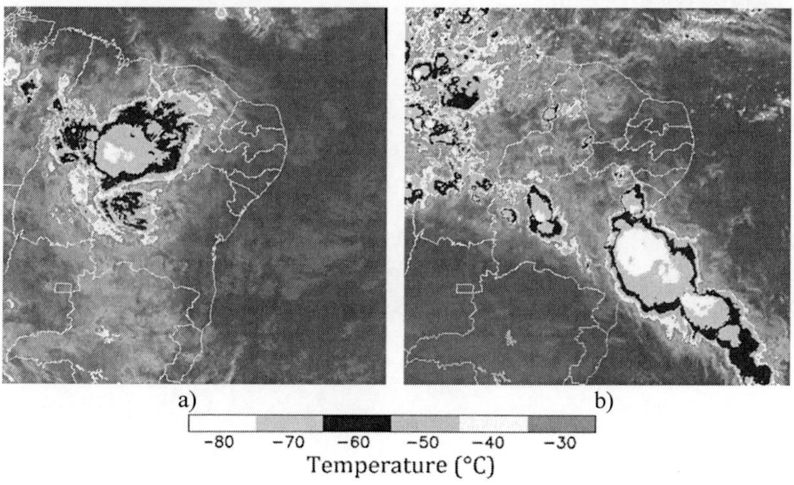

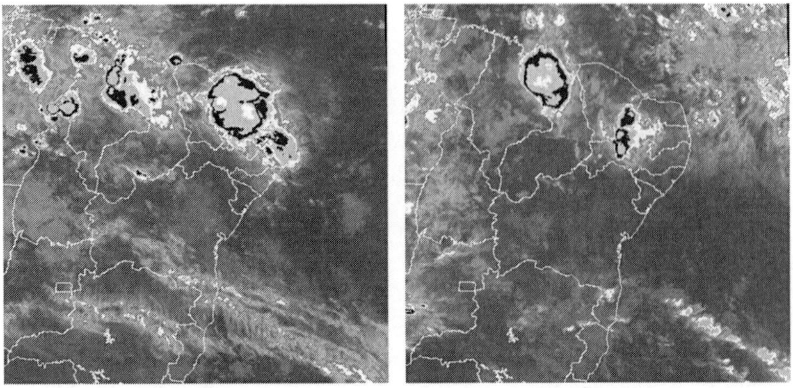

c) d)

Figure II-6-2 – GOES-12 satellite enhanced IR images on: April 08, 2010, 04 UTC (a); GOES-13 satellite enhanced IR images on: January 19, 2013, 10 UTC (b); April 20, 2013, 05 UTC (c) and March 25, 2015, 05 UTC (d).

The MCC life cycle can be divided into four stages (Fedorova, 2008, according to the description of Veltishev, 1988): ***Genesis, Development, Maturity and Dissipation.***

***1 - Genesis*:** At this stage (Figure II-6-3) favorable conditions for instability are created, such as: 1) the confluence of an air current at low levels into a weak baric trough, 2) the supply of moisture and warm air into the lower troposphere (up to 750 hPa), 3) a low levels jet stream promote advection of moisture and warm air and 4) convective instability associated with the influence of the relief. At middle levels, the entrainment of fresh air from the surroundings leads to a strong lifting (with a maximum at 700 hPa) with cold air gusts near the surface of the boundary layer. This stage is characterized by the absence of high horizontal temperature gradients. Separate meso-β convective systems approach each other and form a meso-α system.

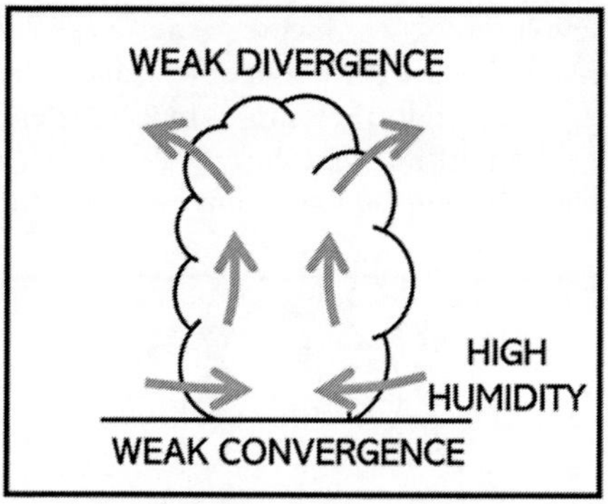

Figure II-6-3 – Schematic representation of the MCC genesis stage.
Source: Adapted from Fedorova (2008)

2 - Development: During this stage (Figure II-6-4) advection of anomalously hot air occurs at low levels. This advection causes an increase in lifting at the middle levels (at 700 hPa), as well as air saturation in these layers, characterized by convergence at low levels and divergence at high levels. Evaporation of precipitation creates high pressure at the surface.

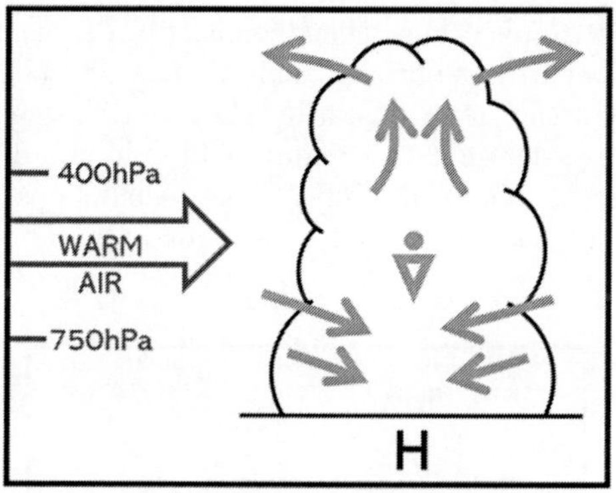

Figure II-6-4 – Schematic representation of the MCC development stage.
Source: Adapted from Fedorova (2008)

3 - Maturity: At the maturity stage (Figure II-6-5), convective processes reach their maximum development. As a result, at high levels of the troposphere and low stratosphere, a divergence of the air flow and a ridge at the top of the convection are formed. The lifting maximum moves up to higher levels (400-500 hPa).

At the middle levels, in the warm core, a mesoscale cyclonic circulation can form, which intensifies the confluence flow and subsidence below this level. This sinking causes high pressure, which creates an air divergence that blocks hot advection, lifting, and thus weakens the MCC. This stage is characterized by a large lifting area and heavy precipitation.

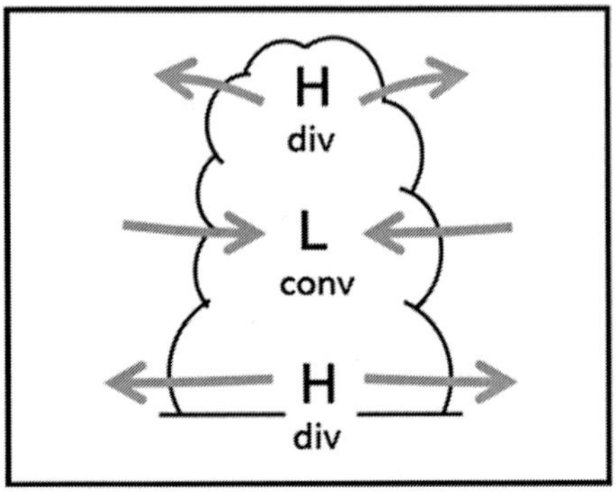

Figure II-6-5 – Schematic representation of the MCC maturity stage.
Source: Adapted from Fedorova (2008)

4 - Dissipation: This occurs when the instability loses its energy supply (warm and humid current in the low and middle troposphere), interrupting the development associated with earlier phases. This stage (Figure II-6-6) occurs when the system enters a region with cold and dry air advection, where there is no source of instability. The system stops receiving power and becomes disorganized, obtaining a chaotic appearance on satellite IR images.

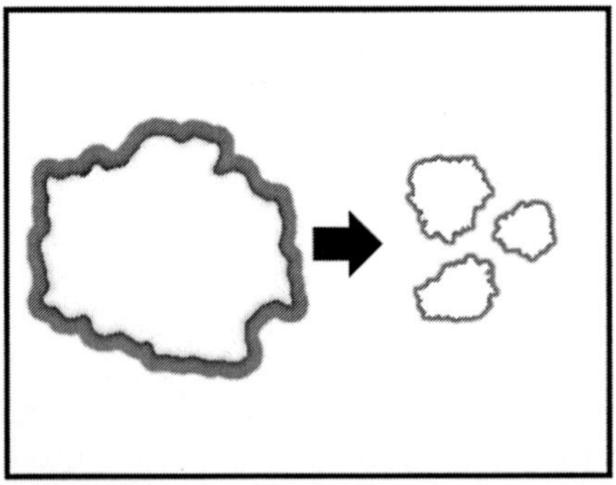

Figure II-6-6 – Schematic horizontal representation of CCM dissipation stage.
Source: Adapted from Fedorova (2008)

3. MCC Seasonality

The information below is a combination of results obtained from the studies by Fedorova et al. (2009), Albuquerque (2011), Milhahn Junior (2013); Lyra (2018), Fedorova et al. (2019) and Lyra et al. (2020). 132 MCCs have been documented in the NEB between 1999 and 2020.

The annual distribution of MCCs in the BNE can be seen in Figure II-6-7. The largest number of cases was recorded in 2002 (14 cases) then in 2008 (12 cases were detected). In general, among the 21 years (1999 - 2020), 10 of them had a frequency below the average, estimated at 6 cases per year. In 1999, the lowest frequency was recorded, when only one case was detected.

35.6% of cases were recorded in El Niño years (19.3% weak, 12.4% moderate and 3.9% strong) and with a larger number of episodes in summer months. On the other hand, 40.3% of the cases were observed in La Niña years (19.4% weak and 20.9% moderate).

In general, it can be noted that a slightly higher MCC frequency developed in ENSO negative phase (La Niña). Also, the neutral Sea Surface Temperature (SST) has a positive effect on the MCC formation. Despite this, some years are poorly characterized with respect to the variation in SST anomalies. For example, the lowest MCC frequency was detected in 2016 (La Niña year). However, the monthly anomaly on where this system was formed had a positive variation. That is, the spread of these values for the month is more significant than when considering the whole year. Therefore, this information becomes important for simplifying the means of studies and forecast these systems.

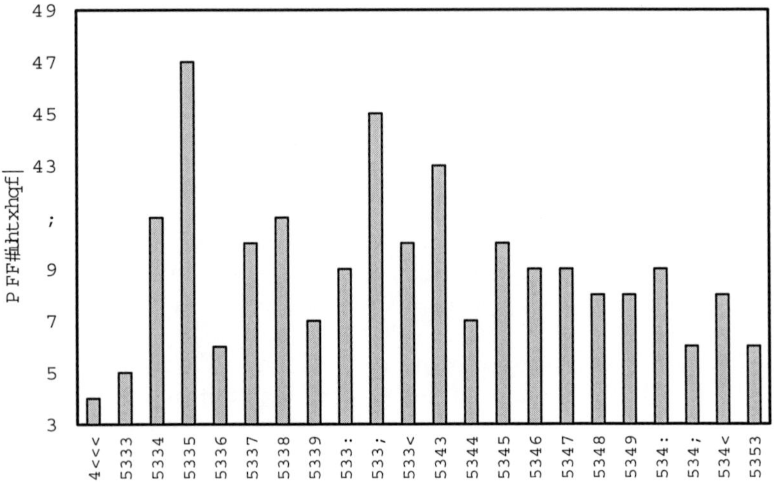

Figure II-6-7 – Annual frequency of MCC in the BNE from 1999 to 2020. Source: Lyra et al. (2022)

The monthly frequency of MCC for 21 years is shown in Figure II-6-8. The systems developed most often in March (31.8%), then in January and February (19.7% and 15.9%, respectively). Between July and October, the system was not identified. The predominance of MCC in the summer and autumn seasons is due to the action of synoptic systems, which are more common in the region during this period.

During the summer months, the convective potential of the Intertropical Convergence Zone (ITCZ, Chapter I-1) in the BNE is quite characteristic, as this system is located south of its climatological position due to the effect of the Atlantic Dipole. As well, the presence of Disturbances in the Trade Winds (Chapter I-3) and the Upper Tropospheric Cyclonic Vortex (UTCV, Chapter I-4) can be mentioned. These systems, together with breezes, increase moisture transport, contributing to increased convection over the region.

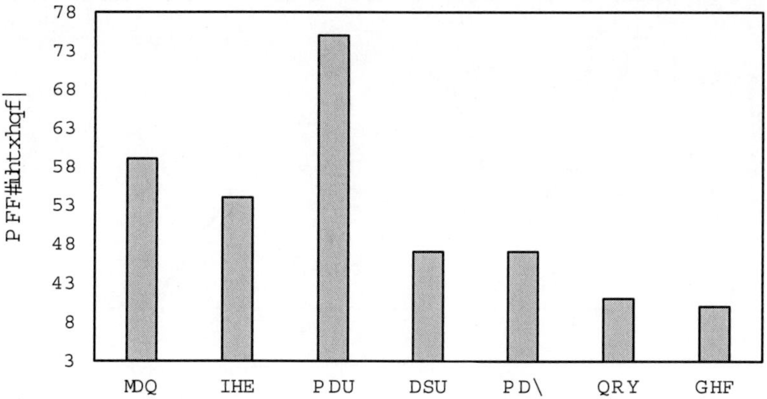

Figure II-6-8 – Monthly frequency of MCC in the BNE from 1999 to 2020. Source: Lyra et al. (2022)

4. MCC Characteristics

The time used to present the data was Universal Time Coordinated (UTC = Brazilian time zone + 3 hours).

The overall life cycle of the MCC in the BNE is shown in Figure II-6-9. Most of the MCCs (over 50%) lasted from 6 to 9 hours, some systems were observed up to 15 hours. MCCs of longer duration (more than 10 hours) did not have a specific period of the year and developed (appeared) from December to April (Lyra, 2018). On average, MCCs in the BNE lasted 8 hours slightly less than the 12 hours in the North American systems analyzed by Maddox (1980). However, these systems have values closer to those found by Blamey and Reason (2012) in the southern region of the African continent, where 53% of MCCs lasted between 6 and 9 hours.

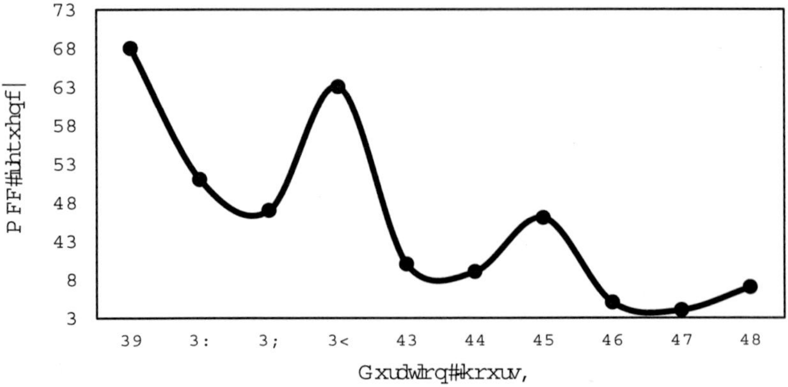

Figure II-6-9 – Duration of the full life cycle of MCCs from 1999 to 2020. Source: Lyra et al. (2022)

Unlike extratropical systems, the first convective explosions in the tropical region (Northern Brazil) do not occur only at night.

In the BNE, half of the cases began to develop predominantly between 00 – 03 UTC, which corresponds to 21 – 00 local time (Figure II-6-10). Many cases began to develop after 19 UTC, with an average of 3 to 4 cases during each hour until 23 UTC. A small number of cases (26.4%) started between 04 and 18 UTC. The MCC reached its maximum extension between night and morning (23 - 07 UTC), mostly around 06 UTC (Lyra, 2018).

On average, it took about 4 hours for MCC to develop from genesis to full size in the BNE. In some analyzed cases, the systems reached their maximum development in just 1 hour, while others took up to 9 hours to reach their peak development. The time for MCCs to reach maximum development was longer compared to the subtropical systems in southern Africa (3 hours) analyzed by Blamey and Reason (2012).

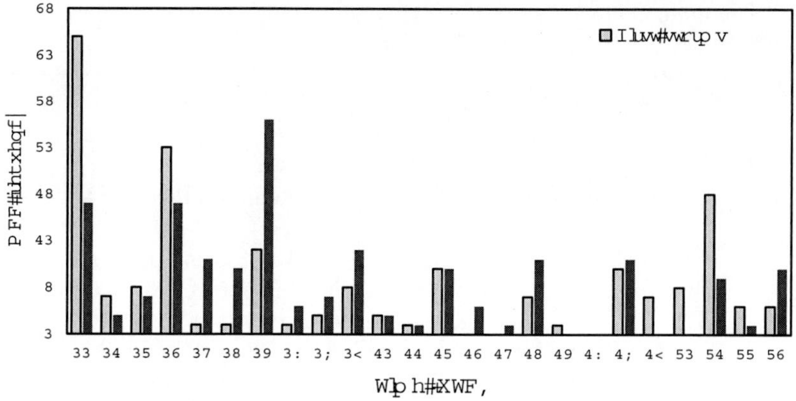

Figure II-6-10 – First storms and time to reach maximum MCC size between 1999 and 2020.
Source: Lyra et al. (2022)

5. Synoptic systems associated with the MCC development

In order for MCCs to reach their physical characteristics and complete their life cycle in the NEB, special synoptic systems are needed.

They are on a larger scale than the MCC (Synoptic Scale) and are usually associated with heavy precipitation. The analysis of various synoptic fields is important for identifying the systems that make up the atmosphere during an MCC.

The synoptic scale systems that influenced the formation and development of this meteorological phenomenon in the period from 1999 to 2020 will be described below. These systems were identified as a result of research by the scientific group of the Laboratory of Physical and Synoptic Meteorology of the Federal University of Alagoas. The main results are described by Cordeiro et al. (2018), Lyra (2018), Fedorova et al. (2019), and Lyra et al. (2020, 2022).

The relationship of these systems was presented using meteorological maps at different levels of the troposphere, formed according to ERA-Interim reanalysis data and from satellite highlighted images of GOES series (12, 13 and 16) in the infrared channel. The construction of these maps at the principal levels facilitates the identification of synoptic and mesoscale systems or processes, since the action of troughs (ridges), cyclones (anticyclones) and the confluence (difluence) of the winds at each level can be seen from the circulation found in them.

The main characteristics of the streamline fields for systems associated with the MCC will be presented, which can be seen in the examples bellow in Figure II-6-11:

1) **Intertropical Convergence Zone** (ITCZ, Chapter I-1) - Wind confluence at 1000 hPa level (Figures II-6-11a and II-6-11b);
2) **Upper Tropospheric Cyclonic Vortex** (UTCV, Chapter I-4) – Cyclonic circulation at 200 hPa level (Figures II-6-11c and II-6-11d);
3) **Frontal Extremity** (FE, Chapter I-7) – Confluence in the trough of the baroclinic cyclone at 1000 hPa level (Figures II-6-11e and II-6-11f);
4) **Trade Winds** (Chapter I-3) - Currents from the South Atlantic subtropical anticyclone at 1000 hPa level (Figures II-6-11g and II-6-11h);
5) **Wave Disturbance in Trade Winds** (WDTW, Chapter I-3) - Trough in the Trade winds at 1000 hPa level (Figures II-6-11i and II-6-11j);

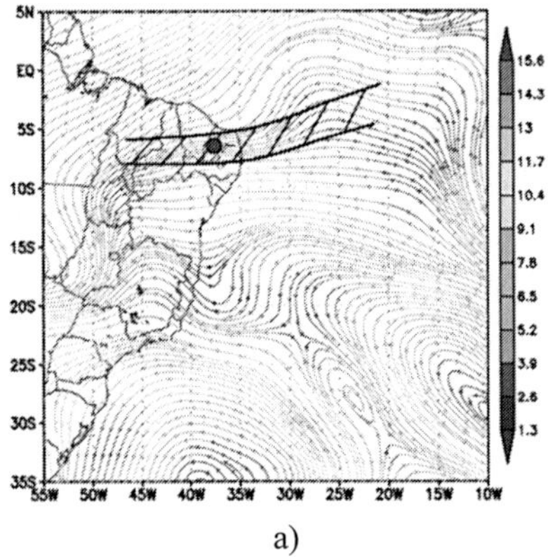

a)

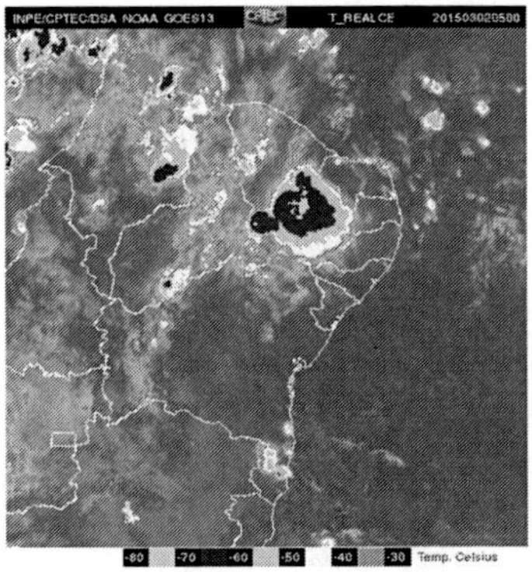

b)

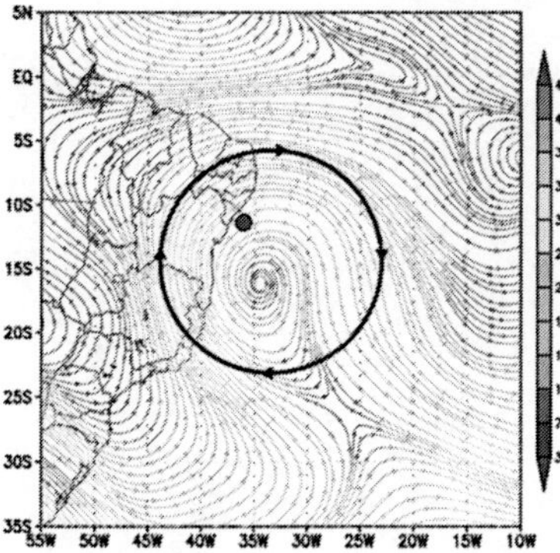
5N
EQ
5S
10S
15S
20S
25S
30S
35S
55W 50W 45W 40W 35W 30W 25W 20W 15W 10W
45.6
41.8
38
34.2
30.4
26.6
22.8
19
15.2
11.4
7.6
3.8

c)

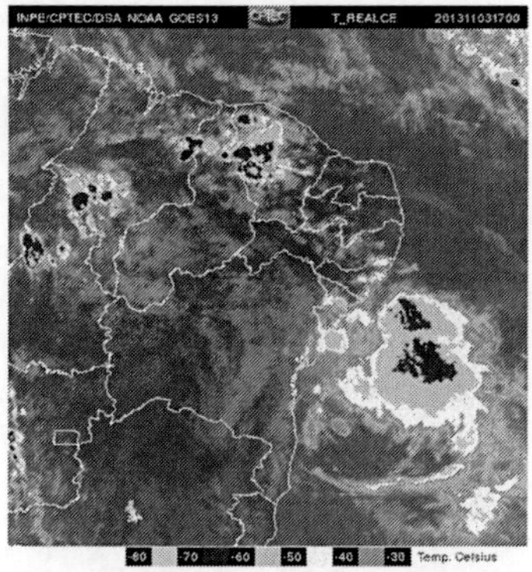
INPE/CPTEC/DSA NOAA GOES13
T_REALCE
201311031700
-80 -70 -60 -50 -40 -30 Temp. Celsius

d)

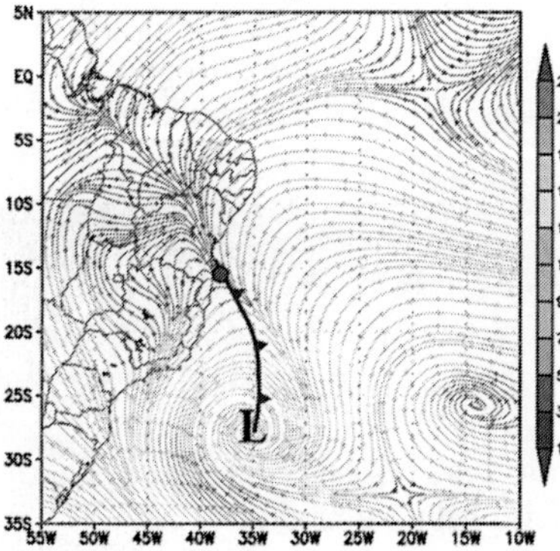
5N
EQ
5S
10S
15S
20S
25S
30S
35S
55W 50W 45W 40W 35W 30W 25W 20W 15W 10W
L
22.8
20.9
19
17.1
15.2
11.4
9.5
7.6
5.7
3.8
1.9

e)

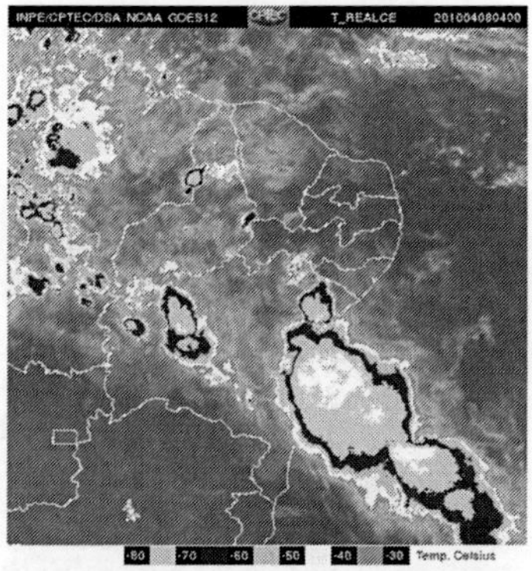
INPE/CPTEC/DSA NOAA GOES12
T_REALCE
201004080400
-80 -70 -60 -50 -40 -30 Temp. Celsius

f)

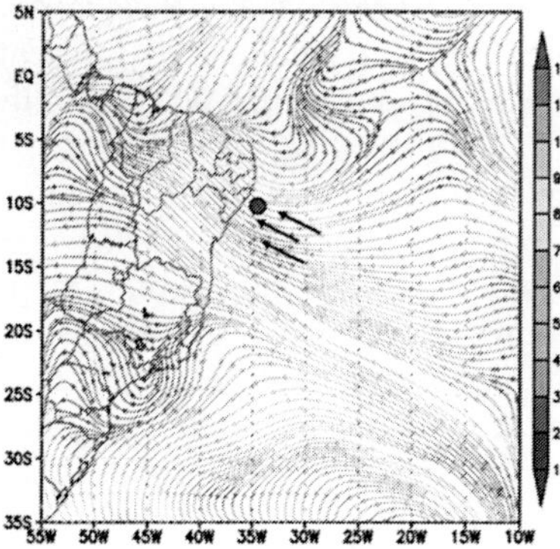
5N
EQ
5S
10S
15S
20S
25S
30S
35S
55W 50W 45W 40W 35W 30W 25W 20W 15W 10W
13.2
12.1
11
9.9
8.8
7.7
6.6
5.5
4.4
3.3
2.2
1.1

g)

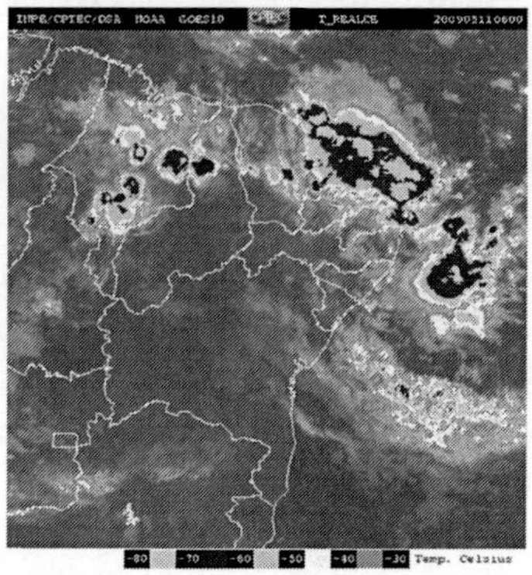
T_REALCE
-80 -70 -60 -50 -40 -30 Temp. Celsius

h)

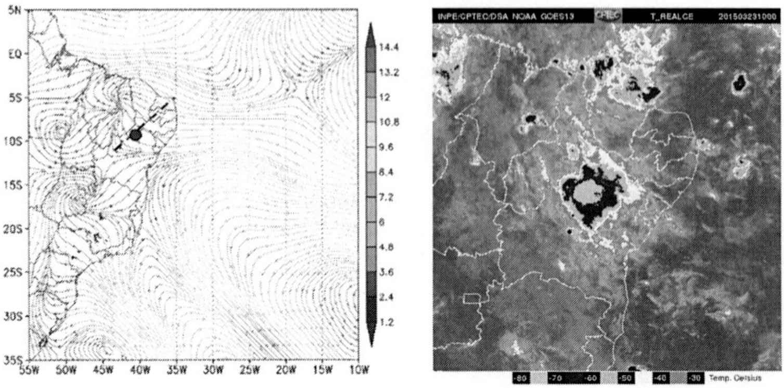

i) j)

Figure II-6-11 – Examples of synoptic systems associated with the MCC development in the BNE. Streamlines (left) and satellite images (right). ITCZ (a) and (b); UTCV (c) and (d); FE (e) and (f); Trade Winds (g) and (h); WDTW (i) and (j).
Source: Adapted from Fedorova et al. (2019)

All systems were considered independently of each other, so in most events more than one system operated at different levels of the atmosphere. Therefore, the total number of synoptic systems analyzed (189) was greater than the total number of reported cases of MCC (132).

According to the database of synoptic systems associated with the MCC in the BNE (Fedorova et al., 2019), the highest percentage of occurrence of mesoscale systems is caused by the ITCZ, 74 cases (57%). This is followed by the presence of UTCV, 50 cases (38%), 17 of which are intense, reaching both high and middle levels. On the other hand, Frontal Extremity presented 35 cases (26%), Trade Winds 30 cases (24%), and finally WDTWs only 5 detected cases (3%).

The great ITCZ influence on the MCC development in the first months of the year, according to Molion and Bernardo (2002), is due to its position further south, which can reach 4° S latitude. Its annual migration is directly linked to the MCC formation. Thus, about 68% of ITCZ events were observed in March and April. In January, February and May, this system did not have a high frequency of occurrence. At the same time, in the remaining months of the year, no cases associated with ITCZ were identified.

In winter, meridional temperature gradients increase in the Southern Hemisphere, so the frontal extremities can reach the BNE more often (Fedorova et al., 2016). Despite this, no MCCs were found on the Frontal Extremities in winter. A higher frequency of occurrence of MCCs associated with this system was observed during the warm period of the year, in December and from February to April. The highest frequency of MCC was observed in March (36.4%; Lyra, 2018).

UTCV is a synoptic scale system in which MCCs developed throughout the warm season, except for April. The largest highlight of this system occurred between November and March, mainly in February (27.3%; Lyra, 2018), when, according to the classical literature (Kousky and Gan, 1981), there is a greater penetration of these systems on the continent.

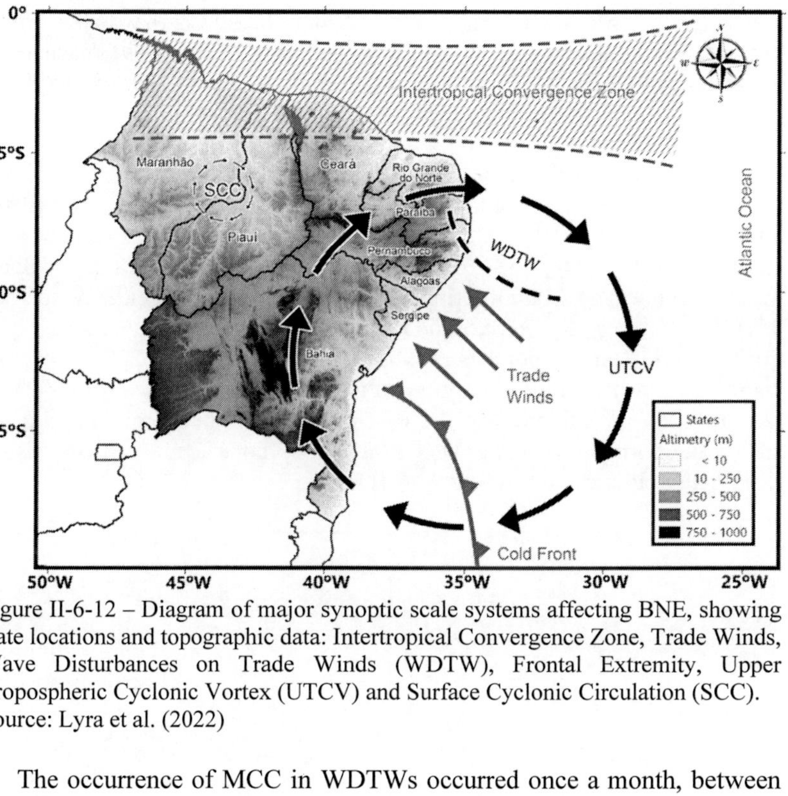

Figure II-6-12 – Diagram of major synoptic scale systems affecting BNE, showing state locations and topographic data: Intertropical Convergence Zone, Trade Winds, Wave Disturbances on Trade Winds (WDTW), Frontal Extremity, Upper Tropospheric Cyclonic Vortex (UTCV) and Surface Cyclonic Circulation (SCC). Source: Lyra et al. (2022)

The occurrence of MCC in WDTWs occurred once a month, between February and April. On the other hand, the Trade Wind pattern showed a larger monthly variability from January to November, demonstrating the absence of a definite pattern (Lyra, 2018).

Figure II-6-12 presents a diagram of the major synoptic scale systems affecting the Northeastern region of Brazil.

6. MCC Locations

The information in this subsection was taken from Lyra et al. (2020), where the authors analyzed 65 MCCs over 10 years (2008 - 2017). More often MCC developed over the continent (54 events) than over the ocean (11 events). This distribution was similar to the results of studies conducted in South America by Velasco and Fritsch (1987) and in the southern part of the African continent by Laing and Fritsch (1997).

Of all the MCCs formed over the ocean, 70% were larger than 100,000 km². The largest of them had an area of about 262,300 km², which is one of the greatest values recorded in those years. All MCC that have reached the largest area in their development, for better visualization, were divided into groups by size with an interval of 25,000 km². The position of the core of each MCC in the group is shown in Figure II-6-13.

The convective potential in MCC core was quite typical for ITCZ. It is noted (Lyra et al., 2019) that in 2014 the MCC formed as a result of the ITCZ cloudiness joining with a frontal extremity, which led to heavy rains in the Maranhao state. The formation of some MCCs over the state of Bahia was influenced by frontal peripheries. The presence of Trade Winds and UTCVs on the eastern coast of the BNE was also quite characteristic, which further increased the number of MCCs (Lyra et al., 2016). On the other hand, other systems that were present during the winter, such as the WDTW, despite a direct influence on the east coast pluviometric regime, did not affect the formation of the MCCs, which is consistent with the results obtained by Pontes da Silva et al. (2011).

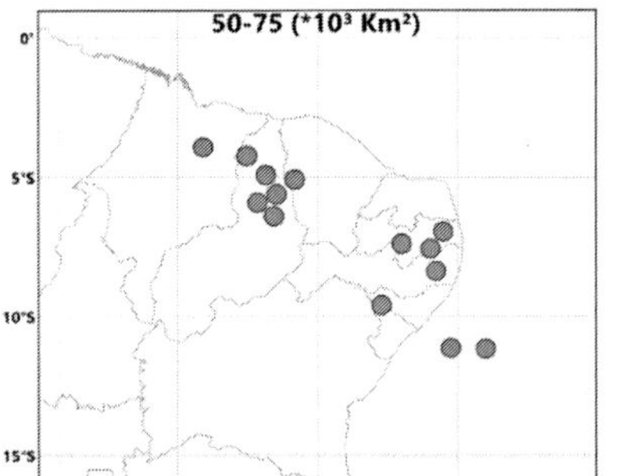

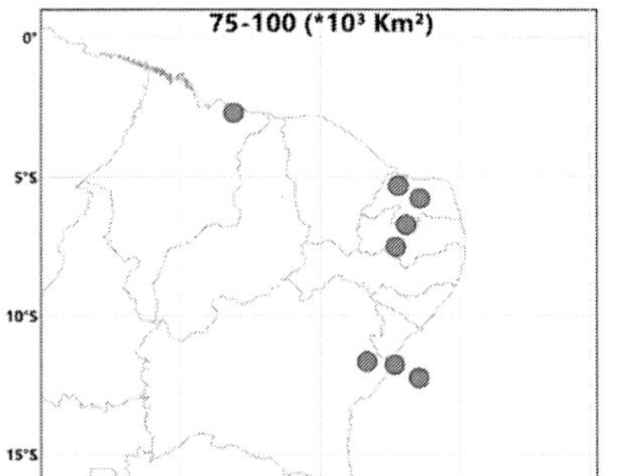

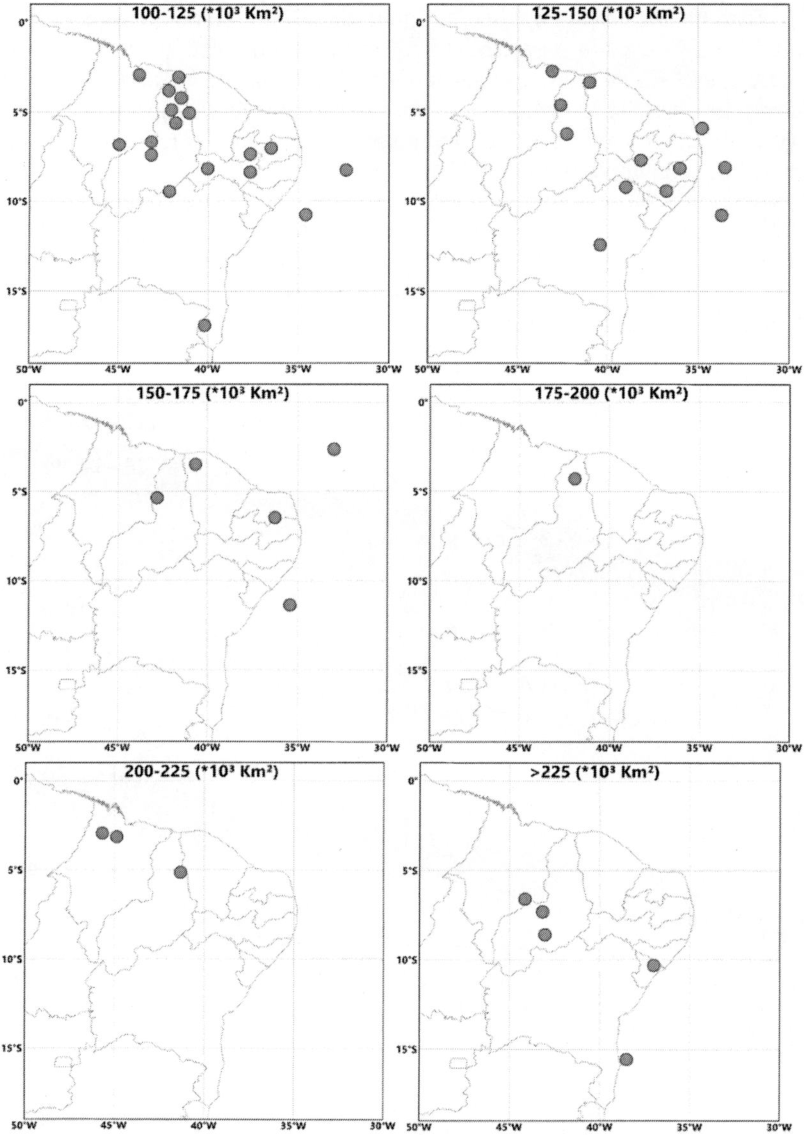

Figure II-6-13 – Locations of the MCC centers according to their maximum extension areas: 50-75 · 10^3 km² (a); 75-100 · 10^3 km² (b); 100-125·10^3 km² (c); 125-150 · 10^3 km² (d); 150-175 · 10^3 km² (e); 175-200 · 10^3 km² (f); 200-225 · 10^3 km² (g) and > 225·10^3 km² (h). Source: Lyra et al. (2020)

Lyra et al. (2020) applied to BNE the Blamey and Reason (2012) methodology used in southern Africa to represent the spatial distribution of MCCs at their maximum extension stage through their respective cloud coverage areas. This methodology was applied to represent specific areas of MCC development in the BNE and display results similar to Figure II-6-13 in a single image in Figure II-6-14.

The two most active regions are mainly concentrated between the borders of the states of Maranhão and Piauí, where the presence of an MCC with larger area can also be identified. The eastern coast of the BNE and its semi-arid region were also favorable for the MCC formation, however, they had smaller extension areas.

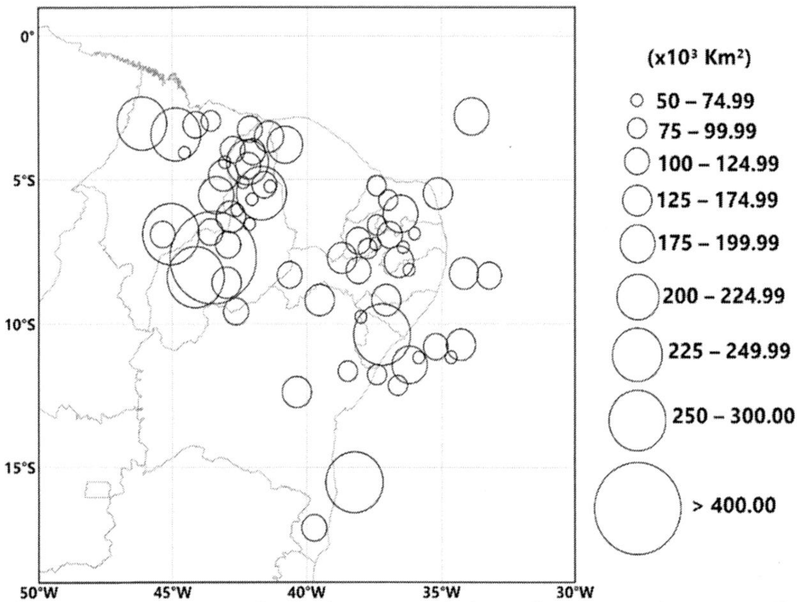

Figure II-6-14 – MCCs distribution across their maximum extension areas. The circles represent the approximate cloud cover of each system.
Source: Lyra et al. (2020)

7. MCC relationship with precipitation and other adverse phenomena

7.1. Precipitation

Precipitation is the main adverse phenomenon associated with MCCs as a result of the intense convection caused by these meteorological systems. Precipitation varies in intensity depending on the MCC location and their life cycle. However, in any case, this can cause great socio-economic damage to the population. It should be noted that precipitation information is not available for all analyzed events, since sometimes the system operated in an area without a meteorological station, while other meteorological systems were located over the ocean. Figure II-6-15 shows maximum precipitation (mm/24h, from the INMET database) for each of the 22 MCC events, as well as climatological data (30-year data series, 1981 - 2010) for each event and region.

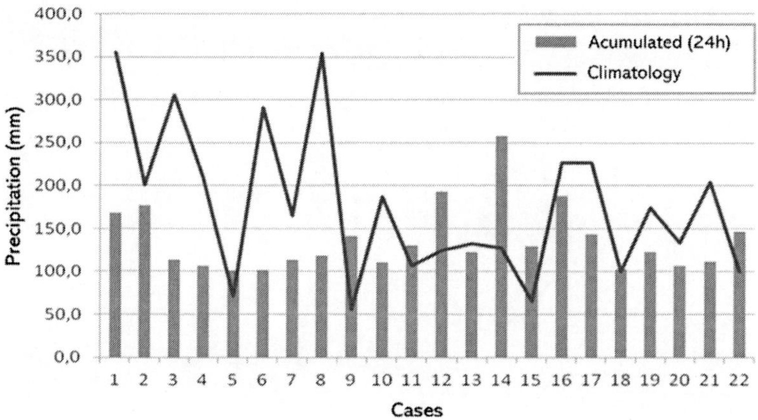

Figure II-6-15 – Precipitation (mm/24h) for some MCC events (gray bar) and Climatology (black line).
Source: Fedorova et al. (2019)

Climatological data showed higher MCC-related precipitation values for the first 08 events. For the remaining 13 events, the relationship between the value of precipitation and the presence of MSS is not so obvious.

7.2. Thunderstorms

The frequency of thunderstorms associated with MCC in the period from 1999 to 2017 in BNE was determined and analyzed in several studies by the research group of the Laboratory of Physical and Synoptic Meteorology of the Federal University of Alagoas (Cordeiro et al., 2018; Lyra et al., 2018; Fedorova, et al., 2019). The term "thunderstorm" will be used to refer to multiple episodes of lightning. That is, each event described will be equivalent to a day characterized by the occurrence of a series of lightning discharges related to MCC cases.

The occurrence of thunderstorms was detected using data from the Tropical Rainfall Measuring Mission (TRMM) satellite, a joint project between the National Aeronautics and Space Administration (NASA) and the Japan Aerospace Exploration Agency (JAXA), together with the Meteorological Aerodrome Report data (METAR) obtained through the REDEMET software, from the Department of Airspace Control (DECEA) and from the GOES-13 satellite images (Geostationary Operational Environmental Satellite) in the Enhanced Infrared spectral channel (Figure II-6-16).

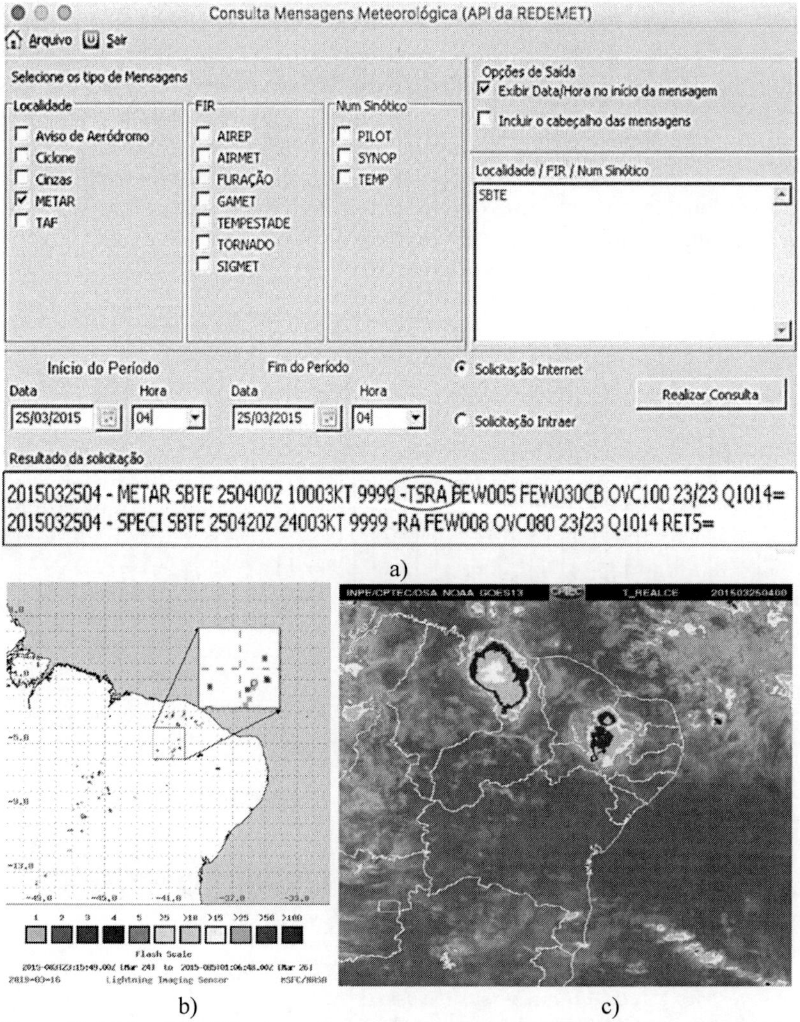

Figure II-6-16 – Identification of atmospheric electric discharges obtained by the following data: METAR via REDEMET Software (a), TRMM Satellite (b) and GOES-13 Satellite in the highlighted IR channel (c).
Source: Fedorova et al. (2019)

Among the 132 MCCs identified in the BNE, thunderstorms were observed for 122 days (94% of events). The formation of this adverse phenomenon occurred mainly in the northern part of the region, mainly in Piauí and Maranhão states.

The annual distribution of MCC-related thunderstorms in the BNE between 1999 and 2020 can be seen more clearly in Figure II-6-17. DECEA and the TRMM satellite recorded about 119 thunderstorms. During this period, the largest number of thunderstorm days (12 days a year) was recorded in 2002 and 2008, followed by 10 days a year in 2005 and 2010.

The monthly average was approximately 1 thunderstorm for all months except the warm months of November to May, where the monthly average was 2 thunderstorms. Some stabilization of the occurrence of this adverse phenomenon has been observed since 2011.

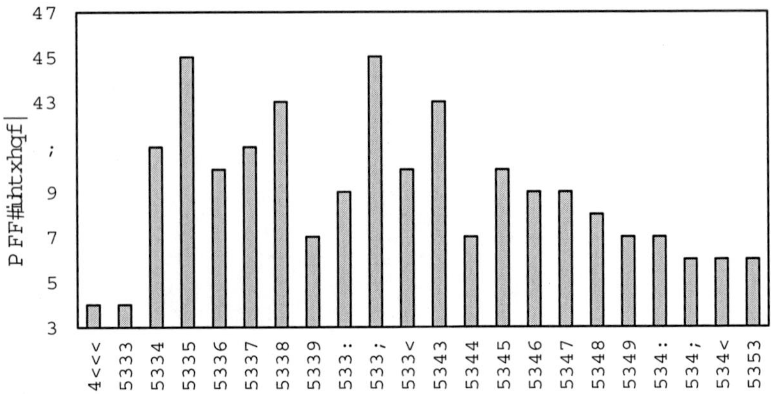

Figure II-6-17 – Annual distribution of thunderstorms in the BNE associated with MCC between 1999 and 2020.
Source: REDEMET/DECEA and TRMM/NASA

8. Simulated and forecasted profiles in MCC

The Forecasted vertical profiles calculated by Lyra (2016) for the 6 MCCs, that developed between 2013 and 2015 are presented in Figure II-6-18 below. The forecast quality was assessed by comparing the temperature (T) and dew point (Td) curves, as well as the instability indices of the forecasted profiles, with those of the simulated profiles.

Methods for elaboration *simulated* and *forecasted* vertical profiles of T and Td are described in Chapter II-2. In general, the forecasted profiles were satisfactory in relation to the T curve up to 48 hours before the formation of the MCC. Regarding the Td curve, a deviation of more than 2°C was observed in most cases. Satisfactory results were only at certain levels, preferably in the middle troposphere.

It should be noted that knowledge of T at lower levels is the most important factor for estimating the intensity of instability and thus for forecasting convective precipitation.

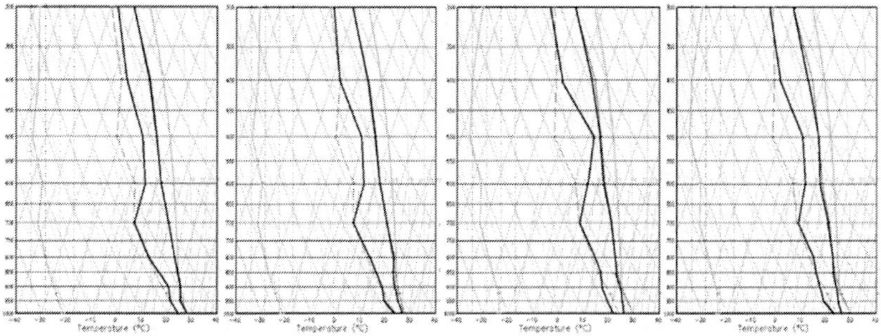

12h 24h 36h 48h

a)

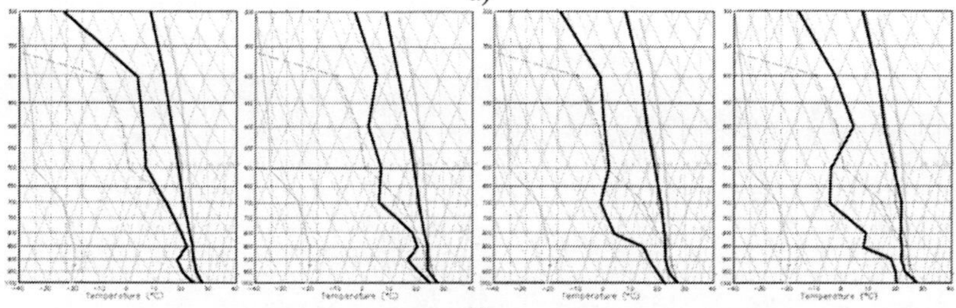

b)

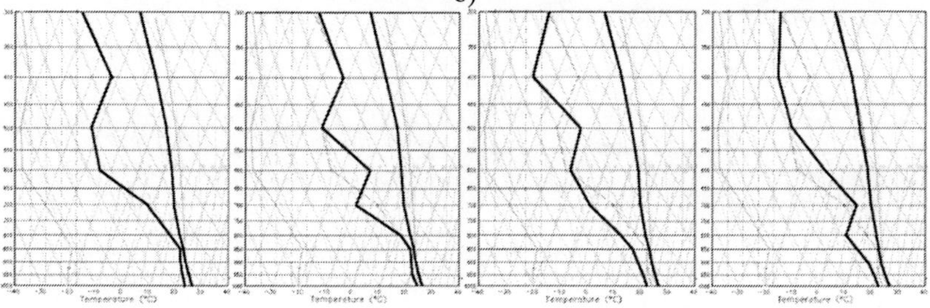

c)

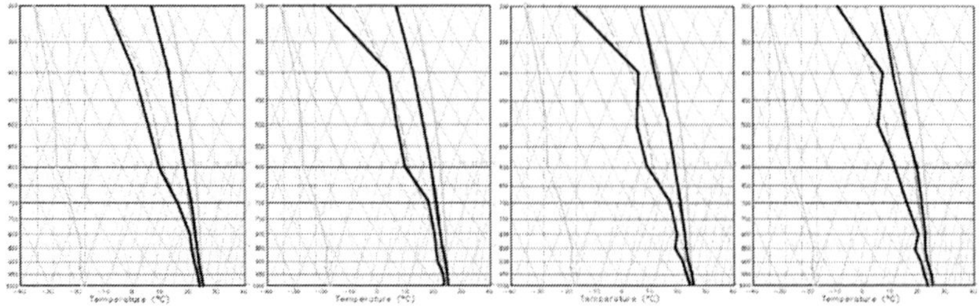

d)

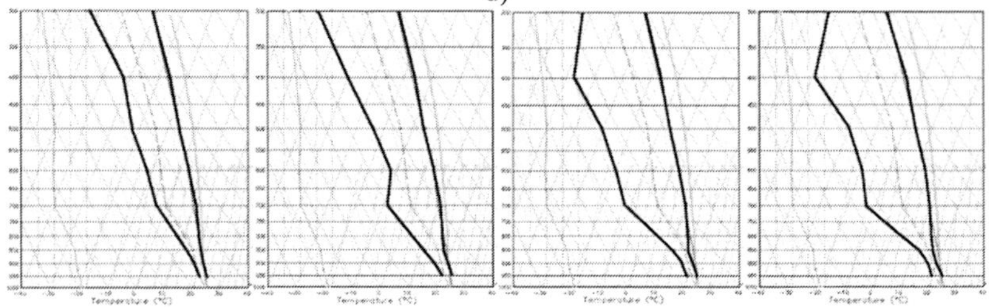

e)

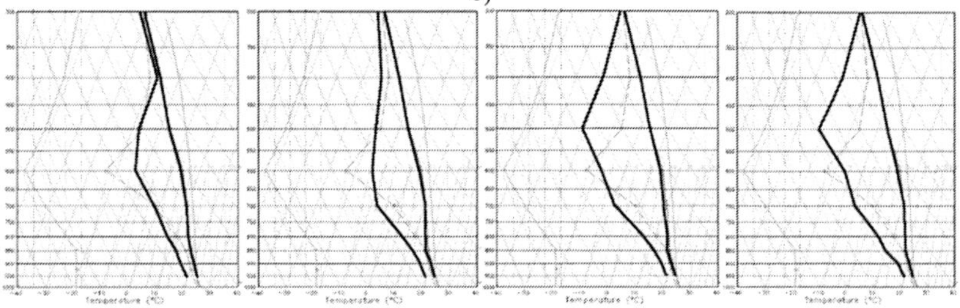

f)

Figure II-6-18 – Forecast profiles developed 12, 24, 36 and 48 hours ahead for MCC in 2013 (a, b); 2014 (c, d) and 2015 (e, f).
Source: Lyra (2016)

9. MCC trajectories

The information below is based on a study by Lyra et al. (2022) for the period from 2008 to 2017, where 65 MCC were analyzed. According to the authors, MCCs in the BNE have a lower frequency and greater complexity of their displacement trajectories compared to subtropical MCCs. Systems associated with the ITCZ were mainly concentrated between 9° and 2° S (Lyra et al., 2022). Developing systems were observed in two longitudinal regions: in the semiarid region, between Rio Grande do Norte and Pernambuco states (37° to 35°W); and in the Piauí state (45° to 42°W).

Trajectories presented in Figure II-6-19a shows the effect of ITCZ. A characteristic feature of this synoptic system is the flow from East to West, in the initial and final hours of MCC development.

The trajectories of MCCs associated with UTCV (Figure II-6-19b) presented the largest variation among all the studied systems. Air current at high levels plays a more important role in displacing these MCCs. This synoptic system has four different development mechanisms (Kousky and Gan, 1981; Ramirez et al., 1999; Fedorova, 2008) and affected the displacement of the MCC. Only four of the MCCs that formed between the states of Piauí and Maranhão and developed in the presence of UTCV (a developmental mechanism proposed by Kousky and Gan, 1981) moved parallel to the vortex current. Forty-three percent of the MCCs associated with these systems followed a vortex circulation pattern: clockwise.

Six of ten MCCs that developed at a Cold Front Extremity (CFE) had different trajectories (Figure II-6-19c). Also, an event with the largest displacement distance among all MCC occurred at the CFE.

The trajectories of three events associated with WDTW (Figure II-6-19e) developed in different regions and in different directions. This shows the great complexity in forecasting these MCC trajectories. Four events under the influence of the trade winds (Figure II-6-19d) moved in different directions at the beginning of their development and to the west at the end of their existence.

The MCCs developed within the weak surface cyclonic circulation (Figure II-6-19f) moved to the west, especially at the dissipation stage. In the beginning they went from south to northwest. The westward displacement and their travel distance were similar to the MCCs boosted by ITCZ.

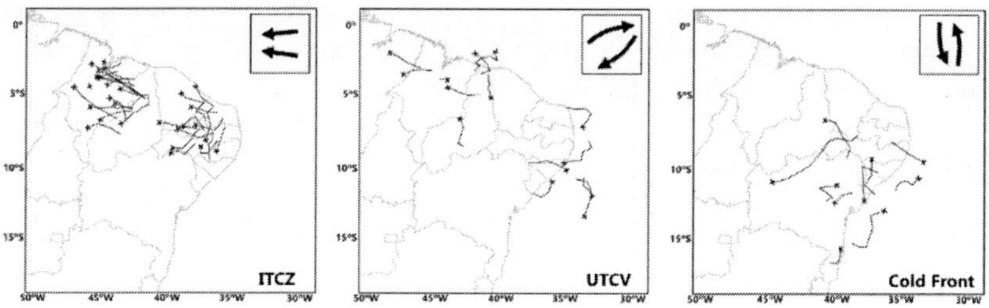

a) b) c)

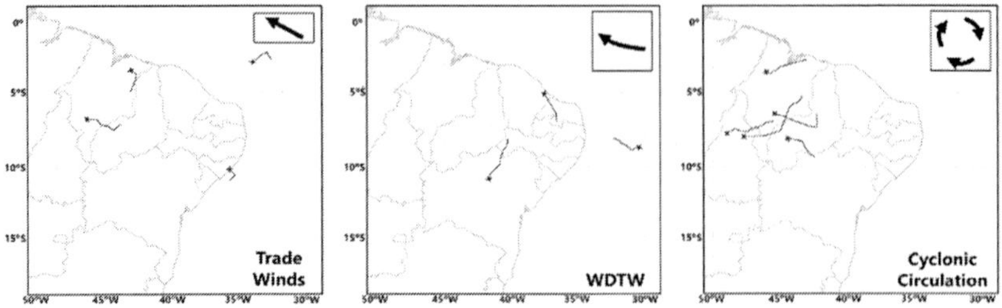

d) e) f)

Figure II-6-19 – Trajectories (lines) and final locations (x) of MCCs associated with ITCZ (a); CF (b); UTCV (c); Trade Winds (d); WDTW (e); surface cyclonic circulation (f). The control current is presented in the upper right corner at 1000 hPa (a); 925 hPa (b); 300 hPa (c); 925 hPa (d); 850 hPa (e); 1000 hPa (f).
Source: Lyra et al. (2022)

Lyra et al. (2022) divided the study area into 3 different sectors, since the BNE has a large area (1,554,257.00 km²): Region N1 consisted of the Bahia and Sergipe states (Figure II-6-20a); Region N2 included Alagoas, Pernambuco, Paraiba and Rio Grande do Norte states (Figure II-6-20b); Region N3 consisted of the states of Ceará, Piauí and Maranhão (Figure II-6-20c). MCCs developing over the adjacent ocean were included in the corresponding region.

Region N1 (Figure II-6-20a) was characterized by many events that developed over the ocean but did not provide a unified pattern of the trajectory direction. It is also important to emphasize that the MCCs that formed over the Bahia state developed relatively close to the coast and did not cross the mountainous region of the Diamantina Plateau.

Two types of trajectories were identified in the N2 region (Figure II-6-20b). The MCCs that developed near the coast (northeastern) did not cross higher regions like the MCCs in the Bahia state (southeastern). The systems formed in the hinterland (mainly 40° W) moved westward. The influence of the local topography on the MCC trajectories in the N2 region was not found.

In the region N3 (Figure II-6-20c) there was little orographic influence because of a less elevated and more uniform topography compared to the other two regions. The predominance of ITCZ and low altitude determine the pattern of the trajectories from west to east in this area.

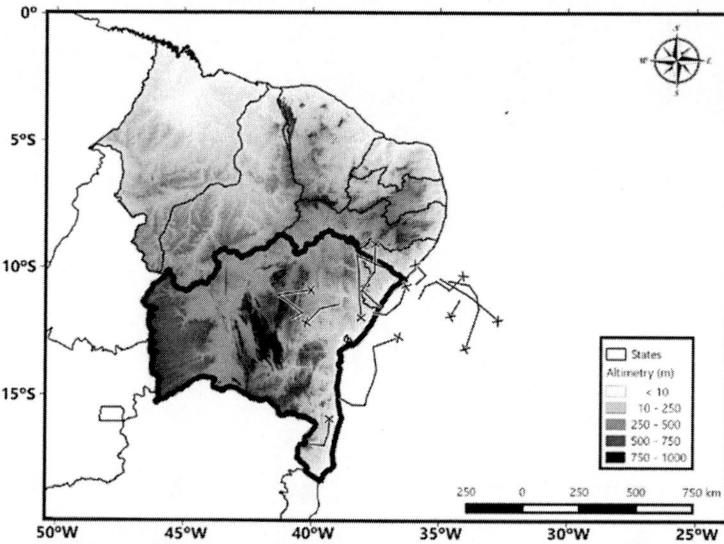
0°
5°S
10°S
15°S
50°W
45°W
40°W
35°W
30°W
25°W
States
Altimetry (m)
< 10
10 - 250
250 - 500
500 - 750
750 - 1000
250 0 250 500 750 km

a)

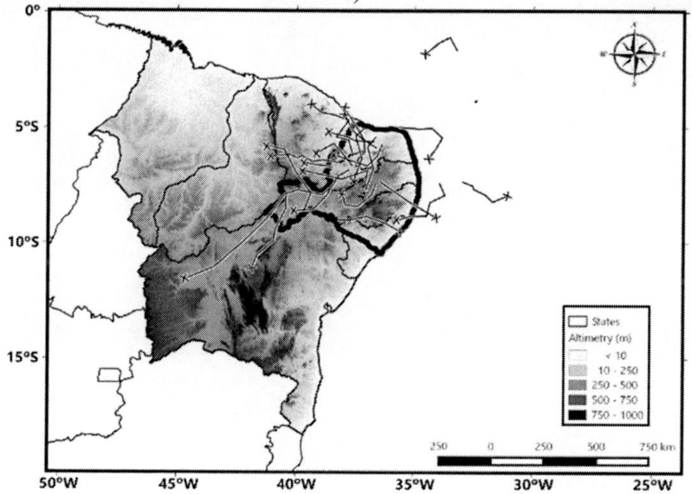
0°
5°S
10°S
15°S
50°W
45°W
40°W
35°W
30°W
25°W
States
Altimetry (m)
< 10
10 - 250
250 - 500
500 - 750
750 - 1000
250 0 250 500 750 km

b)

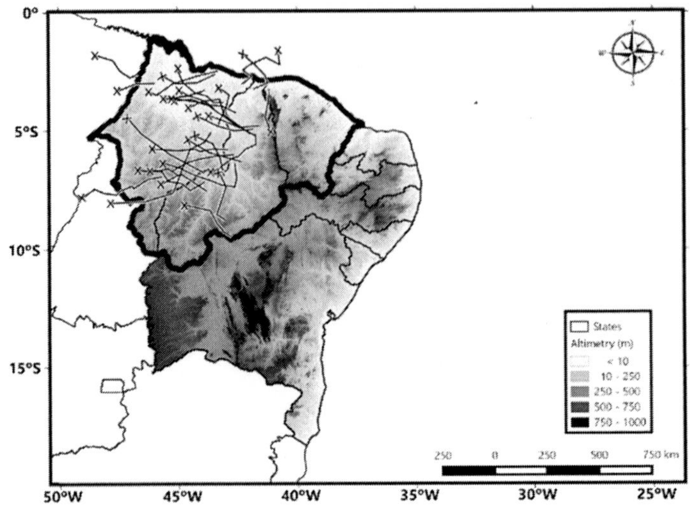

c)

Figure II-6-20 – MCC trajectories (lines) and final locations (x) over topographic maps for each sub-region of the Brazilian Northeast. N1 a), N2 b) and N3 c). The thicker line marks the sub-regions.
Source: Lyra et al. (2022)

10. Relief Influence on the MCCs

The relief of the BNE (Figure II-6-21) is characterized by the Borborema Plateau in the east, the Parnaiba River Basin in the west, and some uplands such as the plateau in the south. Plains are located along the coast.

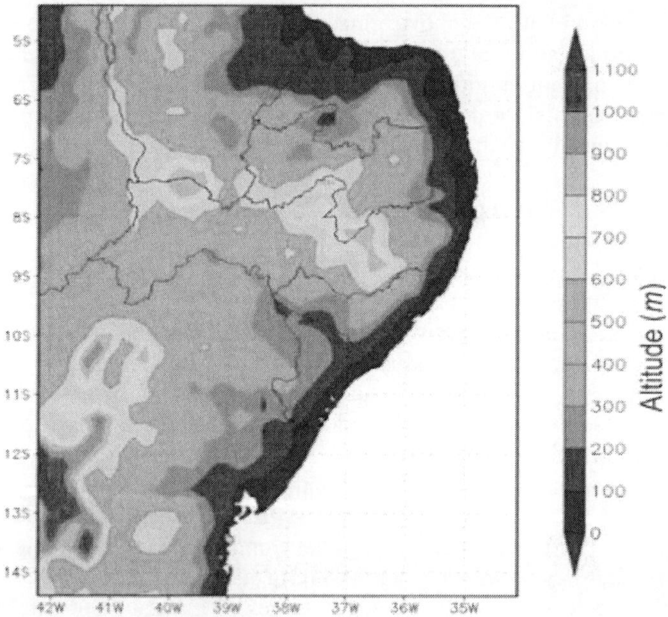

Figure II-6-21 – Relief of the BNE.
Source: Fedorova et al. (2019)

Tracing the relief profile for hundreds of kilometers from the east coast to the interior of the BNE, one can see the change in altitude from sea level in the coastal plain to elevations up to 1000 meters high. Such an elevation contributes to the occurrence of orographic precipitation processes due to its proximity to a warm ocean and being in a region where southeast trade winds prevail almost all year round. The greatest amount of precipitation falls on the windward side of the mountains, in the Atlantic Forests of the coastal plain, while on the leeward side, the climate is semi-arid.

It follows from atmospheric thermodynamics that diurnal temperature fluctuations associated with a daily change in the balance of solar energy are a factor contributing to a change in the instability of an air parcel.

Surface air temperature data from 80 MCC events that took place in the period from 1999 to 2009 showed no significant variations in the diurnal temperature cycle for MCCs of varying intensity. At the initial point of the trajectory, the differences between the surface temperature and the maximum daily temperature, on average, did not exceed 4ºC, and the relative air humidity averaged about 78%.

In connection with the above, since the coast of the BNE is not subject to large thermal amplitudes due to its proximity to the equator and the presence of constant humidity, the convective process due to an increase in daily temperature is not the main factor contributing to the development of atmospheric instability in the region.

10.1. Mountain influence

An air parcel moving along a mountain may undergo changes in its instability. During the lifting, its lower layers are compressed in the adiabatic cooling process, resulting in a decrease in the height of the Lifting Condensation Level (LCL) and positively affects the convective potential (CAPE). Conversely, the descending air parcel undergoes adiabatic expansion, resulting in its heating and drying, which negatively affects the convective potential.

Thermodynamic indices indicate an increase in instability during ascending MCC trajectories and a decrease in descending trajectories from the mountain, regardless of the stage of development. However, at the stages of genesis and development (stages I-II), instability increases even more, and it also decreases even more at the stages of maturity and dissipation (stages III-IV). To better illustrate the influence of orography on the MCC development, let's take an example.

10.2. Example of MCC event

The morphological characteristics of the MCC development stages are presented below based on satellite images (Figure II-6-22).

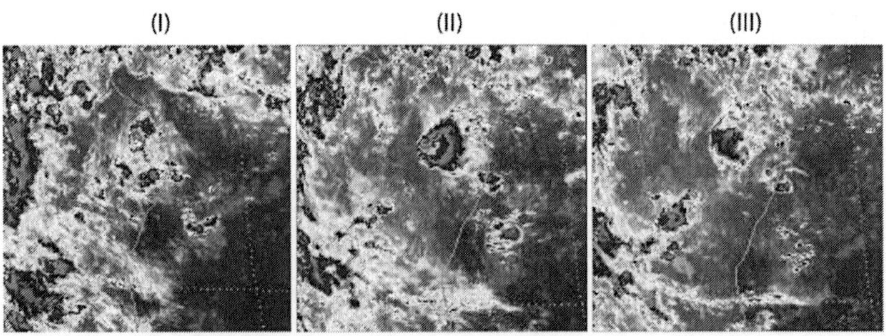

Figure II-6-22 – METEOSAT 9 satellite images at three stages of MCC development on March 25, 2008.
Source: NOAA/GIBBS (KNAPP, 2008)

The distinguished region of cloud tops with temperatures below –32°C covers an area of about 200,000 km², has an approximately circular eccentricity with a length of 300 and 250 km of the major and minor axes, respectively.

With such a development of the MCC (Figure 23 I, II and Table 1) and its displacement from a height of 200 m to 800 m, the following effects were observed: a decrease in the LCL, an increase in the moisture layer and CAPE (from 652 to 1104 J /kg). During the dissipation of the MCC and its descent from a height of 800 m to 500 m (Figure 23 II, III and Table 1), a decrease in the moisture layer and CAPE (from 1104 to 788 J/kg) was recorded.

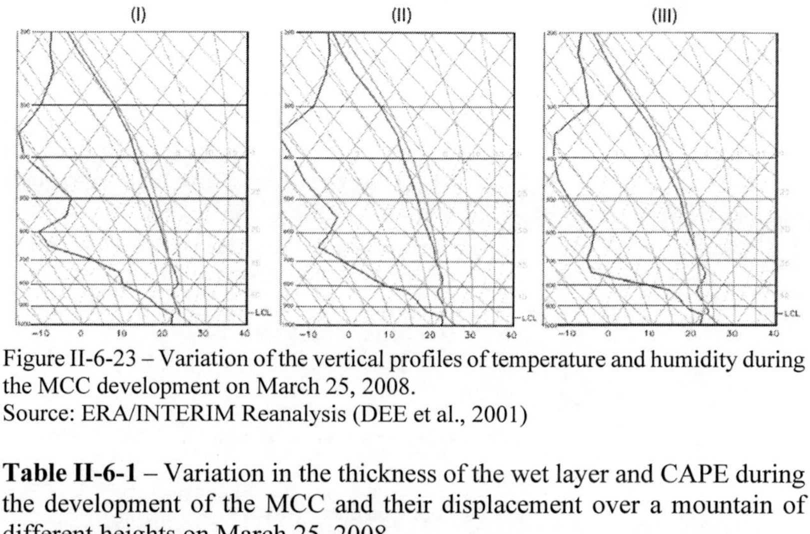

Figure II-6-23 – Variation of the vertical profiles of temperature and humidity during the MCC development on March 25, 2008.
Source: ERA/INTERIM Reanalysis (DEE et al., 2001)

Table II-6-1 – Variation in the thickness of the wet layer and CAPE during the development of the MCC and their displacement over a mountain of different heights on March 25, 2008.

Stage/height	(I)	(II)	(III)
CAPE	652 J/Kg	1104 J/Kg	788 J/Kg
Wet layer thickness	120 hPa	150 hPa	140 hPa
height	200 m	800 m	500 m

The thickness of the wet layer (T–Td ≤ 6°C) increased during the ascending phase of the MCC displacement. This increase was accompanied by an increase in the convective potential (CAPE), which almost doubled, creating instability due to orography. The opposite situation occurred at a downward displacement with the decrease in instability by 316 J/kg.

11. Final considerations and recommendations for MCC forecasting in the BNE

This section provides a summary of the results and recommendations for MCC forecasting. Figure II-6-24 shows a Decision Tree, which is described below.

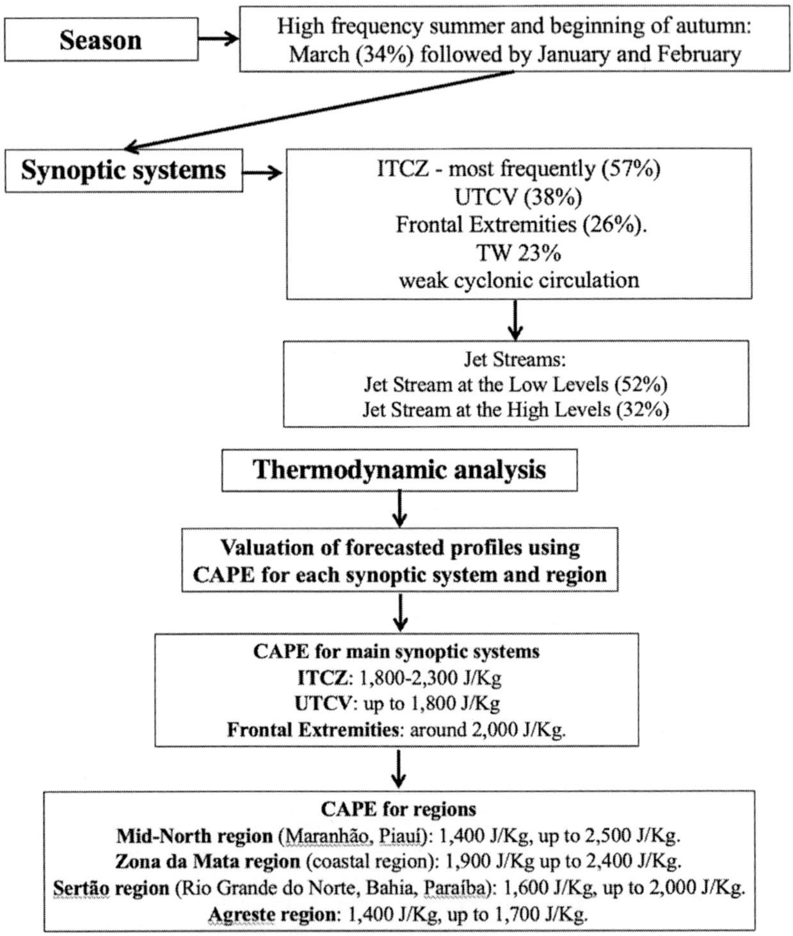

Figure II-6-24 – Decision Tree for MCC forecasting
Source: Authors

11.1 Seasons

The highest MCC frequency was found in March (34%) followed by January and February (16% and 15%, respectively). In the summer, 65% of the cases occurred.

11.2 MCC development time period

Most MCCs (more than 50%) lasted 6 to 8 hours, but some cases up to 15 hours. Longer MCCs (more than 10 hours) were more often observed from December to April. Frequently, MCCs in the BNE form between 21 and 00 UTC, with the highest formation peak at 00 UTC. Another large part of the cases began to develop after 19 UTC, averaging 3 to 4 cases per hour until 23 UTC. MCCs that developed from 00 to 03 UTC reach their maximum extension between 23 and 07 UTC, most often at 06 UTC. On average, these MCCs took about 4 hours to develop from genesis to maximum size.

11.3 MCC Sizes

The average MCC size was about 120,000 km^2, most often between 50,000 and 150,000 km^2. It was not possible to find special months with the largest CCM sizes.

11.4 Continent and adjacent ocean

Predominantly MCCs develop over continental regions (~84% of cases). Of the 68 MCCs, 57 formed over the continent and the rest over the ocean.

11.5 Synoptic systems associated with MCC

Most often, MCCs were encountered in the ITCZ region (57% of cases). Then, MCC were observed in the UTCV vicinity (38%) and Frontal Extremities (26%). On the trade winds, which blow almost all year round in the study region, the MCC formed in 23% of cases. Under trade wind disturbances (Wave Disturbances of Trade Winds - WDTW), MCCs rarely appeared (2% of events). The rest of the MCC formed in a diffuse field, i.e., with weak cyclonic circulation or weak troughs (4% of the total).

11.6 Jet Streams at low and high levels

MCC were associated with some type of Jet Stream (JS) at high and low levels in 62% of cases. 32% of MCCs were associated with Jet Stream at High Levels (JSHL) and 52% with Jet Stream at Low Levels (JSLL). JSHL and JSLL were observed simultaneously in 22% of the MCCs. Most often (in 41% of cases) MCCs were located at the entrance and on the left side of the JSHL.

11.7 Precipitation and thunderstorm

Precipitation more than 100 mm in 24 hours was registered in all MCC cases. The maximum amount of precipitation reached 258 mm/24h. Over the past 10 years, the amount of precipitation has significantly exceeded the historical averages monthly norms.

Thunderstorms were recorded in almost all MCC events (94%). Thunderstorms were more frequent from January to March. On average, about 2 events were recorded in March. There was one event each in January and February. The number of cases recorded in March 2001 and 2008 was 6 and 7 events, respectively. ITCZ, UTCV, and Frontal Extremities were synoptic scale systems favoring the development of strong convective processes.

11.8 Forecast of MCC precipitation and adverse phenomena in the BNE

The forecast of precipitation and adverse phenomena is based on Forecasted Profiles (Section 8). From these profiles, CAPE values and instability indices should be calculated. A conclusion about the possible formation or absence of adverse phenomena must be made based on typical values for each environmental region and synoptic system. The instability characteristics, average and maximum CAPE values for MCCs in the BNE will be presented below:

1) Values of instability indices for the BNE regions:

Mid-North region (Maranhão and Piauí states): strong instability with CAPE = 1,440 J/kg and maximum CAPE = 2,460 J/kg.

Zona da Mata region (coast between Rio Grande do Norte and Bahia): strong instability with CAPE = 1,880 J/kg and maximum CAPE = 2,380 J/kg.

Sertão region (Rio Grande do Norte, Bahia and Paraíba states): strong instability, with CAPE = 1,600 J/kg and maximum CAPE = 2,020 J/kg.

Agreste region: moderate instability, with CAPE = 1,430 J/kg and maximum CAPE = 1,650 J/kg.

2) The values of instability indices in the MCC associated with different synoptic systems are as follows:
   - MCCs in the Mid-North and Zona da Mata regions are closely related to ITCZ and UTCV migration.
   - Under the ITCZ influence, CAPE values can vary from 1,878 J/kg to 2,300 J/kg.
   - Under the UTCV influence, CAPE values can reach up to 1,800 J/kg.
   - Under the Frontal Extremities influence, the highest CAPE values were around 2,000 J/kg.

11.9 MCC trajectories

MCCs influenced by ***ITCZ*** have east-west trajectories throughout their existence in all events. These MCCs are moving along the predominant current.

MCCs at the ***frontal extremity*** move in the opposite direction to the main current. This displacement was not real, since a new convection formed in this direction and it seemed that the entire MCC was moving.

MCCs associated with WDTW and UTCV, represented a large variation of trajectories.

The influence of the relief on the MCC trajectory was found in different regions of the BNE. In the area between the Bahia, Sergipe and Alagoas states, the MCCs formed near the coast and did not cross the mountainous region of Bahia. Further north, between the Pernambuco and Rio Grande do Norte states, systems developed near the coast and dispersed significantly to the east, while those that formed in the semiarid region moved to the west.

12. References

Arlene G. Laing, J. Michael Fritsch, “The global population of Mesoscale Convective Complexes”, *Royal Meteorological Society* 123: 389-405, 1997.

Albuquerque Carolina Silva Miranda. “*Desenvolvimento e Trajetórias dos Complexos Convectivos de Mesoescala no Estado de Alagoas Entre os*

Anos de 1999 e 2009". M. Sc. Teses, Universidade Federal de Alagoas, Maceió, Brazil, 2011.

Blamey Ross and Chris J. C. Reason, *"Mesoscale Convective Complex Over Southern Africa"*, *Journal of Climate* 25: 753-766, (2012).

Cordeiro Edwans Silva, Natalia Fedorova, Vladimir Levit, "*Análise sinótica e termodinâmica dos eventos com trovoadas para o estado de Alagoas no período de 15 anos (1998-2012)"*, *Revista Brasileira de Meteorologia* 33, no. 4: 685-694, (2018).

Dee Dick P. S. M. Uppala, Adrian J. Simmons, Paul Berrisford et al, "*The ERA-Interim reanalysis: configuration and performance of the data assimilation system"*, *Quarterly Journal of Royal Meteorological Society* 137: 553-597, (2001).

Fedorova Natalia. "*Sinótica IV: Sistemas e processos sinóticos atuantes na América do Sul"*. Maceió: Ed. Universitária/ UFAL, 2008.

Fedorova Natalia,Vladimir Levit, Márcio Henrique dos Santos Silveira, Bruce Francisco Pontes Da Silva and Avtandil G. Amiranashvili, *"Mesoscale Convective Complexes on the Northeastern Coast of Brazil"*, *Journal of the Georgian Geophysical Society* 13: 36-49, (2009).

Fedorova Natalia, Vladimir Levit and C. D. Cruz, *"On Frontal Zone Analysis in the Tropical Region of the Northeast Brazil"*, *Pure and Applied Geophysics* 173: 1403–1421, (2016).

Fedorova Natalia, Bruce Francisco Pontes Da Silva and Vladimir Levit. "*Complexos Convectivos de Mesoescala sobre o Nordeste do Brasil"*. Maceió: Clube dos Autores, 2019.

Kousky Vernon E., Manoel Alonso Gan, *"Upper Tropospheric cyclonic vortex in the subtropical South Atlantic"*, *Tellus* 3: 538-551, (1981).

Kenneth R. Knapp, *"Scientific data stewardship of International Satellite Cloud Climatology Project B1 global geostationary observations"*, *Journal of Applied Remote Sensing* 2, (2008).

Lyra Matheus José Arruda. "*Processos de formação e deslocamento dos Complexos Convectivos de Mesoescala no Nordeste Brasileiro"*. Teses, Universidade Federal de Alagoas, Maceió, Brazil, 2016.

Lyra Matheus José Arruda, Natalia Fedorova, Vladimir Levit, Yasmin Uchoa Silva and Hugo Alves Braga. *"Sistemas Sinóticos associados a formação dos Complexos Convectivos de Mesoescala no Nordeste Brasileiro entre 2013 e 2015". In: Brazilian Meteorology Congress* 19, (2016).

Lyra Matheus José Arruda, "*Influência dos processos físicos e sinóticos nas trajetórias dos Complexos Convectivos de Mesoescala no Nordeste Brasileiro entre 2008 e 2015"*. M. Sc. Teses, Universidade Federal de Alagoas, Maceió, Brazil. 2018.

Lyra Matheus José Arruda, Lucas Carvalho Vieira Cavalcante, Vladimir Levit and Natalia Fedorova, *"Complexos Convectivos de Mesoescala sobre o Nordeste do Brasil e fenômenos adversos associados". Revista Ibero-Americana de Ciências Ambientais* 9, no. 3: 95-103, (2018).

Lyra Matheus José Arruda, Lucas Carvalho Vieira Cavalcante, Vladimir Levit and Natalia Fedorova, *"Ligação entre Extremidade Frontal e Zona de Convergência Intertropical sobre a Região Nordeste do Brasil", Anuário do Instituto de Geociências* 42, no. 1: 413-424, (2019).

Lyra Matheus José Arruda, Natalia Fedorova, Vladimir Levit and Ismael Guidson Farias de Freitas, *"Características dos Complexos Convectivos de Mesoescala do Nordeste Brasileiro", Revista Brasileira de Meteorologia* 35: 727-734, (2020).

Lyra Matheus José Arruda, Natalia Fedorova and Vladimir Levit, *"Mesoscale Convective Complexes over Northeastern Brazil" Journal of South America Earth Sciences* 118: 1-7, (2022).

Maddox Robert A., *"Mesoscale Convective Complexes", Bulletin of the American Meteorological Society* 61, no. 11: 1374-1387, (1980).

Milhahn Júnior Aribert Walter. *"Correntes de jato e sistemas sinóticos associados aos Complexos Convectivos de Mesoescala no Nordeste Brasileiro".* Completion of course work (Meteorology) – Federal University of Alagoas, Maceió, Brazil, 2013.

Milhahn Júnior Aribert Walter. "*Uma ferramenta para análise automática dos Complexos Convectivos de Mesoescala no Nordeste do Brasil".* Master's Dissertation (Meteorology) – Federal University of Alagoas, Maceió, Brazil. 2016.

Molion Luiz Carlos Baldicero and Sérgio de Oliveira Bernardo, *"Uma Revisão da Dinâmica das Chuvas no Nordeste Brasileiro", Revista Brasileira de Meteorologia* 17, no. 1: 1-10, (2002).

Pontes da Silva Bruce Francisco, Natalia Fedorova, Vladimir Levit, Bruno Miranda de Brito, A. G. Santos and Clebson do Carmo Raimundo. "*Mecanismos associados às precipitações intensas observadas entre os dias 13 e 14 de janeiro de 2005 no Sertão do Estado de Alagoas*". *Brazilian Meteorology Congress* 11, 2008.

Pontes Da Silva Bruce Francisco, Natalia Fedorova, Vladimir Levit, Bruno Miranda de Brito and Anatoly Peresetsky, *"Sistemas sinóticos associados às precipitações intensas no Estado de Alagoas", Revista Brasileira de Meteorologia* 26, no. 3: 295- 310, (2011).

Ramirez Maria C. V., Mary T. Kayano, Nelson J. Ferreira, *"Statistical analysis of upper tropospheric vortices in the vicinity of northeast Brazil during the 1980-1989 period", Atmósfera* 12: 75-88, (1999).

Silveira Márcio Henrique dos Santos. "*Complexos Convectivos de Mesoescala do Estado de Alagoas*". M. Sc. Teses, Universidade Federal de Alagoas, Maceió, Brazil, 2010.

Velasco Ines, J. Michael Fritsch, *"Mesoscale Convective Complexes in the Americas." Journal of Geophysical Research: Atmospheres* 92, no. 20: 9591–9613, (1987).

Veltishev Nikoly Fedorovich "*Mesometeorology and short-term forecasting*". World Meteorological Organization, Geneva, Switzerland, 1988.

CONCLUSION

The book contains information about the physical process of formation and forecasting of adverse meteorological phenomena in the tropical region of the BNE. This information includes, firstly, the theorical foundations of physical processes, and secondly, the Guide to Short-Range Weather Forecasting. Therefore, this book can be used to:

- aid students in their education
- carry out operational weather forecasting in the BNE
- verify forecasting methods for other tropical regions
- aid the development of automatic methods for short-term weather forecasting

It is important to note that the development of automatic methods depends on technical progress and has its own characteristics in each center with different computing equipment. At the same time, the theorical foundations and empirical methods do not change quickly. Therefore, the data in this book can be used to develop an automatic method for short-term forecasting of such phenomena as heavy and light rain, thunderstorms, low visibility, haze, fog.

The book describes in detail synoptic-scale systems, which are of great importance for weather forecasting in the BNE, recently described for the first time, such as: Brazilian Northeast Jet Stream and Middle Tropospheric Cyclonic Vortexes. In addition, the need for a detailed study of the interaction of meteorological systems for the weather formation in the BNE is emphasized.

The interaction of synoptic-scale systems between the two hemispheres in the Atlantic region, as well as between tropical and extra-tropical regions, has not been sufficiently studied for use in operational practice. This requires additional research.

List of Figures

Introduction

Part I
Synoptic and Mesoscale Systems Affecting the Weather Formation in Northeast Brazil

Chapter I-1 Intertropical convergence zone

Chapter I-2 South Atlantic Subtropical High

Chapter I-3 Disturbances in the Trade Winds

Chapter I-4 Upper Tropospheric Cyclonic Vortex in the Brazilian Northeast

Chapter I.5 Middle Tropospheric Cyclonic Vortexes

Chapter I.6 Brazilian Northeast Jet Stream

Chapter I.7 Frontal zones in the tropical region

Chapter I.8 Impact of a Northern Hemisphere Tropical cyclone on the BNE

Chapter I.9 Synoptic systems analysis using satellite data

Part II
Short Term Forecasting in the BNE

Chapter II.1 Temperature and maximum temperature forecasting

Chapter II.2 Elaboration of a forecasted profile

Chapter II.3 Precipitation forecasting

Chapter II-4 Thunderstorms forecasting

Chapter II.5 Short term forecasting of weak rain, low visibility, fog and mist in the BNE

Chapter II.6 Mesoscale Convective Complexes in the Northeastern Brazil

List of Tables

Part I

Synoptic And Mesoscale Systems Affecting the Weather Formation in Northeast Brazil

Chapter I-3 Disturbances in the Trade Winds

Chapter I-4 Upper Tropospheric Cyclonic Vortex in the Brazilian Northeast

Chapter I.5 Middle Tropospheric Cyclonic Vortexes

Chapter I.6 Brazilian Northeast Jet Stream

Chapter I.7 Frontal zones in the tropical region

Chapter I.8 Impact of a Northern Hemisphere Tropical cyclone on the BNE

Chapter I.9 Synoptic systems analysis using satellite data

Part II
Short Term Forecasting in the BNE

Chapter II.1 Temperature and maximum temperature forecasting

Chapter II-4 Thunderstorms forecasting

Chapter II.6 Mesoscale Convective Complexes in the Northeastern Brazil

LIST OF CHAPTER AUTHORS

Bruce Francisco Pontes da Silva
Instituto Capixaba de Pesquisa, Assisteencia Teecnica e Extensao Rural, Meteorologist

Bruno Miranda de Brito
Instituto Nacional de Pesquisas Espaciais, INPE, Meteorologist

Davidson Lima de Melo
Federal University of Alagoas, Brazil, Meteorologist

Edwans Silva Cordeiro
Somar Meteorologia - São Paulo – SOMAR, Meteorologist

Joao Pedro Gonçalves Nobre
Instituto Nacional de Pesquisas Espaciais, INPE, Ph.D. in progress in Meteorology

Lucas Carvalho Vieira Cavalcante
Universidade Federal de Campina Grande, UFCG, Ph.D. in progress in Meteorology

Marcio Henrique dos Santos Silveira
Infraero – INFRAERO, Guarulhos, Meteorologist

Matheus José Arruda Lyra,
Federal University of Alagoas, Brazil, Meteorologist, Dr.